PRINCIPLES FOR ELECTRIC POWER POLICY

Technology Futures, Inc.,
and
Scientific Foresight, Inc.

Q

Quorum Books
Westport, Connecticut • London, England

Library of Congress Cataloging in Publication Data

Main entry under title:

Principles for electric power policy.

Bibliography: p.
Includes index.
1. Electric utilities—Government policy—United States.
I. Technology Futures, Inc. II. Scientific Foresight, Inc.
HD9685.U5P75 1984 333.79'32 84-4692
ISBN 0-89930-095-2 (lib. bdg.)

Library of Congress Catalog Card Number: 84-4692
ISBN: 0-89930-095-2

First published in 1984 by Quorum Books

Greenwood Press
A division of Congressional Information Service, Inc.
88 Post Road West, Westport, Connecticut 06881

Printed in the United States of America

10 9 8 7 6 5 4 3 2 1

Disclaimer

This material is based upon work supported by the
National Science Foundation under Grant No. PRA-8019781.
Any opinions, findings, conclusions, or recommendations
expressed in this publication are those of the authors
and do not necessarily reflect the views of the Foundation.

PRINCIPLES FOR ELECTRIC POWER POLICY

Recent Titles from QUORUM BOOKS

Business Strategy for the Political Arena
Frank Shipper and Marianne M. Jennings

Socio-Economic Accounting
Ahmed Belkaoui

Corporate Spin-Offs: Strategy for the 1980s
Ronald J. Kudla and Thomas H. McInish

Disaster Management: Warning Response and Community Relocation
Ronald W. Perry and Alvin H. Mushkatel

The Savings and Loan Industry: Current Problems and Possible Solutions
Walter J. Woerheide

Mechatronics
Mick McLean, Editor

Establishing and Building Employee Assistance Programs
Donald W. Myers

The Adversary Economy: Business Responses to Changing Government Requirements
Alfred A. Marcus

Microeconomic Concepts for Attorneys: A Reference Guide
Wayne C. Curtis

Beyond Dumping: New Strategies for Controlling Toxic Contamination
Bruce Piasecki, Editor

Payments in the Financial Services Industry of the 1980s: Conference Proceedings
Federal Reserve Bank of Atlanta, Sponsor

Japanese Business Law and the Legal System
Elliott J. Hahn

YOUTHJOBS: Toward a Private/Public Partnership
David Bresnick

State Government Export Promotion: An Exporter's Guide
Alan R. Posner

Contents

Exhibits vii

Tables ix

Preface xi

Acknowledgments xv

Assessment Team and Review Panel xvii

Abbreviations xix

PART ONE: INTRODUCTION **3**

I. Assessment Overview 5

A. Background 5

B. Assessment Purpose 6

C. Six Alternative Sets of Assumptions 7

D. Conclusions 8

E. Methodology 16

F. Book Organization 17

PART TWO: ANALYSES **19**

II. Six Potential Futures 21

III. Potentials for Regret: Policy/Future Mismatches 53

IV. Principles for Electric Power Policy 80

PART THREE: BACKGROUND 119

V. Finance 121

VI. Health and the Environment 147

VII. Power Systems Reliability 166

VIII. Demography 193

IX. Research, Development, and Demonstration Policy 209

X. Historical Perspective on Electrification 230

PART FOUR: PANELISTS' VIEWS 249

Expert Panelists' Individual Comments: Disagreements and Elaborations 251

Appendix A. Research Design and Methodology 263

Appendix B. Expert Opinion Survey—Final Results 287

Appendix C. Example Statistical Descriptions of Six Potential Futures 427

Appendix D. Data Sheets for Exhibits in Chapter V 435

Index 443

Exhibits

II.1 Real per Capita GNP Trends, 1980-2010 31

II.2 Energy Intensity of Real Gross National Product, 1980-2010 32

II.3 Annual Energy Consumption, 1980-2010 33

II.4 Electricity's Share of Domestic Energy Market, 1980-2010 34

II.5 Issue Importance by Scenario 50

IV.1 Energy for Electricity Production, 1950-1980 88

IV.2 U.S. Energy Consumption and Real GNP, 1950-1980 89

IV.3 Industrial Electricity Consumption, 1950-1982 94

V.1 Real Electricity Prices, 1935-1980 122

V.2 Comparison of General Inflation and Electricity Price Inflation, 1935-1980 123

V.3 Utility Plant Utilization, 1930-1980 124

V.4 Total Electricity Supply, 1935-1980 125

V.5 Utility Fuel Conversion Rate, 1930-1980 126

V.6 Real Utility Fossil Fuel Cost, 1954-1980 127

V.7 Power Plant Construction Cost Inflation: Materials and Labor, 1954-1980 128

V.8 Utility Real Return on Equity, 1950-1980 131

VI.1 Water Consumption, by Use, 1960-1975 150

VI.2 Water Withdrawal, by Use, 1950-1975 151

IX.1 Trends in Technology and Cost of Electric Energy, 1880-1980 212

IX.2 R&D: Distribution of Total Estimated Annual Cost to Utilities, Manufacturers, and Government 213

Tables

I.1	Scenario Rates of Change	8
I.2	Scenario Point Values	9
II.1	Side-by-Side Scenario Comparison	36
III.1	Summary of Stakeholder Analyses	76
VII.1	Attacks Against Energy Facilities, 1970-1980	170
VII.2	Incidents of Energy-Related Terrorism, 1968-1978	171
VII.3	Worst Bulk Power Interruptions Due to Technical Failure, 1974-1979	175
VII.4	Reactor Accident Consequences	177
VII.5	Worst Power System Interruptions, 1974-1979	190
IX.1	Major Research and Development Needs in the Field of Electric Power	216
IX.2	National Need: Energy, 1981-1985	221
IX.3	Joint Projects with DOE at Risk Because of Changing Federal Funding Policies	222

Preface

The presence of electric power in contemporary American life is pervasive. In the one hundred years since the financial elite of San Francisco and New York first enjoyed electricity, access to electric power has become tantamount to an American birthright. As witnessed, for example, by the history of the Tennessee Valley Authority, this evolution has been very much influenced by government policy. By the early 1980s, electric power was so interweaved with American life that more than a third of all conventionally measured energy consumption was for electricity generation.

The versatility of electricity argues that its use will continue to expand for the foreseeable future. Indications abound that the future U.S. economy will be dominated by electricity-intensive activities, such as information exchange and high value-added manufacturing. Conservation efforts notwithstanding, the proliferation of robots and a host of other precisely controllable electricity-based industrial machinery, such as arc furnaces, induction heaters, and electro-magnetic curing and drying processes, will likely continue to spur electricity demand growth. Furthermore, electricity is appealing because most of its end uses are environmentally benign. (It is also a critical input to such pollution control technologies as residential air purifiers and industrial small particulate precipitators.) Certainly, electricity is the most feasible means by which to realize the energy content of the nation's most plentiful conventional fuels, coal and uranium.

Thus, electric power is not only a major element of life today, it promises to be an even larger element in the future. Government policy, without doubt, will act repeatedly to shape the role electricity plays in our future. How should government act? What are the guiding principles that, if adhered to, are likely to produce what posterity will judge as "wise" policy regarding electric power?

One could argue that these are superfluous questions, questions of only academic interest. After all, more than a hundred years of electricity and government policy interaction already lies behind us, and the "future" will arrive not instantaneously, but in an incremental fashion that presumably leaves time for policy adaptation. Thus, regardless of the answers to these questions, this line of thought holds that the system will get along in the future as, and because, it has gotten along in the past.

This thinking, at a minimum, underestimates the historical sophistication of the electric power policy process. That process has been asking these questions since, at least, the 1930s. Furthermore, this thinking does not acknowledge, though it should, that increased societal-electricity integration can lead to problems that are qualitatively different from those encountered in the past. For example, if electricity is only a minor input to production, an industry and its workers may be able to remain competitive in the face of a steep electricity price increase. If electricity is a large input, however, the consequences could be quite dire and extend well beyond the immediately affected industry and its workers.

This thinking also discounts the unprecedented uncertainties that have recently entered the electric power arena. Prior to the mid-1970s, electric power policy makers worked in a rather stable environment. Since then, however, the national rate of electricity demand growth has dropped 50 percent. Oil- and gas-fired generation, once touted, has become a financial liability for both utilities and ratepayers, as has the construction of nuclear generating facilities. Coal-based power generation has come under attack for its contribution to "acid rain," atmospheric carbon dioxide buildup, and potentially cancer-causing small particulate emissions. National and state energy legislation has promoted utility diversification and encouraged electricity generation by non-utilities. Though electricity price has been regulated almost since the inception of commercial electricity service, various schemes for electricity price deregulation are now being seriously discussed. These and other changes have sown uncertainties that show no signs of dissipating for, at least, the remainder of this decade.

Because the already large role of electricity in U.S. society is probably increasing, the potential for future regret because of missteps in government electric power policy is also increasing. Furthermore, heightened uncertainty about factors that are important to the shaping of electric power policy heightens the probability that policy missteps will occur. This convergence of circumstances led the National Science Foundation to fund, in mid-1981, a two-year, multi-disciplinary technology assessment of alternative U.S. electric power futures.

That assessment unearthed and organized information, insights, and understanding that we believe can be of practical utility to much of U.S. society and, especially, to anyone who would influence government electric power policy. The assessment paid particular attention to the long-term consequences of more near-term policy expediencies. By so doing, it illuminated the increasingly important and complex considerations at play in the shaping of the nation's electric power policy.

This book closely parallels the official report of that assessment. It presents the assessment's data, methodology, and conclusions. We commend it to you in the belief that it can further the cause of prudent long-term and short-term electric power policy.

John H. Vanston, Jr.
President
Technology Futures, Inc.

David O. Frederick
President
Scientific Foresight, Inc.

Acknowledgments

This book is the result of a long-running project, one involving a rather large staff and an extended network of formal contributors. A great many people could legitimately be acknowledged for their help. The following paragraphs acknowledge a number of these, but, lamentably, also omit mention of numerous others. For these omissions the authors apologize and request understanding.

Conventional form acknowledges last those people whose practical contributions are absolute pre-requisites to the existence of a document such as this. These are the people who type multiple drafts of each page, label and re-label dozens of figures, and maintain a storage system from which almost anyone can retrieve the current version of anything. This is detailed work and, in a semi-automated office, really transcends its conventional "clerical" label. Deb Robison, Robin Corbett, Kathy Miller-King, and Helen Mary Vanston were these people for this project—and are sincerely thanked for their efforts. Though last in time, the efforts of our publisher, particularly those of Lynn Taylor and Susan Baker, were by no means last in importance and are deeply appreciated.

The quality of the underlying project has been materially improved by advice from analysts at the National Science Foundation (NSF). They have gained valuable insights through past work with similar projects. Joshua Menkes, Frank Huband, and, particularly, G. Patrick Johnson did a commendable job of offering wisdom without imposing direction.

Many subject matter authorities gave freely of their time and insights. Among these were numerous people at the Electric Power Research Institute. Foremost among these were Oliver Yu, one of the assessment's twelve expert panelists, and Walter Esselman. Claude Anderson, Sherman Feher, René Males, Kathy Miller, Bud Nelson, Lynn Sagan, Phil Schmidt, Sam

Schurr, Milton Searle, Chauncey Starr, Walter Weyzen, and Orin Zimmerman also made valuable contributions. Similarly, various members of the university community provided insights that are appreciated. In addition to members of the assessment's expert panel, the contributions of Bob Sullivan (University of Florida); Faye Duchin and Paul Christensen (New York University); Evan Vlachos (Colorado State University); Todd LaPorte (University of California at Berkeley); and Martin Baughman, Kenneth Land, and Marlin Blisset (University of Texas at Austin) stand out. Several people in private industry or with private consulting groups also donated time and insight; Terry Day (Exxon Corporation), Charles Komanoff (Komanoff Energy Associates), and Ian Wilson (SRI International) were particularly helpful. The contributions of David Snyder (Snyder Family Enterprises) and William Renfro (Policy Analysis Company, Inc.) were very valuable.

Much of the data and theory on which the project drew were supplied by the individuals who participated in its very time-consuming expert opinion survey. They are listed individually in Appendix B and their contributions are acknowledged here.

Members of this assessment's Issue Review Panel read and critiqued the entire project report as it was submitted to NSF. They previously reviewed a number of the working documents. They did this for no money and very little fame, but they do enjoy the appreciation of the assessment's staff.

Finally and most importantly, the staff acknowledges the contributions of the individuals who participated on this project's panel of experts. They provided, by far, the greatest part of the intellectual fodder for the project, and they did so under circumstances requiring considerable patience and understanding. The authors are sincerely and lastingly grateful to each.

Assessment Team and Review Panel

ASSESSMENT EXPERT PANEL

Dr. Sanford Berg	University of Florida
Dr. Richard D. Brown	The MITRE Corporation
Dr. David Howard Davis	International Energy Associates, Ltd. and Gettysburg College
Dr. Gordon A. Enk	Gordon A. Enk & Associates, Inc.
Dr. Kirby Holte	Southern California Edison Company
Mr. Alvin Kaufman	U.S. Congressional Research Service
Dr. Steve Murdock	Texas A&M University
Dr. Grant Thompson	Conservation Foundation
Dr. John Wardwell	Washington State University
Mr. David Wood	Massachusetts Institute of Technology
Dr. Oliver Yu	Electric Power Research Institute
Mr. René Zentner	Shell Oil Company and University of Houston Law Center

ASSESSMENT STAFF

Dr. John H. Vanston, Jr. Co-Principal Investigator and Project Director	Technology Futures, Inc.
Mr. David O. Frederick Co-Principal Investigator and Deputy Project Director	Scientific Foresight, Inc.
Dr. Ken Roberts Co-Principal Investigator	Southwestern University
Mr. Patrick Drew	Technology Futures, Inc.
Dr. Parker Frisbie	University of Texas at Austin

Mr. Rick Lowerre	Henry, Lowerre, and Mason
Dr. Dudley Poston	University of Texas at Austin
Ms. Donna C. L. Prestwood	Technology Futures, Inc.
Ms. Carolyn Vanston	Technology Futures, Inc.
Dr. Peter Zandan	Technology Futures, Inc.
Mr. Georg Zappler	Technology Futures, Inc.

ISSUE REVIEW PANEL

Dr. Jack Allison	Oklahoma State University
Dr. Doug Bauer	Edison Electric Institute
Dr. Raphael Kasper	National Research Council
Mr. Blair Ross	American Electric Power Service Corp.

Abbreviations

AEC	Atomic Energy Commission
AFUDC	allowance for funds used during construction
BEA	Bureau of Economic Analysis
BPA	Bonneville Power Administration
COMSAT	Communication Satellite Corporation
CONAES	Committee on Nuclear and Alternative Energy Systems
CWIP	construction work in progress
DOE	Department of Energy
ECAR	East Central Area Reliability Coordinating Agreement
EIA	Energy Information Administration
EMP	electromagnetic pulse
EPRI	Electric Power Research Institute
ERA	Economic Regulatory Administration
ERC	Electric Research Council
ERCOT	Electric Reliability Council of Texas
ERDA	Energy Research and Development Administration
FPC	Federal Power Commission (later Federal Energy Regulatory Commission)
GAO	General Accounting Office
HELCO	Hartford Electric Light Company of Connecticut
LWRs	light water reactors
MAAC	Mid-Atlantic Area Council
MAIN	Mid-American Interpool Network
MARCA	Mid-Continent Area Reliability Coordinating Agreement
MHD	magnetohydrodynamics
NEPOOL	Northeast Power Pool
NERC	National Electricity Reliability Council
NPCC	Northeast Power Coordinating Council
NPS-FPC	National Power Survey of the Federal Power Commission

NSF	National Science Foundation
PSCs	public service commissions
PUCs	public utility commissions
PURPA	Public Utility Regulatory Policy Act
REA	Rural Electrification Administration
SERC	Southeastern Electric Reliability Council
SERI	Solar Energy Research Institute
SMSAs	Standard Metropolitan Statistical Areas
SNG	synthetic natural gas
SWPP	Southwest Power Pool
TVA	Tennessee Valley Authority
UHV	ultra high voltage
USGS	United States Geological Survey
WPPSS	Washington Public Power Supply System
WSCC	Western Systems Coordinating Council

PRINCIPLES FOR ELECTRIC POWER POLICY

Part One

INTRODUCTION

This part presents an overview of:

- The reasons for undertaking the technology assessment
- The scenarios on which the assessment relied
- The nine electric power policy principles through which the conclusions of the assessment are presented
- The methodology by which the assessment was conducted
- The organization of the remainder of the book.

Chapter I is, by design, rather terse. It covers the high points of those aspects of the assessment that will be relevant to most readers, and it is also a guide by which the person who would read only selected sections of the book can decide which sections to read. Factual statements in this chapter do not cite underlying data sources; these are cited in subsequent chapters. Also, this chapter does not reflect the elaborations or dissents expressed by individual members of the assessment's panel of experts; these are noted in subsequent chapters and are presented in Part Four.

CHAPTER I

Assessment Overview

A. BACKGROUND

Since the opening of the Edison Electric Company's Pearl Street Station in New York City just over one hundred years ago, the role of electric power in the life of the American citizen has steadily grown in both scope and importance. Technological developments in electric lighting, fractional horsepower motors, computers, and telecommunications, to name a few, have continuously reshaped American family life. Moreover, ever increasing use of electric power in commercial and industrial activities has served to increase productivity and improve working conditions.

With this growing role, electricity and electricity generation have accounted for a progressively larger share of this nation's total energy consumption. Today, more than one-third of all conventionally measured energy consumption is fuel for electricity generation. Notwithstanding the slackening of electric power demand growth since the 1973 Arab oil embargo, the versatility of electricity argues that its role, relative to those of energy resources, will continue to grow over the next few decades. Factors supporting this conclusion include a growing population with expanded expectations of comfort and convenience and with an increasing desire for power sources that are clean at the point of use. Further, new industrial and commercial uses for electricity continue to emerge. It is expected that industrial and commercial electricity conservation efforts will be more than offset as robots, information manipulation, high-technology manufacturing, and other electricity-intensive technologies proliferate.* Moreover, electric

*Examples of emerging electricity-intensive technologies are electric arc furnaces, induction heating, laser beam torches, electrically conducting flywheels, and microwave production processes. These new technologies allow finer process control, higher conversion efficiencies, and greater potential for automation than many conventional technologies. The MITRE Corporation citation at Chapter IV, Exhibit IV.3, discusses these and other electricity-intensive technologies in detail.

power could come to play a larger role in transportation, as through increased U.S. use of electric cars, trains, and buses. One of the most compelling reasons to expect increased dependence on electric power is that electricity is the most practical way to utilize the nation's most abundant conventional energy sources, coal and uranium, as well as many of the proposed new energy systems, such as solar, fusion, and ocean thermal power.

Because of the critical role electric power plays and will apparently increasingly play in both our personal and professional lives, it is essential that government policies promote the reliable supply of electricity at acceptable societal, environmental, and monetary costs. Broadly stated, the assessment summarized in this overview was undertaken to develop and to organize insights that support that policy objective.

B. ASSESSMENT PURPOSE

Historically, governments at all levels have sought to influence electric power production, transmission, and distribution. This influence has been realized through regulation of, for example, electricity price, utility organizational structures, and environmental and health standards. As national dependence on electric power increases, the importance and influence of government electric power policy are also expected to increase. The assessment, therefore, focuses on those policies within the provinces of the federal government or state public utility commissions. The assessment examines a 30-year period beginning in 1980, but pays particular attention to policies that might be adopted during the 1980s and have long-term ramifications.

Increased electrification will have important consequences both for the country as a whole and for various groups within the country. For example, one would expect the economic consequences of power interruptions and irregularities to be greater as the importance of electricity to the nation's economic activities becomes greater. Similarly, one would expect increased electrification to bring with it increased competition for certain natural resources, such as land, water, air quality, and fuel. Though increased competition for natural resources will affect all of society, some segments, such as agricultural water users and heavy industry, will be disproportionately affected. The assessment, therefore, addresses the more important potential consequences of electrification for the nation as a whole and also addresses consequences of special interest to particular groups within the nation.

Government policies affecting electric power typically have long-term impacts, as witnessed by, for example, the long technologically useful lives of (government licensed) power plants. Furthermore, policies that directly influence electric power can have important indirect influences over other aspects of life in the United States. Electric power policies, for example, affect electricity availability, quality, and price. These, in turn, can affect economic growth, particularly at the regional level. Policy support for electricity generation options affects prospects for U.S. energy independence and technological leadership.

The assessment is concerned with policies, such as those just mentioned, that would directly affect the future of electrification and indirectly affect other aspects of society as well. Conversely, it also examines "non-electric" policies that could have indirect, but important, effects on electrification. U.S.-Canadian agreements for the control of acid rain are an example of this latter type of policy.

C. SIX ALTERNATIVE SETS OF ASSUMPTIONS

Many technologies could potentially supply future electricity demand. No one technology is so obviously superior that the others can safely be ignored. Moreover, the absolute and relative sizes of future energy and electricity demand are highly uncertain. The consequences of future electrification and the policies pertinent to that electrification are sensitive to the technologies by which electricity is supplied, the absolute and relative amounts of electricity supplied, and a multitude of societal variables, most of which could take many potential future values.

Because of these uncertainties, the analyses of the assessment are conducted within the contexts of six alternative sets of assumptions (scenarios) about how the nation's 30-year future might evolve. One set of assumptions is exceedingly "middle of the road." The other sets are designed to bound the plausible extremes within which electric power policy might be formed.

The central scenario is the "Average Future" Scenario (Scenario A). It postulates total energy and electricity demands near the averages of those postulated by the other scenarios. It assumes a limited but growing role for nuclear generation and a dominant role for coal-based generation.

Three of the bounding scenarios postulate high total energy demand growth. Two of these postulate that a high percentage of that demand will be supplied by electricity. One of the high energy, high-electricity demand scenarios is the "Nuclear Resurgence" Scenario (Scenario B). It postulates a future in which society's disenchantment with coal grows and in which generation from nuclear energy is increasingly acceptable. The other high energy, high electricity demand scenario is the "Mega-plant" Scenario (Scenario C). It postulates a future in which electricity supply from neither nuclear nor coal resources is considered acceptable for the long term. Here, the nation embarks on a program to supply electricity by unconventional, large-capacity sources, exemplified by solar power satellites. The "Small Coal Plants" Scenario (Scenario D) is the third high energy demand scenario. It postulates a future in which oil prices rise very moderately, the trend to electrification slows, and relatively small coal-based facilities are the preferred technology for electricity generation.

The two other bounding scenarios are low-energy demand scenarios. One postulates relatively high and the other relatively low electricity demand.

The low energy, high electricity demand scenario is the "Post-industrial" Scenario (Scenario E). Here, business activity is dominated by services and very high technology manufacturing. This scenario also postulates significant advances in distributed electric power technology, exemplified by terrestrial solar photovoltaics. The low energy, low electricity demand scenario is the "Economic Malaise" Scenario (Scenario F). In this future, a series of unfortunate circumstances and policies lead to low, oscillating economic growth and very modest electricity demand growth. Power supply in this scenario continues to be dominated by plants similar to those in use today (i.e., conventional, large coal-fired units).

Tables I.1 and I.2 present abbreviated statistical descriptions of the scenarios. Exhibits II.1 through II.4 in Chapter II present selected graphical descriptions of the scenarios.

Table I.1
Scenario Rates of Change

		Average Annual Future Growth Rates (%)					
Descriptor	**1951–1980**	*A*	*B*	*C*	*D*	*E*	*F*
Real GNP	3.44	2.20	2.70	3.00	2.40	3.00	1.50
Population	1.33	.65	.83	.83	.71	.33	.33
Per Capita RGNP	2.10	1.54	1.85	2.16	1.65	2.66	1.16
BTU/RGNP Ratio	−.66	−1.25	−.87	−1.16	−.58	−2.74	−1.30
Energy Demand	2.75	.92	1.81	1.81	1.81	.18	.18
Electricity Demand	6.23	2.03	3.54	3.57	2.61	1.88	.63

D. CONCLUSIONS

The substantive conclusions of the assessment are presented as two underlying principles and seven synthesized principles. Summarized, these are as follows:

Underlying Principles

1. *The electric power system and electric power policies are inexorably interrelated with most public policies and many facets of society.* The nation's electric power system is often affected as much by general government policies (e.g., those involving anti-trust, defense, or occupational health) as by policies that are more strictly "electric" (e.g., those involving utility financial accounting or transmission line corridors). Furthermore, because electricity is tightly interwoven with American life, few policies are so specifically targeted to electric power that they do not also affect other elements of society.

Table I.2
Scenario Point Values

Descriptor	1980	2010 Descriptor Values *A*	*B*	*C*	*D*	*E*	*F*
RGNP (billion 1980$)	2,633	5,050	5,850	6,400	5,325	6,400	4,110
Population (millions)	226.5	275	290	290	280	250	250
Energy Demand (quadrillion BTU)	76	100	130	130	130	80	80
Net Electricity Demand (billion kwh)	2,381	4,350	6,760	6,825	5,160	4,160	2,875
Non-utility Generation (% of net electricity demand)	3.5	4.5	2.0	4.5	12.0	18.0	6.5
Electricity Intensity of Energy Demand (% of energy demand)	33	46	55	55	42	55	38

This principle highlights the need for institutionalizing methods to anticipate possible societal changes and to devise and, especially, to test contingency policies for a wide range of possible circumstances. Most of the promising methods, for example, cross-impact analysis, are significantly less well developed than are more conventional planning methods, such as econometric modeling. Institutionalization of these less well developed forecasting methods is important if they are to be improved over time and more frequently produce useful information for policy makers. A necessary adjunct to these methods will be structured efforts to transfer information and ideas between the electric power industry and other industries and research disciplines.

2. *Electric power policies typically have long-term implications and are difficult to modify.* Major electric power decisions have historically involved complex technological questions, large expenditures of funds, and implications experienced over decades. Once initiated, these programs have often developed self-reinforcing momenta and had long-term effects on capital markets, building and equipment contractors, labor forces, and power consumers. Program modifications or terminations have typically been costly and caused wide-spread distress. The long lives of power plants, the multiple-generation consequences of nuclear and hydrocarbon wastes, and the length of the technology development cycle ensure that electric power policies of the 1980s will affect life for decades to come.

Given that government policies in this area tend to have long-term implications and tend to be difficult to modify, policy makers have only limited options for constraining potential future regrets. In general, the nation will benefit from conscious efforts to maintain policy flexibility. This implies support of diverse experimental programs. It also implies a premium on policies that allow incremental decision making. Certainly, the policy-making process will be enhanced by explicit acknowledgment of future uncertainties. The development and consideration of alternative scenarios is a proven means to this latter end. The policy process will also be well served by increased understanding about the nature of technological innovation. Though research in this area currently abounds, little of it has been particularized to the observed evolution of electric power technologies.

Synthesized Principles

3. *The importance of electric power to U.S. society will continue to grow in the foreseeable future.* Between 1950 and 1983, the fraction of U.S. energy consumption attributable to electricity generation and use rose from about 17.5 percent to about 34 percent. There are compelling reasons to believe this fraction will continue to rise in the future. Furthermore, most credible views of the future contemplate national electricity reliance that is not only quantitatively large but, also, functionally wide-spread.

The signal message here is that high-quality, efficiently priced electricity is and will continue to be essential to the well-being of the nation. The excess generation capacities of the early 1980s should not dull awareness of this. Underestimation of future demand for electricity will surely lead to stop-gap actions that will be costly in monetary terms and, potentially, dangerous in terms of environmental and health risks.

Emphasis on adequate supply obviously entails the risk of overcapacity. Therefore, efforts leading to an expressed definition of ''adequate'' supply and to a consensus about how to distribute across society the costs of meeting that supply standard deserve more support than they have thus far received. These efforts should involve, perhaps in dominant roles, segments of society beyond the electric utility industry and its regulators. An important element of these efforts will be improved understanding of the relationships between electric power and non-monetary ''quality of life'' measures.

4. *Projections this decade of the long-term size and distribution of electric power demand will be increasingly uncertain.* Traditionally, estimates of future electric power demand have been based on projections of economic growth and on the relationship between economic growth and electric power requirements. Long-term projections of both of these factors are unusually uncertain. In the latter case, this is due primarily to questions about future electricity prices, electricity price/use elasticities, and the changing economic structure of the country. Technological and sociological factors also obscure future electricity demand projections. While robotics and high-technology manufacturing tend to spur demand growth, new elec-

tricity technologies, more efficient than their predecessors, tend to slow demand growth. Population growth patterns strongly affect regional electricity demand. Implementation of time-of-use electricity pricing is widely expected and will alter historical demand relationships in ways that are not yet clear.

This principle reinforces the contention that "appropriate" electric power policies necessarily run the risk of developing capacity that may prove unneeded solely to meet demand. The need to assume this risk should be explicitly acknowledged as a criterion for policy formulation. With explicit acknowledgment of this criterion, the decision, in fact, to assume the risk at some level will be perceived as prudent by conscientious observers. If the risk assumption is perceived as prudent, the potential for subsequent disrespect of the regulatory system will be lessened.

This principle also obviously calls for efforts to lessen uncertainty about future electricity demand. This is an imposing mandate. A more manageable subtask, though, is detailed analysis of how different national economic structures would affect electricity demand. (Models capable of manipulating the Commerce Department's detailed input/output tables in response to assumptions about societal and technological evolution would seem to be useful tools). Also, experimentation with load management techniques and time-of-use rate schedules should be encouraged, because this will provide hard data on the demand-reduction and demand-shifting potentials of these innovations. Finally, continuing encouragement of small power producers and cogenerators is prudent. Though the economic justification for these "unconventional" sources of supply is currently questionable, they may be needed in the future and will not likely exist then in sufficient numbers without interim regulatory encouragement.

5. *The range of practical electric power generation sources may expand dramatically over the next three decades, and the relative attractiveness of traditional sources may change significantly.* Although it is difficult to conceive of major changes in the electricity generation mix over the next decade, there could be important shifts in later years that should be anticipated during the 1980s. Certainly, large research investments were made in unconventional electric power technologies in the 1960s and 1970s. Some of these investments will yield technological returns. Public perceptions will continue to be a dominant influence affecting the practicality of any generation option. The generation source presently most subject to public resistance is nuclear power and, therefore, its future role is the most fraught with uncertainty. It is possible, however, that public awareness of environmental and health problems associated with coal-based power generation could also foster aggressive opposition to this supply technology. Public perceptions are already extremely difficult to anticipate; they will likely become more so as information access becomes more common and international.

The foregoing argues for maintaining multiple generation options; it, therefore, also argues for formal support of a broad spectrum of generation technologies. Because almost all authorities anticipate a dominant future role for coal generation, special research attention should continue to focus on coal technologies. Were perceived difficulties with that resource to increase its environmental and health protection costs markedly, nuclear generation is the most likely base load alternative. Therefore, continued efforts to improve nuclear fission reactors and waste-handling technologies are warranted, regardless of wide-spread disenchantment with nuclear power. The pace of technological evolution, the very high costs of new and embedded generating facilities, and the increased influence of public perceptions call for reevaluation of utility capital recovery methods. Means of cushioning the impacts of sudden devaluation of electricity supply assets should be explored.

6. *The roles, structures, and procedures of electric utilities could change significantly in the relatively near future.* Revolutionary restructuring of the electric power industry is highly unlikely. Nonetheless, changes in the conservative, highly regulated, and monopolistic character of privately owned electric utilities are taking place, and the pace of these changes will accelerate in coming years. Recent federal acts not only define more broad energy service roles for private utilities, they also encourage expansion of non-utility power production. The development of new types of generating technologies could also bring fundamental change to the electric utility industry. Very large capacity units could result in new relationships between utilities and the federal government; very small scale, distributed units could change utility-customer relationships and beget major new electric power institutions.

Electric utility change may originate within the industry, largely in the absence of regulatory direction, or it may occur in response to regulatory direction. In the former case, policy makers must reach an early accord as to what best serves the public interest in areas not currently subject to regulation. Most possibilities for technology-induced change are not so near in time that they, in fact, press for regulatory attention. The time is ripe, however, for policy accord on power brokerage services and, certainly, on the relationships between utilities and other organizations that might offer financial or energy advisory services. Furthermore, as witnessed by events in telecommunication regulation, private enterprise can quickly respond to technological change, and this can radically foreshorten conventional regulatory planning horizons. Where policy induces utility change, special policy attention should focus on allaying the uncertainties of the transition period. Finally, utility entry into new business areas calls for attention to possible conflicts between conventional electric utility regulators and other regulators whose jurisdictions encompass, or could encompass, the new utility activities.

The effects on electric utility institutions of small power producer and electricity cogenerator proliferation could be significant. There is a conceptual predisposition to characterize, particularly, the small power producer as "small" in the corporate sense. In fact, it could be a giant energy or aerospace company or even an electric utility.

7. *The means, extents, and purposes of electric utility regulation could change significantly in the relatively near future.* As the societal, technological, and institutional transformations discussed in the previous four principles come to pass, they will create stresses that will drive change in electric utility regulation. The geographical focus of regulatory jurisdiction could shift from the state level to, depending on events, the international, national, or more local level. The subject matter jurisdictions of regulators will shift as the roles of utilities change. Changed technological and economic realities will continue to encourage experimentation with various degrees of electricity price deregulation.

The institutions most affected by electric power policies are changing, the substance about which policies are formulated is changing, and the means for development and implementation of policies are also changing. With these important variables simultaneously in transition, the risk is high that regulatory actions will lead to unintended results. It is important, therefore, that regulatory changes be incremental and, preferably, preceded by pilot programs. Caution is particularly called for in the matter of electricity price deregulation because that is something with which the nation has no modern experience. In considering varying degrees of price deregulation, the debate should clearly acknowledge that the choice is not between imperfect regulatory pricing, on the one hand, and some type of perfect free-market pricing, on the other. The choice is between pricing mechanisms that are imperfect.

8. *It will be increasingly difficult for electric power policy to strike acceptable balances among considerations of efficiency, equity, and risk.* Efficiency (optimum allocation of resources), equity (fairness), and risk (exposure to loss or peril) are fundamental human concerns of obvious importance to the future of electric power policy. The achievement of an acceptable balance among these concerns has always been difficult. The remainder of this century promises to be a time during which public concern with, particularly, environmental and health risks will remain high, and rising power costs will limit further compromise of efficiency or equity. This circumstance, combined with changing value systems and an increasingly diverse and complex society, will make striking the policy balance progressively more difficult in the future.

Certainly, the electric power policy process will not grow less political. There is, therefore, the need to adapt the process further to a political environment. The public is now more conscious than in the past of values that can be only poorly quantified. Therefore, the regulatory process must come to in fact and in appearance more explicitly honor not only that which is measurable, but also that which, though intangible, is valued. More effective use of citizen advisory boards, the electoral process, and new telecommunications opportunities are modifications to the policy process that merit exploration.

In a very political regulatory environment, it will be difficult to form long-term views of electric power issues. Therefore, the policy process will be well served by institutionalization of formal mechanisms to understand and to weigh future impacts of current policy compromises. An example of such a mechanism might be the creation of posts for guardians *ad litem* to represent future stakeholders in electric power decisions.

The assessment was undertaken on the assumption that what happens in the long term is important. This assumption militates for less regulatory reliance on maintenance of low current electricity prices. For example, construction work in progress (CWIP) should be more "in" than "out" of rate bases, and schemes to diminish or delay utility returns on equipment in service should be critically evaluated.

9. *Broadly based, long-term research, development, and demonstration (RD&D) programs are the single most efficient means of achieving flexibility in electric power policies and decisions.* RD&D is the only mechanism by which to pretest future technological options and analyze the opportunities or problems these new options might present. RD&D can, in fact, create policy opportunities that would not otherwise exist. Particularly, the "research" element of this triad offers inexpensive insurance against the costly consequences of limited future options.

Although direct governmental involvement in the electric power field has waxed and waned in the past, the overall trend has been increased government involvement. In light of this and the increasing importance of electricity to all of society, the federal government should bear the primary responsibility for definition of the electric power research agenda. The assessment argues that the first step in defining this agenda should be to establish research priorities based on functional objectives (e.g., expansion of current generating capabilities), not on technologies or social science disciplines; then decisions could be made about the level of support that should be accorded the meeting of each objective. Next, particular technological and social science research areas could be targeted. Finally, means for providing support to each of these areas could be selected. A procedure of this nature should result in RD&D programs that are more balanced, better understood, and less subject to shifting public moods than have been past policies.

Many of the unknowns confronting electric power policy makers are not and will not be technological in nature. Analyses throughout the assessment highlighted the magnitude and importance of social science unknowns.

There is, for example, obvious value for electric power policy in increased wisdom about the balancing of monetary and non-monetary costs, or about relating the measure of risk and the perception of risk. These are among the crucial and relatively inexpensive RD&D tasks that have nothing to do with hardware.

A bit of humility is also in order. The imperfection of human foresight must be acknowledged. The unknowns are too numerous and sophisticated to permit only eager pursuit of "right" solutions. There is needed a willingness to pursue potentially "wrong" solutions also. Failure to make this acknowledgment will lead to continuing shifts in RD&D goals, emphases, and funding levels. The assessment's evaluation of historical RD&D policy shifts is that they were too numerous and, by and large, inefficient, costly, and counterproductive.

The first eight principles present the policy maker with a taxing dilemma. The first two indicate that electric power decisions tend to have long-term impacts, are difficult to modify, and have societal effects far beyond the electric power area. Principle 3 stresses that the importance of wise electric power policy decisions is increasing, while Principles 4-7 point up the inordinate uncertainties that will characterize, at least, the next decade. Finally, Principle 8 describes how the options practically available to the decision maker are becoming fewer.

In short, the policy maker is called upon to make fair, long-term decisions that have sufficient flexibility to accommodate unexpected changes without unreasonable disturbance or cost. To accomplish this end, the policy maker must project as accurately as possible future sociological, economic, political, and technological developments; analyze with perception the interrelationships among these developments and between them and electric power policies; ensure the availability at reasonable costs of alternate technological choices; contrive means for deferring irreversible decisions without undue costs and risks; and frame means for changing policies, if circumstances require, with minimal societal disruption. Fortunately, Principle 9 offers hope that these ends may be met.

Overall, the strongest consensus among the assessment participants was that well-conceived and executed RD&D programs, in both technological and non-technological areas, provide the most efficient means of ensuring the fairness and flexibility that will be essential to effective future electric power policies. Assessment participants and most experts in the electric power field stress the necessity for RD&D policies that are clearly defined, generally stable, and supported at levels that are predictable and reasonably consistent. Paradoxically, policies of this nature will be necessary if the nation is to confront successfully a future that promises to be mercurial, diverse, and surprising.

E. METHODOLOGY

There were two central elements to the methodology used in this assessment. One element was conceptual: the use of six alternative scenarios to test the long-term implications of near-term electric power policy decisions under different assumptions about the nation's long-term future. The other element was organizational: the combination of the efforts of a small, interdisciplinary core staff and a panel of twelve highly skilled professionals in a manner that enhanced the effectiveness of each group. The core staff provided the continued efforts required for sustained progress; the experts on the panel provided wisdom that can only come from prolonged experience in a subject area. The efforts of the two groups were coordinated through telephone calls, letters, personal visits, and, most importantly, four workshops.

The first task of the core staff, after the selection of the expert panel, was the preparation of a "normal political response" or "base" scenario. To accomplish this, the staff chose approximately 75 factors—technological, social, political, economic—that would both drive the demand for electric power and constrain the manner by which that demand could be met. A three-round survey involving approximately 125 knowledgeable people was conducted. It asked how and why each factor would change over the next three decades. The mean projections of this survey were used as a foundation for the base scenario, a predecessor of the "Average Future" Scenario.

At the first workshop in January 1982, the expert panel examined two variations of the base scenario and identified issues associated with the developments described there. In addition, they suggested other feasible developments that should be included in alternative scenarios and identified subject areas where they felt more detailed analyses were needed. These suggestions served as bases for development of the five "non-base" scenarios. Specific research was begun in each of six major subject areas identified by the panel: finance; environmental health and safety; power system security; demography; technology research, development, and demonstration; and the relationship between electric power and the overall quality of American life. The results of this research are presented in the six "Background" chapters (Chapters V-X) of this book. (Adequate data on the electric power-quality of life question were not found. A historical overview of electric power, rather than the electric power-quality of life research, is presented in Chapter X.)

In July and August 1982, two more workshops were held, each involving approximately half of the expert panel. The work of the core staff to that time, specifically drafts of the "Background" chapters and the alternative scenarios, was reviewed by the experts and additional issues were identified and discussed. The panel and staff also attempted to identify the overarching ideas that had emerged from their analyses to date. This effort provided the foundation for the nine principles that constitute the assessment conclusions.

Prior to the final meeting with selected members of the expert panel in November 1982, the staff prepared a new side-by-side comparison of the six scenarios and a draft analysis of the implications of the nation's following a policy set designed for a given future, in the event a different future evolved. The impacts of such policy/future mismatches on 20 different groups (stakeholders) within society were evaluated and policy modifications for alleviating adverse effects were suggested. At the final meeting, panel members reviewed this work and suggested improvements.

During the winter and early spring of 1983, the staff prepared final drafts of each of the chapters for the assessment final report. Those chapters were, then, reviewed by each member of the expert panel. Panel members who found statements in the drafts with which they disagreed, or to which they felt elaborations were desirable, prepared comments to those ends. Panelists' comments are referenced at appropriate places in this text.

Finally, members of an issue review board, who collectively served as a surrogate for the body of "electric power policy makers," critiqued the report on the pertinence of the issues it discussed. These critiques were part of the "research validation" process and are available from the corporate authors and the National Science Foundation.

Regarding the two basic elements of this assessment's methodology, the use of alternative scenarios and a core staff/expert panel assessment team, both worked well. The alternative scenario approach provided detailed contexts within which to test realistically available policy options under a variety of disparate assumptions about the future. This made it possible to identify non-obvious electric power policy issues and implications. Comparisons among scenarios highlighted the matters on which policy flexibility is particularly important and helped to focus thinking about the ways in which specific types of flexibility might be obtained.

The continuing interest and involvement of expert panel members over the nearly two-year period of the assessment contributed immensely to the progress of the assessment. These members, by virtue of their regular professional involvement with subject matters pertinent to the assessment, brought to the assessment both factual knowledge and theoretical insights that would have been very difficult, if not impossible, to obtain from literature or other secondary sources. The daily reliance on a core staff and intermittent intense reliance on the expert panel members made it possible to draw effectively and efficiently on high-quality knowledge and insight.

F. BOOK ORGANIZATION

This book is identical in most respects to the assessment final report. Each is divided into four parts, of which this Introduction is Part One. Part

Two presents the basic analyses of the assessment in three chapters. Chapter II describes in detail the six scenarios employed in the assessment, together with the major issues associated with each. Chapter III examines the impacts of implementing policies based on anticipation of a future similar to that described by the "Average Future" Scenario, were a different future to unfold; it also suggests policies for limiting the consequences of such policy/future mismatches. Chapter IV discusses the basic findings of the assessment, each of which is accompanied by a listing of questions for further analysis. Part Three presents the six "Background" chapters developed during the assessment. Part Four presents individual comments of expert panel members.

In addition to these parts, three of the assessment report's eight appendices are reproduced in this book. Appendix A describes the assessment methodology in detail; Appendix B presents the final results of the assessment's expert opinion survey; and Appendix C offers examples of statistical descriptions for each of the assessment's six scenarios. Appendix D reproduces data sheets for the exhibits in Chapter V.

Part Two

ANALYSES

This part consists of three chapters (Chapters II-IV) that document the majority of the analyses conducted in the course of the assessment. Conceptually, the analyses of Chapters II and III are inputs to Chapter IV, "Principles for Electric Power Policy." Chapter IV is the heart of this book.

Chapter II, "Six Potential Futures," presents and elaborates upon the assessment's six scenarios. Each of the scenarios is a caricature of a possible future and was developed to highlight particular types of potential electric power policy issues. Therefore, none of the scenarios should be understood as a forecast of the way an individual or organization thinks the future will in fact unfold. Pains were taken to construct plausible scenarios. Certainly, however, the set of values associated with the variables of each scenario is not an exclusive set. For example, the "Economic Malaise" Scenario is here characterized by, among other things, low population growth and inappropriate industrial policy. A condition of economic malaise that would highlight the same electric power policy issues could be hypothesized consistent with higher population growth and international instability that supersedes industrial policy.

Chapter III, "Potentials for Regret: Policy/Future Mismatches," explores the consequences for society of electric power policy that anticipates a future that does not, in fact, come to pass. The chapter examines the risks attending policies that future hindsight may reveal to have been inappropriate. The chapter starts with the electric power policy set for the "Average Future" Scenario. It then explores, for each of the other five scenarios, the dislocations that should be expected were "Average Future" electric power policy and the future described by other scenarios to be juxtaposed. For example, the chapter examines the dislocations to be expected were "Average Future" policies favoring coal-based electricity generation to be implemented in a future where distributed solar photovoltaic generation suddenly proves viable. One output from analyses such as this is an understanding of areas in which policy flexibility is particularly important.

Chapter IV draws information from the analyses of the two preceding chapters, as well as the six chapters of Part Three and the appendices. It presents, in the form of nine principles, one integration of this information. There are doubtless other ways this information could have been integrated and presented, but Chapter IV offers policy makers the practical strength of a manageable number of concepts around which to organize thought about particular, perhaps currently unforeseen, electric power issues.

There are no references by which one can validate the truth of statements about the future. Footnotes are, therefore, used rather sparingly in the chapters of this part. In this part and the next, parenthetical references to dissents or elaborations offered by individuals on the assessment's twelve-member panel of experts occasionally appear. References are in the form: "(panelist's initials: chapter-comment number)," for example, (AK:IV-1) indicates Alvin Kaufman's first comment in Chapter IV. Panelists' comments are reproduced in Part Four of this book.

CHAPTER II

Six Potential Futures

A. OVERVIEW

This chapter presents six scenarios of the future. Each scenario covers a period of 30 years beginning in 1980. These scenarios form the contexts within which most of the policy analysis of this assessment was conducted.

It was a preliminary hypothesis of this assessment that the future means and extent of electrification in this country were subject to a great deal of uncertainty. Research and analysis early in the assessment bore out this hypothesis. There are uncertainties about how much electricity will be produced in the future, how much energy will be carried by electricity relative to other energy forms, the technologies by which electricity will be produced, the political acceptability of various technologically feasible means of electricity production, and the end uses to which electricity will be put.

The decision to use scenarios as the vehicles for analyses followed from the high level of uncertainty associated with future electrification. In the absence of assumptions about the uncertainties, analyses of electric power policy would have been too vague to be useful to real-world decision makers. Scenarios were developed to make explicit the contexts in which future policies might exist.

Material concerning the scenarios is presented in four parts. Immediately following this overview is a description of the general process by which the scenarios were developed; this part includes an explanation of the means by which the quantitative assumptions within each scenario were made. The next part presents a brief narrative description of each scenario. It also conveys the gists of the scenarios through tabular and graphical comparisons of key quantitative variables. The next part is a side-by-side comparison of each of the scenarios. The comparison is presented in a matrix format, in which the rows describe the variables for which values are changed among scenarios and in which columns correspond to individual scenarios. The final part is a narrative summary of the policy issues that were found to be particularly important within the context of each scenario.

The appendices of this book also bear on the subject matter of this chapter. Appendix A presents a more detailed explanation of the methodology by which the entire assessment was conducted and thereby sheds additional light on the development of the scenarios. Appendix B presents the complete results of an expert opinion survey from which data for the scenarios were drawn. Appendix C presents example statistics of selected data (principally relating to the electricity generation mix) for each scenario. A fourth appendix, included in the report submitted to the National Science Foundation but omitted from this book, presented two highly detailed scenarios that were developed early in the assessment and that were precursors to the six scenarios described in this chapter.

B. SCENARIO DEVELOPMENT

The development of the six scenarios described in this chapter began with the listing of approximately 200 descriptors by which one could characterize a society. These descriptors were weighted and the listing winnowed to a group of approximately 75 "critical" factors. The principal weighting criterion was the ability of the descriptor to say something about the facets of society with which electric power and electric power policy might interact. Descriptors that were subject to quantification were somewhat more likely than non-quantifiable descriptors to become critical factors. For example, the average number of leisure hours available to an individual per week became a critical factor, while the emphasis placed by the average individual on material success, as opposed to self-actualization, was a society descriptor that did not become a critical factor.

Most critical factors were next reduced ("operationalized") to quantifiable measures so that precise opinions about their future values could be ascertained. Though in most cases this was rather easily done, as for electricity generation mix and persons living in poverty, it was not feasible to reduce each critical factor to a quantifiable measure. The degree of automation of work is an example of a critical factor for which no acceptable quantitative measure was developed. For some critical factors, such as citizens' support of environmental and conservation movements, imperfect surrogate (quantitative) measures were developed.

Historic data series were developed for quantifiable critical factors and experts were surveyed about the values that these might come to take over the next 30 years. Experts' responses to this survey guided the assessment determination of future values for scenario descriptors. Generally, the responses bound the values that the assessment assumed for the purposes of its scenarios. The survey, because it allowed experts to offer their narrative opinions, also helped to reveal the theories ("models") on which experts based their opinions about the future.

A total of 121 experts agreed to participate in the survey. They were divided into four expertise areas: economics, energy, environment and health, and sociology, with an emphasis on demography. The survey inquired about 58 of the critical factors. It was conducted in three rounds, the first of which was by mail in September 1981. Of the 121 participants, 79 responded to this round. Round 2 was conducted by telephone in October 1981 and was directed to those respondents who had made projections substantially different from the norm on any question. The third round was conducted by mail in November and December 1981. This round presented participants with brief statements of the opinions of other participants and with tabulations of the projections from round 1. During round 3, the experts had opportunities to rebut opinions expressed during round 1 and, if they desired, to change their own previously expressed projections. Appendix B presents the results of this survey and historical data on most of the 58 critical scenario factors on which the survey made inquiries.

Principally on the basis of responses to this survey, two rather detailed 30-year scenarios were developed. These served as policy discussion vehicles for the assessment staff and the assessment's twelve-member expert panel. One scenario was a relatively high energy demand scenario, 116 quadrillion BTU in 2010. The other was a relatively low energy demand scenario, 82 quadrillion BTU in 2010. In general, the values specified for the critical factors in each scenario were approximately a standard deviation either side of the mean responses from the expert opinion survey. This guideline occasionally yielded values that were inappropriate in light of the theories expressed by survey respondents and linkages commonly used in energy/econometric models, such as those of the Department of Energy, the Committee on Nuclear and Alternative Energy Systems (CONAES) study, Exxon Corporation, and the Solar Energy Research Institute (EIA, 1981; CONAES, 1979; Exxon, 1980; SERI, 1981). Inappropriate values were respecified within the ranges of responses to the expert opinion survey.

After protracted discussions and analyses of these two scenarios, six alternative scenarios were developed. These six are the scenarios described in this chapter. They were developed to focus analysis on important electric policy issues that had arisen during consideration of the two detailed scenarios. These six alternative scenarios are, therefore, emphatically not projections of the way the future will be. Five are caricatures of ways the future might be, each developed to encompass extremes from the universe of potential electric power issues. The sixth scenario is a caricature only in that it is "average" in all respects. It is the "middle of the road" view of the future.

The six alternative scenarios are not as quantitatively specified as were their two predecessor scenarios. Where the alternative scenarios specify quantitative values, those values fall within the ranges bounded by responses to the expert opinion survey. Many of the variables specified in the alternative scenarios are not quantitative in nature; assumptions about these qualitative variables (e.g., focus of electric power RD&D efforts, national policy on coal wastes, and regulatory treatment of power plant construction costs) were developed by the assessment's staff and twelve-member panel of experts (SM:II-1).

C. NARRATIVE SCENARIO DESCRIPTIONS

Three of the six scenarios postulate "high" energy consumption; 130 quadrillion BTU annually by 2010. (The 1982 figure was 71 quadrillion BTU, down from 79 quadrillion BTU in 1979.) Two of the high energy consumption scenarios also postulate "high" electricity intensity in U.S. society, with 55 percent of 2010 energy consumption being imported electricity and primary fuels input to generate electricity. The other high energy consumption scenario postulates 42 percent electricity intensity. (Electricity intensity has historically increased annually; it stood at 34 percent in 1982.) Two other scenarios postulate "low" energy consumption, 80 quadrillion BTU annually in 2010. One of these also postulates high electricity intensity and the other, 38 percent electricity intensity. The remaining scenario postulates 2010 energy consumption and electricity intensity measures near the averages of those for the other scenarios, 100 quadrillion BTU energy consumption and 46 percent electricity intensity.

The following paragraphs briefly describe each scenario. The side-by-side comparison in Table II.1 (at the end of this section) presents a more detailed description of each.

"Average Future" Scenario (Scenario A). In the "Average Future" Scenario there are no significant changes from economic and life-style patterns that characterized life in the United States in the mid to late 1970s. The

Scenario A (Average Future)

		Average Annual Growth Rate (%)					
Descriptor	**1951–1980**	*A*	*B*	*C*	*D*	*E*	*F*
RGNP	3.44	2.20	2.70	3.00	2.40	3.00	1.50
Population	1.33	.65	.83	.83	.71	.33	.33
Per Capita RGNP	2.10	1.54	1.85	2.16	1.65	2.66	1.16
BTU/RGNP	−.66	−1.25	−.87	−1.16	−.58	−2.74	−1.30
Energy Consumption	2.75	.92	1.81	1.81	1.81	.18	.18
Electricity Consumption	6.23	2.03	3.54	3.57	2.61	1.88	.63

Power System Characterization: Rather large coal plants.

services sector of the economy expands, but conventional manufacturing remains the base of the nation's economy. The energy intensity of the economy decreases about 30 percent over the 30-year period. The dominant policy influence continues to be national policy, and that is normally formulated only in response to problems that have become so aggravated that the public demands government action.

For the 30-year period: real GNP growth averages 2.2 percent annually, approximately two-thirds its rate during the 1950-1980 era; U.S. population growth averages .7 percent annually, approximately one-half its 1950-1980 rate; and total energy consumption growth averages .9 percent annually, approximately one-third its 1950-1980 rate. Oil prices increase approximately 1.6 percent annually in real terms. The growth in demand for electricity falls to about one-third of its 1950-1980 average but still averages 2 percent annually, twice the rate of growth in total energy demand. Large (e.g., 600-800 megawatts) coal-based generating plants continue to dominate the nation's electricity supply system in the year 2010. The generation of electricity from nuclear power increases modestly over the 30-year period to constitute approximately 20 percent of all electricity generation in 2010.

"Nuclear Resurgence" Scenario (Scenario B). The "Nuclear Resurgence" Scenario is a high energy consumption, high electricity intensity scenario. In the "Nuclear Resurgence" Scenario, the nation experiences a vigorous revitalization of its industrial sectors. This leads to an economy that is more energy intensive than that of the "Average Future" Scenario, though energy intensity still drops 23 percent over the 30-year period. Increased public reservations about the environmental and health consequences associated with coal conversion and increased public acceptance of, if not enthusiasm for, nuclear waste containment and storage techniques lead the nation to growing reliance on nuclear energy.

For the 30-year period: real GNP growth averages 2.7 percent annually, a higher rate than in the "Average Future" Scenario and about 80 percent of its rate during the 1950-1980 era; population growth averages .8 percent

Scenario B (Nuclear Resurgence)

		Average Annual Growth Rate (%)					
Descriptor	**1951-1980**	*A*	*B*	*C*	*D*	*E*	*F*
RGNP	3.44	2.20	2.70	3.00	2.40	3.00	1.50
Population	1.33	.65	.83	.83	.71	.33	.33
Per Capita RGNP	2.10	1.54	1.85	2.16	1.65	2.66	1.16
BTU/RGNP	−.66	−1.25	−.87	−1.16	−.58	−2.74	−1.30
Energy Consumption	2.75	.92	1.81	1.81	1.81	.18	.18
Electricity Consumption	6.23	2.03	3.54	3.57	2.61	1.88	.63

Power System Characterization: Much expanded nuclear generation.

annually, again, a higher rate than in the "Average Future" Scenario and about two-thirds of its rate during the 1950-1980 era; total energy demand growth averages 1.8 percent annually, twice its rate of growth in the "Average Future" Scenario and about two-thirds its rate during the 1950-1980 era. Real oil prices increase at approximately 1.9 percent annually, a slightly greater rate of increase than in the "Average Future" Scenario. Electricity demand growth averages 3.5 percent annually, a much higher rate than in the "Average Future" Scenario but, still, little more than one-half its rate during the 1950-1980 era. In the "Nuclear Resurgence" Scenario, nuclear generation of electricity comprises approximately 40 percent of all electricity generation by 2010.

"Mega-plant" Scenario (Scenario C). The "Mega-plant" Scenario is another high energy consumption, high electricity intensity scenario. In the "Mega-plant" Scenario, the nation experiences a re-vitalization of the industrial sectors of its economy and very strong growth in the information services and high-technology manufacturing sectors. The net result of this by 2010 is an economy that is more robust but only slightly more energy intensive than that postulated for the "Average Future" Scenario. The energy intensity of the economy declines 29 percent over 30 years. In the "Mega-plant" Scenario, strong public opposition develops to both coal and nuclear fuels. This opposition is based on concerns about the environmental and health risks associated with these fuels. In response to this opposition, the nation's leaders adopt a strong, long-term national commitment to research programs aimed at the development of new, more environmentally benign, power generation sources. By 2010, three five-gigawatt (GW) solar power satellite stations are undergoing successful demonstration, indicating that plants like these will be the predominant power sources in the more distant future.

For the 30-year period: real GNP growth averages 3 percent annually, a higher rate than in either of the scenarios discussed thus far and approxi-

Scenario C (Mega-plant)

		Average Annual Growth Rate (%)					
Descriptor	**1951–1980**	*A*	*B*	*C*	*D*	*E*	*F*
RGNP	3.44	2.20	2.70	3.00	2.40	3.00	1.50
Population	1.33	.65	.83	.83	.71	.33	.33
Per Capita RGNP	2.10	1.54	1.85	2.16	1.65	2.66	1.16
BTU/RGNP	−.66	−1.25	−.87	−1.16	−.58	−2.74	−1.30
Energy Consumption	2.75	.92	1.81	1.81	1.81	.18	.18
Electricity Consumption	6.23	2.03	3.54	3.57	2.61	1.88	.63

Power System Characterization: Five GW + plants coming on line.

mately 90 percent of its rate during the 1950-1980 era; the population growth rate is as postulated for the "Nuclear Resurgence" Scenario; and total energy consumption is also as postulated in the "Nuclear Resurgence" Scenario. Real oil prices increase an average of 2.2 percent annually, a greater rate of increase than in either of the other two scenarios thus far discussed. Growth in electricity demand is approximately as postulated in the "Nuclear Resurgence" Scenario. Despite the environmental and health hazards the public associates with coal fuels in this scenario, those fuels supply more than 50 percent of all electricity generated in 2010. Because the public, generally, perceives the nuclear risk to be greater than the hydrocarbon risk, nuclear power for electricity generation does not expand as it does in the "Nuclear Resurgence" Scenario, though by 2010 it comprises approximately 20 percent of all electricity generation.

"Small Coal Plants" Scenario (Scenario D). The "Small Coal Plants" Scenario is a high energy consumption, low electricity intensity scenario. In the "Small Coal Plants" Scenario, most of the conventional industrial sectors of the nation's economy are re-vitalized. High-technology manufacturing and services sectors also expand, approximately as in the "Average Future" Scenario. Population growth in the "Small Coal Plants" Scenario is highest in the suburban areas, leading, by 2010, to significant geographic dispersion in the population. The structure of the economy, population dispersion, and relatively low oil prices bring about an economy that is more energy intensive than that of any of the other scenarios postulated for this assessment. Even so, the economy is 16 percent less energy intensive in 2010 than in 1980. Though public opposition remains strong to nuclear power, the environmental and health consequences of coal-based power generation prove to be tractable to technological remedies, so wide-spread public opposition to coal-based generation does not develop. Coal plants in 2010, however, are smaller (200-400 megawatts) and more dispersed than were plants in the 1980s.

Scenario D (Small Coal Plants)

		Average Annual Growth Rate (%)					
Descriptor	**1951-1980**	*A*	*B*	*C*	*D*	*E*	*F*
RGNP	3.44	2.20	2.70	3.00	2.40	3.00	1.50
Population	1.33	.65	.83	.83	.71	.33	.33
Per Capita RGNP	2.10	1.54	1.85	2.16	1.65	2.66	1.16
BTU/RGNP	−.66	−1.25	−.87	−1.16	−.58	−2.74	−1.30
Energy Consumption	2.75	.92	1.81	1.81	1.81	.18	.18
Electricity Consumption	6.23	2.03	3.54	3.57	2.61	1.88	.63

Power System Characterization: Small coal plants, increased cogeneration.

For the 30-year period: real GNP growth averages 2.4 percent annually, a rate higher than in the "Average Future" Scenario and lower than in the other scenarios thus far discussed; population growth is slightly higher than in the "Average Future" Scenario; and total energy demand growth is as postulated in the "Nuclear Resurgence" and "Mega-plant" Scenarios. Real oil prices rise modestly in the "Small Coal Plants" Scenario, less than 1 percent annually and significantly less than postulated in any of this assessment's other scenarios. Although electricity's share of the energy market grows more slowly in this scenario than in the "Average Future" Scenario, its absolute rate of growth, 2.6 percent annually, is greater than that postulated for the "Average Future" Scenario because of greater total energy demand. The differential between the rates of growth of electricity demand and total energy demand is lower in this scenario than in any other postulated. Significantly, 12 percent of all electricity generation in 2010 is produced by non-utility generators.

"Post-industrial" Scenario (Scenario E). The "Post-industrial" Scenario is a low energy consumption, high electricity intensity scenario. In the "Post-industrial" Scenario, there are significant changes in the economic and life-style patterns that characterized the United States during the 1970s and early 1980s. By 2010, the national economy is dominated by very high-technology manufacturing and information services. Energy-intensive basic manufacturing and refining industries have largely been exported to developing countries. Historically energy-intensive industries that remain in the country, such as aluminum, iron, and steel industries, rely heavily on relatively low energy materials technologies, such as electric arc furnaces. Robots are thoroughly integrated throughout industry. By 2010, energy consumption in residential and commercial buildings has been substantially reduced by the adjustment of thermostat settings, insulation of hot water heaters, approximately 40 percent efficiency improvements in household

Scenario E (Post-industrial)

		Average Annual Growth Rate (%)					
Descriptor	**1951–1980**	*A*	*B*	*C*	*D*	*E*	*F*
RGNP	3.44	2.20	2.70	3.00	2.40	3.00	1.50
Population	1.33	.65	.83	.83	.71	.33	.33
Per Capita RGNP	2.10	1.54	1.85	2.16	1.65	2.66	1.16
BTU/RGNP	–.66	–1.25	–.87	–1.16	–.58	–2.74	–1.30
Energy Consumption	2.75	.92	1.81	1.81	1.81	.18	.18
Electricity Consumption	6.23	2.03	3.54	3.57	2.61	1.88	.63

Power System Characterization: Distributed generation; much non-utility generation.

and commercial appliances, and extensive adoption of passive energy conservation techniques, such as improved building orientation. These economic and life-style changes result in an economy that, by 2010, is 57 percent less energy intensive than in 1980.

This scenario postulates an important technological breakthrough in solar photovoltaic technology in the early 1990s that makes terrestrial solar electricity generation economically feasible. Thereafter, an increasing share of all electricity generation occurs outside the electric utility industry.

For the 30-year period: real GNP growth averages 3 percent annually, the same rate as is postulated for this assessment's other high economic growth rate scenario, the "Mega-plant" Scenario; population growth averages .3 percent annually, about one-half the growth rate postulated in the "Average Future" Scenario and one-fourth the rate at which population grew during the 1950-1980 era; and overall energy consumption rises very modestly, averaging .2 percent annually. The average energy consumption growth rate in this scenario is approximately 10 percent of the rate postulated for this assessment's three high energy growth scenarios. 2010 electricity generation in this scenario is only slightly lower than in the "Average Future" Scenario, reflecting a high average annual growth rate for electricity relative to the other energy forms. This growth rate is driven by high-precision electro-technologies in industry, telecommunications and computing in the service sectors, and new technologies, such as electrostatic air filtering, in the residential sector. By 2010, coal-based generation still comprises 44 percent of all utility generation, but imports and renewable energy sources account for about 25 percent of utility supply. Of all generation, 18 percent is non-utility, about one-fourth of this generated by residential and commercial customers.

"Economic Malaise" Scenario (Scenario F). The "Economic Malaise" Scenario is a low energy consumption, low electricity intensity scenario. In the "Economic Malaise" Scenario, a series of government policies ultimately

Scenario F (Economic Malaise)

		Average Annual Growth Rate (%)					
Descriptor	**1951–1980**	*A*	*B*	*C*	*D*	*E*	*F*
RGNP	3.44	2.20	2.70	3.00	2.40	3.00	1.50
Population	1.33	.65	.83	.83	.71	.33	.33
Per Capita RGNP	2.10	1.54	1.85	2.16	1.65	2.66	1.16
BTU/RGNP	−.66	−1.25	−.87	−1.16	−.58	−2.74	−1.30
Energy Consumption	2.75	.92	1.81	1.81	1.81	.18	.18
Electricity Consumption	6.23	2.03	3.54	3.57	2.61	1.88	.63

Power System Characterization: Coal plants of early 1980s technology.

prove inappropriate, and the nation experiences a protracted period of low, oscillating economic growth (RB:II-1). Despite initial government attempts to support industrial re-vitalization, this sector of the economy is ultimately unable to compete successfully with foreign industries, for which inputs to production are less expensive than they are in the United States. This circumstance comes to be generally acknowledged in the late 1980s and early 1990s, and government policies switch to support of high-technology manufacturing and service industries. However, the United States is for many years not able to compete effectively in these areas, either, because of superior infrastructures developed by foreign competitors during the 1980s. The nation's problems in the late 1990s and early 2000s are compounded by the impacts of poor U.S. economic performance on economic growth in less-developed countries; slow economic growth in these countries retards expansion of markets for high-technology products and information services. By 2010, the energy intensity of the U.S. economy is about as specified in the "Average Future" Scenario. The nation, at this juncture, is poised for a period of improved economic growth.

For the 30-year period: real GNP growth averages only 1.5 percent annually, the lowest growth rate postulated for any of this assessment's scenarios and less than one-half the rate of economic growth in the 1950-1980 era; population growth is also low, about the Bureau of the Census Series III rate, as assumed in the "Post-industrial" Scenario; and total energy consumption growth averages only .2 percent annually, also, as in the "Post-industrial" Scenario. Real oil prices are largely supported by demand elsewhere in the world and escalate moderately, as specified in the "Average Future" Scenario. In the "Economic Malaise" Scenario, growth in electricity demand averages .63 percent annually, a lower rate than that for any of this assessment's other scenarios and about 10 percent the rate experienced during the 1950-1980 era. Relatively more electricity (6.5 percent) is produced by non-utility generators in 2010 than was the case in 1980. This generation is from industrial generators, whose electric power activities largely reflect concerns about the quality and reliability of utility electricity supply. Few new utility electric power plants come on line after the early 1990s, and these are generally coal-fired plants similar to those in use in the early 1980s.

Selected Graphical Comparisons. The following four exhibits present graphical scenario comparisons of real per capita GNP, energy consumption per dollar of real GNP, total energy consumption and the electricity intensity of total energy consumption (i.e., electricity's share of the energy consumption market).

Exhibit II.1 reflects the real per capita GNP trends for each of the six scenarios. From 1950 through 1980, real per capita GNP growth averaged 2.1 percent. The solid square symbol shows the 2010 value of that ratio were

Exhibit II.1
Real per Capita GNP Trends, 1980–2010

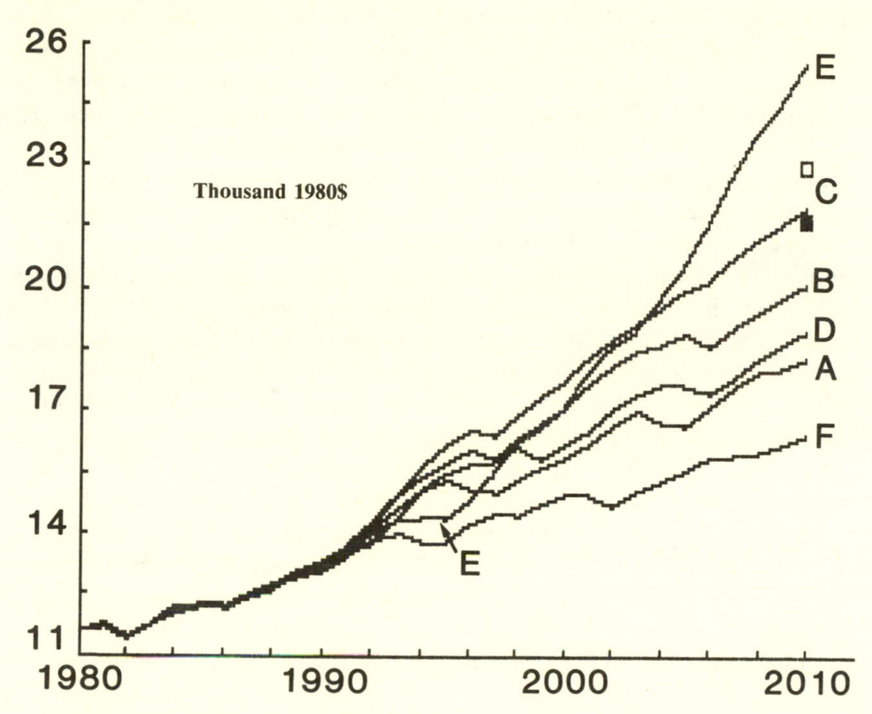

it to grow from 1980 at the 1950-1980 rate. The hollow square symbol reflects the result of growth at the average rate (2.3 percent) for the 1950-1973 period.

The "Post-industrial" and "Mega-plant" Scenarios postulate the most prosperous future societies (by this measure). The average growth rate for real gross national product is the same in each of these scenarios, but population growth is considerably lower in the "Post-industrial" Scenario. Per capita GNP is, therefore, greater in the "Post-industrial" Scenario. The onset of prosperity, however, is somewhat delayed in the "Post-industrial" Scenario. Through 1995, society in this scenario is only slightly more prosperous than society in the "Economic Malaise" Scenario. Once the economic structural changes postulated for the "Post-industrial" Scenario begin to mature in the late 1990s, economic growth in that scenario becomes much more vigorous.

Note that the "Economic Malaise" Scenario, a low population growth scenario, would appear even less desirable if its absolute level of RGNP growth were compared to that of other scenarios.

Exhibit II.2 reflects the energy intensity of real gross national product for each of the six scenarios. This ratio continues its historic decline in each of the scenarios. The solid square symbol indicates the 2010 value of this ratio were it to decrease during the next 30 years at the rate (– .66 percent) it decreased during the past 30 years. The hollow square symbol reflects the 2010 value of the ratio were its rate of decrease to be that (– .24 percent per year) experienced during the 1950-1973 time period.

For each scenario, except the "Small Coal Plants" Scenario, the rate of decrease in the energy intensity of gross national product is greater than that experienced during the 1950-1980 era. Even in the "Small Coal Plants" Scenario, the energy intensity of gross national product decreases at a rate greater than that experienced prior to the 1973 Arab oil embargo. For each scenario, except the "Post-industrial" Scenario, the period of most rapid decrease in gross national product energy intensity is the decade of the 1980s. The more rapid decrease during this decade reflects implementation of energy conservation plans formulated during the 1970s.

The trend to decreasing energy intensity of gross national product reverses in the "Small Coal Plants" Scenario in the mid-1990s. This is attributable to that scenario's low rate of increase in real oil prices, its assumption of successful re-vitalization of heavy manufacturing, and its

Exhibit II.2
Energy Intensity of Real Gross National Product, 1980–2010

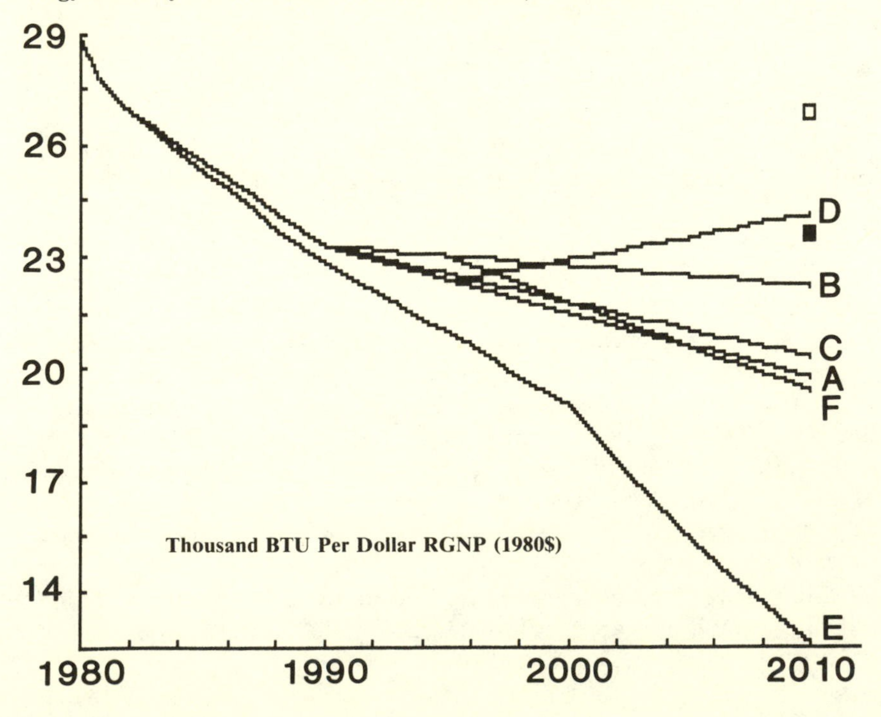

assumption of relatively wide population dispersion. Note that an accelerating rate of decrease in the energy intensity of gross national product occurs after 2000 in the "Post-industrial" Scenario. This reflects the coalescing and maturing of life-style and economic structural changes postulated in that scenario.

Exhibit II.3 reflects patterns of annual energy consumption for each of the six scenarios. Were energy consumption to grow during the 1980-2010 period at the rate it grew during the 1950-1980 period, annual energy consumption in 2010 would equal 171.5 quadrillion BTU (referenced by the solid square symbol). Were it to increase from 1980 at the pre-1973 oil embargo rate, it would in 2010 equal 214.3 quadrillion BTU (far off the scale of the exhibit).

The trends plotted in Exhibit II.3 reflect one of the major themes by which the scenarios were formed. Three scenarios reflect "high" energy consumption, two scenarios reflect "low" energy consumption, and one scenario reflects "average" energy consumption. Even the high energy consumption scenarios postulate a 2010 level much lower than would be produced by historic growth rates. Note that the "Post-industrial" and the "Economic Malaise" Scenarios reflect 30 years of almost no growth in national energy consumption. In the case of the "Post-industrial" Scenario,

Exhibit II.3
Annual Energy Consumption, 1980–2010

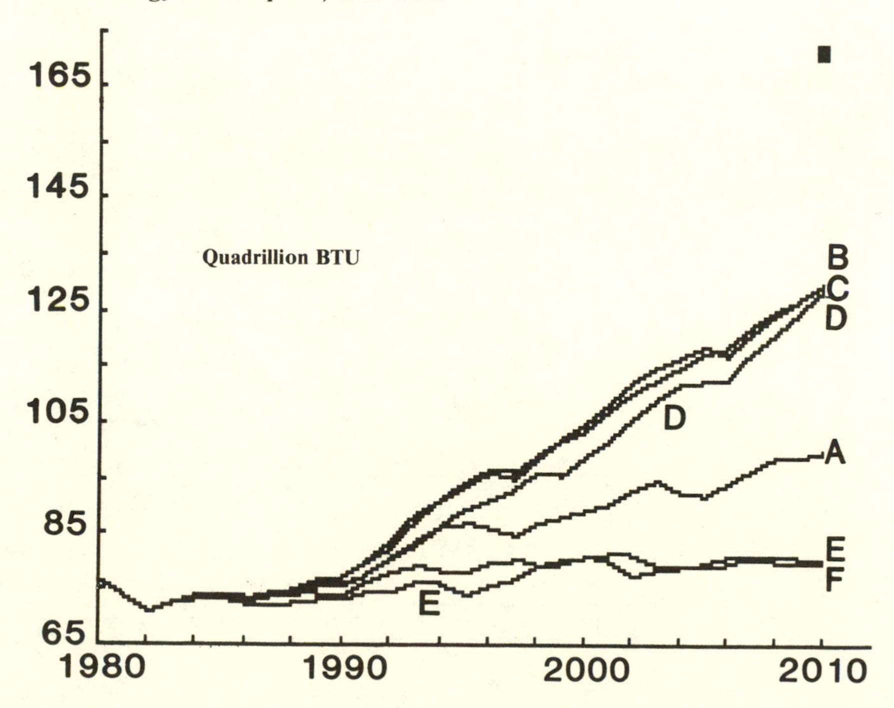

low energy consumption is brought about by life-style and economic structural changes. In the "Economic Malaise" Scenario, of course, low energy consumption results principally from very poor economic growth.

Exhibit II.4 reflects the scenarios' assumptions about the electricity intensity of domestic energy consumption. Electricity intensity is calculated by first summing the energy content of imported electricity and the energy content of all primary fuels input to domestic electricity generation. This sum is then divided by total domestic energy consumption and the resulting fraction is converted to a percentage.

Each scenario postulates an increased electricity intensity in 2010. This intensity is, however, less than if electricity dependence were to continue growing at either the rate of the 1950-1980 era or the rate of the 1950-1973 era (2.16 percent/year or 2.12 percent/year, respectively). The historic rates would imply 2010 electricity intensities at approximately 62 percent.

Electricity's rate of market penetration is postulated to fall off somewhat during the 1980s as electricity prices increase rapidly. In the "Post-industrial" Scenario, the electricity intensity of national energy consumption accelerates noticeably beginning in the late 1990s. This principally reflects the accelerating growth of information and very high technology

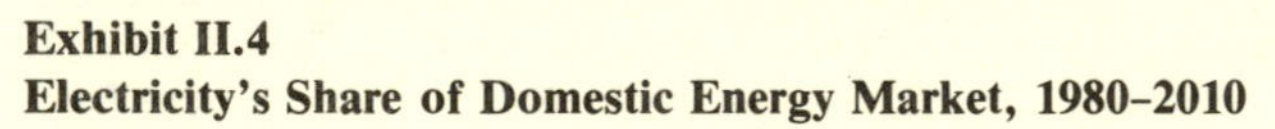

Exhibit II.4
Electricity's Share of Domestic Energy Market, 1980–2010

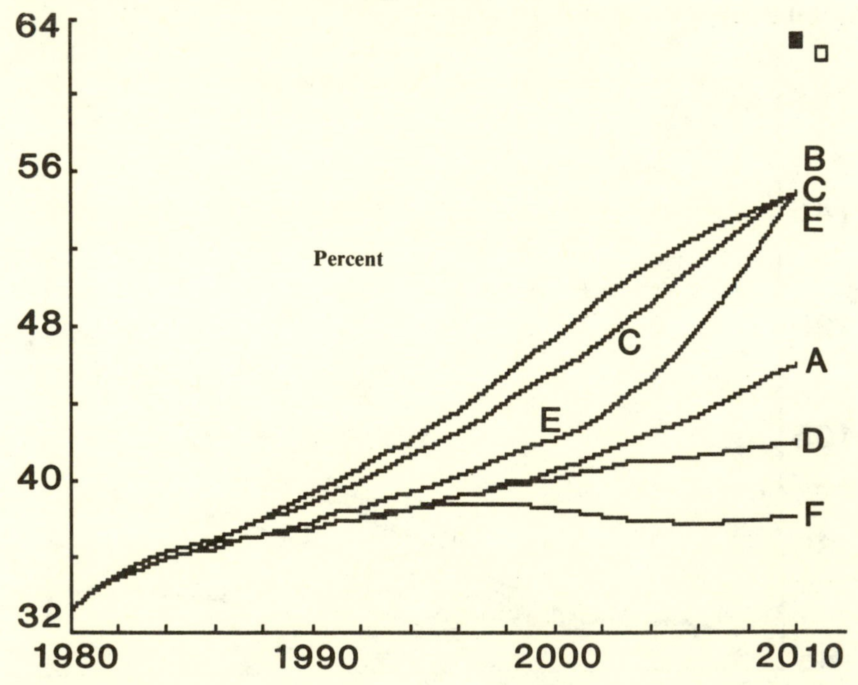

manufacturing activities postulated for that scenario. The electricity intensity actually decreases in the "Economic Malaise" Scenario for a period beginning in the mid-1990s. This reflects the assumption of that scenario that the nation is initially unable to compete effectively in the information and very high technology manufacturing industries and must continue to rely heavily on outmoded manufacturing techniques that are primary energy intensive.

Side-by-side tabular comparisons. The following table, Table II.1, presents a side-by-side comparison of the six scenarios across 28 variables. In some instances, the entries in Table II.1 elaborate on scenario variables that have been compared in the foregoing exhibits. In most instances, however, the table entries allow comparisons among scenarios on the bases of their assumptions regarding policy. The assumptions regarding policy variables, which are specified in rows twelve through 28 of Table II.1, determine scenario-specific policy sets. These are analyzed at length in Chapter III. Issues associated with particular areas of policy (i.e., particular policy variables) are discussed on a scenario-by-scenario basis in the final section (Section D) of this chapter.

D. IMPORTANT SCENARIO ISSUES

This section elaborates on electric power policy issues associated with each of the six assessment scenarios. These issues are addressed by policy variables as specified beginning at row 12 in Table II.1, "Side-by-Side Scenario Comparison." Within the context of any particular scenario, certain issues are more demanding of policy attention than are others; this section focuses on these more demanding issues.

In considering the future importance to the policy maker of one issue relative to another, it is generally essential to describe explicitly the context within which the comparison is to be made. Otherwise, the definitions of the issues are likely to be ambiguous and, therefore, issue comparisons are likely to be meaningless. Consider the issue of electricity price deregulation. This is an important issue in both the "Mega-plant" Scenario and the "Post-industrial" Scenario. In the "Mega-plant" Scenario, however, the facets of the deregulation issue of most interest deal with the wholesale pricing of and fair utility access to power generated by the mega-plants. In the "Post-industrial" Scenario, the price deregulation issues of interest involve more than bulk power sales and more societal parties than electric power institutions. The relative importance a policy maker attaches to the "price deregulation" issue will depend on that policy maker's assumptions about, among other things, the number and financial strengths of the segments of society directly affected by the issue. Because of the need to clearly

Table II.1
Side-by-Side Scenario Comparison

Scenario Variable	Average Future A	Nuclear Resurgence B	Mega-plant C	Small Coal Plants D	Post-industrial E	Economic Malaise F
1. Total U.S. Energy Consumption (quadrillion BTU = Q)	100 Q in 2010. Smooth evolution. .92%/yr.	130 Q in 2010. Highest growth 1990-2005. 1.8%/yr.	130 Q in 2010. Accelerating growth post-1990. 1.8%/yr.	130 Q in 2010. Highest growth 1990-2005. 1.8%/yr.	80 Q in 2010. Growth peaks in mid-1990s. .18%/yr.	80 Q in 2010. Irregular growth rate. .18%/yr.
2. Electrification of U.S. Society (primary fuel input to generate electricity plus imported electricity, divided by total U.S. energy consumption)	46% in 2010. Smooth evolution. 2010 net generation = 4350 bkwh (2.0%/yr).	55% in 2010. Highest growth 1990-2005. 2010 net generation = 6760 bkwh (3.5%/yr).	55% in 2010. Accelerating growth post-1990. 2010 net generation = 6825 bkwh (3.6%/yr).	42% in 2010. Slightly higher growth 2000-2010. 2010 net generation = 5160 bkwh (2.6%/yr).	55% in 2010. Most growth 1980-2000. 2010 net generation = 4160 bkwh (1.9%/yr).	38% in 2010. Most growth 1980-1995. 2010 net generation = 2875 bkwh (.6%/yr).
3. Real Gross National Product Growth Rate (annual compound rate)	2.2% 1980-2010. Highest during 1990s.	2.7% 1980-2010. Highest 1990-2005.	3.0% 1980-2010. Highest 1990-2010.	2.4% 1980-2010. Highest 1990-2005.	3% 1980-2010. Slowly accelerating improvement post-1990.	1.5% 1980-2010. Oscillating 1980-2005. Then, slow improvement.[a]
4. RGNP Energy Intensity (BTU/1980$; assumes expanded service economy)	19,770 BTU/$ in 2010. Steady evolution from 1980.	22,200 BTU/$ in 2010. Re-industrialization economy.	20,340 BTU/$ in 2010. Re-industrialization economy with much expanded service base.	24,240 BTU/$ in 2010. Re-industrialization economy.	12,520 BTU/$ in 2010. Post-industrial economy blooms post-1990. Selective industrial base remains; other industries exported.	19,440 BTU/$ in 2010. Still-born re-industrialization. Still-born post-industrial economy.

5. International Competition and Interdependence (more integrated world economies assumed)	Slow improvement for LDCs. Two or three nations achieve economic parity with United States by 2010.	Moderate improvement for LDCs. United States remains dominant world economic power in 2010.	Moderate improvement for LDCs. United States remains dominant world economic power in 2010.	Slow improvement for LDCs. Two or three nations achieve economic parity with United States by 2010.	Accelerating improvement in economies of LDCs. United States remains dominant world economic power by 2010.[b]	Slow improvement for LDCs. Two or three nations' economies surpass that of United States by 2010.
6. Population Size	275 million in 2010.	290 million in 2010.	290 million in 2010.	280 million in 2010.	250 million in 2010.	250 million in 2010.
7. Population Distribution	Continued migration to South and West. More urban and suburban than 1980.	Continued but slowed migration to South and West. Slightly more urban and suburban than 1980.	Continued migration to South and West. Slightly more urban and suburban than 1980.	Continued migration to South and West. Same urbanization as 1980, more suburban than 1980.	Increasing migration to South and West. Less urban than 1980, more suburban than 1980.	Very little migration to South and West. Urbanization and central city up substantially from 1980.

AFB = Atmospheric Fluidized Bed
PFB = Pressurized Fluidized Bed
PV = Photovoltaic
ROR = Rate of Return
SPS = Solar Power Satellite

[a]Scenario assumption is that this future results principally from policy failure. During 1980s, policy is formulated to aid an industrial future (see, for example, 15F and 28F), which forces beyond policy undermine. During 1990s, policy is formulated to aid post-industrialization. However, the delay has placed the United States at a great competitive disadvantage and, further, has slowed economic progress worldwide.

[b]Scenario assumption is that LDCs prosper because of raw materials and primary manufacturing activities once performed in United States and other developed countries. LDCs, therefore, have economies to support information service and high value added purchases from the United States and other developed countries.

Table II.1 *(continued)*

Scenario Variable	Average Future A	Nuclear Resurgence B	Mega-plant C	Small Coal Plants D	Post-industrial E	Economic Malaise F
8. Oil Price (domestic refiner acquisition cost 1980$/BBL; $33.89 in 1980)	$55 in 2010. Steady escalation from late 1980s (1.6%/yr.).	$60 in 2010. Steady escalation from late 1980s (1.9%/yr.).	$65 in 2010. Steady escalation from late 1980s (2.2%/yr.).	$45 in 2010. Most escalation post-2000 (.95%/yr.).	$65 in 2010. Steady escalation from late 1980s (2.2%/yr.).	$55 in 2010. Steady escalation from late 1980s (1.6%/yr.).
9. Commercially Available Generation Technology (conventional coal and nuclear assumed)	AFB—late 1990s. PFB—Early 2000s. Fuel cells and coal gasifiers—1990. Breeder reactor—2005.	AFB—late 1990s. PFB—late 1990s. Fuel cells and coal gasifiers—1990. Breeder reactor—2000.	AFB—late 1990s. PFB—late 1990s. Fuel cells and coal gasifiers—1990. Breeder reactor—2010. Solar power satellites—2005 + / − .[c]	AFB—mid-1990s. PFB—mid-1990s. Fuel cells and coal gasifiers—1990. Breeder reactor—2010.	AFB—late 1990s. PFB—Early 2000s, Fuel cells and coal gasifiers—2010. Breeder reactor—2010. Solar PV—1993.	AFB—late 1990s. PFB-2005. Fuel cells and coal gasifiers—1990. Breeder reactor—2010.
10. Utility Electricity Price to Industry (1980 cents/kwh).[d]	6 in 2010. Highest escalation pre-1995.	5 in 2010. Highest escalation pre-1995.	7 in 2010. Highest escalation during 1990s.	6 in 2010. Steady escalation 1980-2010.	7 in 2010. Steady escalation 1980-2010.	6.5 in 2010. Price peak in 2000.
11. Social Movements Directly Affecting Electricity (assumes increased recognition of "right" to electricity)	—	Increasing disenchantment with coal in early 1990s.[e] Lessened nuclear resistance.	Increasing disenchantment with coal in early 1990s.[e] Continued nuclear resistance.	Increasing nuclear resistance.	Continued nuclear resistance. Increasing ex-urbanization.	Lessened nuclear resistance. "Right" to electricity particularly strong.

12. Generation Fuel Availability (assumes federal coal slurry eminent domain and 1980s evisceration of Fuel Use Act)	Nuclear fuel reprocessing by U.S. government adopted 1987 + / −.	Nuclear fuel reprocessing by U.S. government adopted 1987 + / −.	Nuclear fuel reprocessing by U.S. government adopted late 1980s.	No nuclear fuel reprocessing allowed in United States.	No nuclear fuel reprocessing allowed in United States.	Nuclear fuel reprocessing by U.S. government adopted 1987 + / −.
13. Power Plant Characterization (slowed phase out of oil and gas assumed)	Smooth rise in coal and nuclear, though both below late 1970s orders. 60% of 1970s nuclear orders delayed (or cancelled).	Strong nuclear rise post-1990. Most coal rise occurs pre-2000.	Coal rises more rapidly than nuclear. Oil and gas phase out reverses in 1990s. 3 SPS (first of many) on line by 2010.	Coal rises rapidly post-1990. Plants are rather small (200-400MW). Slow death of nuclear power. Unconventional sources and imports grow post-1990.	Coal rises until late 1990s. Nuclear peaks in early 1990s. Unconventional sources rise post-1995. Imports peak in early 1990s.	Coal rises until mid-1990s. Nuclear peaks in early 1990s and declines.
14. Non-utility Generation (% of total generation)	4.5% in 2010. Mostly industrial cogeneration.	Insignificant	4.5% in 2010. Mostly industrial cogeneration.	12% in 2010. 8% industrial cogeneration. 4% small power producer.	18% in 2010. 8% industrial cogeneration. 6% small power producer. 4% residential solar.	6.5% in 2010. Mostly industrial cogeneration.

[c]Fusion and SPS research are supported in parallel until mid-1990s, when the SPS election is made.

[d]The spectrum of prices here is derived from responses to Item 50, Appendix B, which was based on one of two EIA electricity price series. Exhibit V.1 of this book is based on the other, longer-running EIA series. Based on this latter series and escalation rates implied by responses to Item 50, Appendix B, scenario 2010 prices would be: $0.085 (A), $0.07 (B), $0.115 (C), $0.085 (D), $0.115 (E), and $0.10 (F).

[e]This comes about because of publicity given credible scientific evidence of long-term coal environment and health hazards. These hazards might be hazards related to acid rain, carbon dioxide buildup, and/or small particulate emissions.

Table II.1 *(continued)*

Scenario Variable	Average Future A	Nuclear Resurgence B	Mega-plant C	Small Coal Plants D	Post-industrial E	Economic Malaise F
15. Conservation Policy	Federal tax incentives allowed to expire in 1980s. Federal policy relies on market forces.	Federal tax incentives allowed to expire in 1980s. Federal policy relies on market forces.	Federal tax incentives allowed to expire in 1980s. Federal policy relies on market forces.	Federal tax incentives allowed to expire in 1980s. Wheeling and "avoided" cost fees encourage non-utility generation.	Federal tax incentives extended indefinitely. Additional cogeneration incentives adopted in early 1990s. Wheeling and "avoided" cost fees encourage non-utility generation.	Federal tax incentives allowed to expire in 1980s. Tax incentives reinstated in early 1990s.
16. Focus of Electric Power R&D Effort	Mostly improvement of present capabilities. Also, breeder reactor[f] and reprocessing. Little emphasis on conservation, new end use technology, or non-hardware research.	Mostly expansion of fission capability. Also emphasis on advanced coal, breeder reactor, and reprocessing. Little emphasis on conservation, end use technology, or non-hardware research.	Mostly development of new capabilities. Also emphasis on fundamental engineering. Some emphasis on end use technology and coal. Little emphasis on breeder reactor, conservation, or non-hardware research.	Mostly expansion of coal capability. No breeder reactor or reprocessing. Little emphasis on end use technology or conservation. Some emphasis on cogeneration and non-hardware research.	Mostly end use technology and conservation. Also emphasis on cogeneration and non-hardware research. No breeder reactor or reprocessing.	Mostly improvement of present coal capabilities. Some emphasis on conservation. No breeder reactor or reprocessing. Little continuity of research.
17. Distribution of R&D Burden	Progressive federal withdrawal, except for breeder reactor and reprocessing.	Continued moderate federal presence.	Very large federal presence.	Progressive federal withdrawal.	Modest increase in federal presence.	Continued moderate federal presence.

18. Nuclear Proliferation Policy	No effective policy.	Effective policy developed by mid-1980s.	No effective policy.	No effective policy.	Effective policy developed in 1990s.	No effective policy.
19. Nuclear Waste Treatment (see also variable 12)	Government-monitored retrievable temporary storage operational by mid-1990s.	Permanent storage operational by 2000.	Government-monitored retrievable temporary storage operational by early 1990s.	Government-monitored retrievable temporary storage operational by 2000.	Government-monitored retrievable temporary storage operational by mid-1990s.	Government-monitored retrievable storage operational by mid-1990s.
20. Coal Waste Policy (assumes no carbon dioxide regulation)	Standards of 1970s are not relaxed. Acid rain regulations by 1990. No new small particle regulations.	Standards of 1970s are not relaxed. Acid rain regulations by 1990. New small particle regulations issued in 1990s.	Standards of 1970s are not relaxed. Acid rain regulations by 1990. New small particle regulations issued in 1990s.	Standards of 1970s are not relaxed. Acid rain regulations by 1990. No new small particle regulations.	Standards of 1970s are not relaxed. Acid rain regulations by 1990. New small particle regulations issued in 1990s.	Standards of 1970s are relaxed in mid-1980s. No acid rain regulations. No new small particle regulations.
21. Utility Regulatory Policy Forum	Mostly state level.	Increasing regional and national regulation.	Increasing regional and national regulation. Some international regulation.	Mostly state level. Increasing regional regulation.	Mostly state, but increasing local regulation.	Mostly state, but increased national regulation.
22. Utility Economic Deregulation Policy	Slow movement to bulk price deregulation.	Very slow movement to bulk price deregulation.	No change in historic policy.	Movement to deregulation of generation.	Accelerating movement to deregulation of genertion.	No change in historic policy.

[f]This scenario contemplates abandonment of the Clinch River project and continuing, but erratic and poorly focused, support of other breeder research.

Table II.1 *(continued)*

Scenario Variable	Average Future A	Nuclear Resurgence B	Mega-plant C	Small Coal Plants D	Post-industrial E	Economic Malaise F
23. Policy on Permissible "Utility" Functions	Slight broadening of scope re conservation activities.	No change in scope.	No change in scope.	Slight broadening of scope re brokering of power.	Significant broadening of scope in 2000s re brokering of power, conservation, and residential generation equipment.	No change in scope.
24. Utility ROR Policy (average)	Higher than during 1970s.	Higher than during 1970s.	Significantly higher than during 1970s.	Higher than during 1970s.	1980-1995: Higher than during 1970s. 1995-2010: Significantly higher.	1980-1990: Higher than during 1970s. 1990-2010: Leveling off, then lower.
25. Regulatory Treatment of Construction Costs	Added to rate base on an ad hoc basis.	General movement to inclusion in rate base as built.	General movement to inclusion in rate base as built.	Added to rate base on an ad hoc basis.	Rarely added to rate base.	1980-1990: Added to rate base on an ad hoc basis. 1990-2010: Rarely added to rate base.
26. Regulatory Treatment of Plant Cancellation Costs	Mostly borne by ratepayers.	—	—	Shared by ratepayers and shareholders.	Shared by ratepayers and shareholders.	Mostly borne by ratepayers.

27. Provision for "Catastrophic" Accident Costs' (property loss and third party liability)	Insurance industry pool.	Insurance industry pool.	Insurance industry pool, with strong federal government coverage of liability in excess of pool.	Insurance industry pool.	Insurance industry pool.	Insurance industry pool.
28. Miscellaneous Federal Policies	Maintenance of selective import tariffs and quotas.	—	—	Increased emphasis on regional environmental data collection and monitoring.	Many policies reflect national commitment to post-industrial economy (e.g., support and retraining of displaced workers). More stringent control of immigration.	Protective tariffs and quotas in 1980s. Increased social support legislation in late 1990s. Increased utility "welfare" responsibilities.

understand the context in which an issue will arise, this section's elaborations are presented on a scenario-by-scenario basis.

The scenarios that postulate the greatest changes in the nation's electricity supply system are the scenarios that raise the largest numbers of demanding electric power policy issues. Exhibit II.5, which concludes this section, presents in tabular form a qualitative, scenario-specific ranking of each of the issues discussed in this section. It reflects substantially more important electric power policy issues associated with the "Mega-plant" Scenario and the "Post-industrial" Scenario than with other scenarios considered. On the other hand, the "Average Future" Scenario, the scenario that postulates a future most similar to today, has associated with it no electric power policy issues as demanding of attention as several that arise within the contexts of other scenarios.

Exhibit II.5 also reflects that issues associated with nuclear power and coal wastes containment are either highly important or moderately important within the contexts of each scenario. Furthermore, unlike the price deregulation issue, the details of these issues are rather consistent across scenarios. Even the nation's evolution to small-scale, decentralized, renewable power sources will not put to rest nuclear power issues. Capital investments already made or, today, ongoing and the containment of nuclear wastes already created ensure that nuclear power issues will continue to confront electric power policy makers for decades to come. Based on current knowledge of the environmental and health implications of coal-based power generation and the large role coal will likely play in future power generation, coal-related environmental and health issues are also likely to be important in any future. As a concession to the ennui of repetition, nuclear material proliferation and coal and nuclear waste containment issues are not discussed anew with each scenario in the following paragraphs.

"Average Future" Scenario (Scenario A). It was previously mentioned that the "Average Future" Scenario has associated with it no issues that, alone, are as pressing as issues associated with other scenarios. This is not to say that electric power policy in the "Average Future" Scenario is unimportant or easily promulgated. This scenario has associated with it far more moderately important issues than any other scenario. Here, electric power decision makers will have to confront a multitude of issues that are important to some segments of society and moderately important to society as a whole, but that never focus widespread and intense debate. For example, the costs of construction cancellation and suspension on nuclear power plants planned during the 1970s will exert upward pressure on electricity prices throughout the 1980s and into the 1990s. At the same time, the capitalized costs of very expensive completed plants will be entering utility rate bases, also exerting upward pressure on electricity prices. Ratepayers will be unhappy about the circumstance of increased electricity rates but will not be

as vocal in this scenario as in others where rates escalate more, power generation health hazards are perceived to be greater, or feasible alternatives exist to utility power generation.

"Nuclear Resurgence" Scenario (Scenario B). Nuclear power issues are obviously the ones of most importance to this scenario. Nuclear fuel reprocessing, nuclear proliferation, and short- and long-term nuclear waste storage must be satisfactorily addressed in this scenario.

Issues related to catastrophic losses of third parties because of events at electric power facilities will also be important. These will be important not because of the fact of major nuclear power plant accidents, which this assessment treats as inconsistent with a nuclear resurgence. Rather, they will be important because of public apprehension (here misapprehension) that the risk of nuclear accidents expands as nuclear generation expands. They will also be made important by this scenario's assumption that the public becomes increasingly aware of health hazards, such as acid rain and small particulate emissions, associated with coal-based electricity generation. Public concern about hydrocarbon-induced illnesses analogous to asbestosis is implied in this scenario. (GE:II-1).

The scenario postulates most environmental and health regulatory responsibility at the state agency level during the 1980s. Nuclear power plant regulation would remain primarily a federal function. The state role will lessen in the 1990s because the long-range air transport of pollutants from combustion of coal (and, to lesser extents, other hydrocarbons) will militate in favor of environmental and health regulation at the regional, multi-state level. Expanded reliance on nuclear power during the 1990s will also expand the general federal regulatory role. Under conditions similar to those of this scenario, there will be protracted policy debate during the late 1980s and early 1990s about regulatory jurisdictions.

Public concern about the environmental and health risks this scenario associates with coal-fired power generation will also intensify policy debate on the wisdom of regulatory encouragement of industrial cogeneration. To the extent that hydrocarbon pollution from electricity cogeneration is (1) greater than from central station power plants or (2) less easily regulated and monitored by environmental agencies, policy favoring cogeneration will draw vocal critics.

Issues that will not be of particular importance in a future like that postulated by the "Nuclear Resurgence" Scenario are also worth noting. In an era when electricity demand growth is significantly greater than it was during the 1970s, policy makers will likely be under little pressure to deal with issues such as the treatment of power plant cancellation costs (they will rarely occur), utility diversification into new business areas (there will be little utility interest or public need), or electricity price deregulation (prices will rise slowly and technological change will not accelerate deregulation).

"Mega-plant" Scenario (Scenario C). This is a scenario with which many

electric power policy issues are associated. In this scenario, regulation of the coal and nuclear fuel cycles will be particularly important and difficult because the scenario assumes mounting public opposition to power generation from either resource. Regulatory hearings and the promulgation of regulations that affect power generation mixes will be contentious. As in the "Nuclear Resurgence" Scenario, the issues of utility insurance for third party losses and regulatory encouragement of industrial cogeneration will be important; the third party insurance issue will actually be more pressing in a future such as the "Mega-plant" Scenario describes because of the strong perceived risks of both nuclear and coal generation—and the absence of short-term alternatives to these technologies.

There are a number of important issues directly related to the mega-plants themselves. During the late 1980s and early to mid-1990s, there will be continuing debate about the levels of funding for the competing mega-plant options and about the length of time during which expensive RD&D will be funded on more than one of the options. Later, international interests will seek a voice in the regulation of the mega-plants. Therefore, the degrees and mechanisms for international regulation will be issues. To the extent that mega-plants are privately funded, the costs of construction work in progress will re-emerge as an important issue. Similarly, treatment of losses associated with potential cancellation of mega-plant construction programs will be an important issue. Provision for losses suffered by third parties because of electric power facility operations will be an important issue, particularly in light of the potential associated with solar power satellites for accidents of international scope. Finally, the pricing of mega-plant power and the access of small utilities to this power will be important issues in this scenario.

"Small Coal Plants" Scenario (Scenario D). Most of the issues of interest in the "Small Coal Plants" Scenario are those that in some manner relate to the scenario's assumptions about distributed coal-based power generation and marked increase in non-utility cogeneration. The increased geographical dispersion of coal-based power plants will raise new debate about the control and monitoring of coal wastes. In particular, the importance of regional environmental regulation and monitoring efforts will increase. Regulatory structures at regional levels are commonly not strong today, so development of these will entail new expenses and affect the sovereign authority of individual states.

This scenario's relatively low oil prices will remove some of the economic incentive for utilities to become less reliant on petroleum-fired generation. Federal and state regulatory policies that weight generation mixes in favor of non-petroleum fuels will be subject to moderate, but wide-spread, opposition.

The "Small Coal Plants" Scenario postulates a significant increase in power production from non-utility sources. This will intensify policy debate about the degree of regulation to which non-utility generators should be subjected. To the extent that non-utility generators are less regulated than are utilities, the issue of utility deregulation will also be important. Furthermore, the control of utility diversification will be important within the context of a future such as this, where electricity demand growth is moderate and a significant percentage of demand is satisfied by sources outside the electric utility industry. Utility diversification will accelerate because of demand for grid management and electricity common carrier services. Also, utilities will be driven to venture beyond their conventional business areas to earn revenue.

A review of this scenario also suggests a number of issues that will be of less importance here than in other scenarios. For example, there will not be much contention over regulatory treatment of construction work in progress; small coal plants can be built rather quickly, and the increase in non-utility electricity generation will decrease the instances of utility power plant construction. Similarly, the treatment of power plant construction cancellation costs will be a low importance issue after the 1980s because the costs will be modest by comparison with today's experiences. Because the scenario assumption is that environmental and health risks potentially associated with coal-based power generation are amenable to technological control, the issue of catastrophic third party losses because of power facility operations will be less important here than in the "Nuclear Resurgence" and "Mega-plant" Scenarios. It will still be non-trivial, however, because a meaningful percentage of the public will remain skeptical about the adequacy of these technological remedies.

"Post-industrial" Scenario (Scenario E). This is another scenario, like the "Mega-plant" Scenario, in which there are many important electric power issues. The "Post-industrial" Scenario postulates a future in which the structure of the electric utility industry will be significantly changed. Therefore, the restructuring of relationships between electric utilities and other elements of society will be important. For example, the proliferation of non-utility generators will lead to competition among utilities and non-utilities in the sale of electric power. Increased electricity generation by residential and commercial power consumers will probably lead to cooperatives or community supply systems existing in parallel with conventional electric utilities and depending on conventional electric utilities for backup power. In each of these instances, regulators will be called upon to balance the interests of the established utility industry and customers especially dependent on it against the interests of new electricity supply parties and their growing networks of customers. The ability of regulators to enforce policy will diminish as the number of alternatives to utility-supplied power increase.

In futures like that postulated by this scenario, there will be accelerated diversification of electric utilities into new business areas, such as energy conservation, management, and brokerage services. This diversification will raise issues about the appropriateness of these activities to regulated utility functions, about the degree of regulation to be imposed on these new activities, and about the regulatory bodies (including some that now have little contact with utilities) that should have jurisdiction over these new business activities.

The distribution of electric power costs in futures like that postulated here will be a contentious issue, particularly during the period of transition from a fully regulated electricity supply system to one that is not fully regulated. Although the treatment of power plant construction costs will not be a significant issue, recovering the costs of plants already constructed will be. Some conventional generating facilities will be outmoded by the solar photovoltaic technologies postulated in this scenario. Some not outmoded will be operated at low capacity factors because of the moderate increase in demand for electric power and the simultaneous increase in non-utility generation. Government actions will transfer from electric utilities to society at large at least some portion of the cost of this obsolete or unnecessary capital investment; these actions will be lightning rods for debate. Similarly, regulatory actions that increase utility rates of return to make them commensurate with this scenario's much riskier utility operating environment will face strong public opposition.

Electric power policies in general will be very important here because this scenario postulates such low energy consumption and such high electricity dependence. The "Post-industrial" Scenario is different from today in many economic and life-style characteristics. Strong government policies will be necessary to reduce the stress of social dislocations during the transition between the two societies. (GE:II-2). These broader government policies will focus on issues such as energy conservation, worker retraining programs, and control of foreign immigration (given a strong U.S. economy and low U.S. birth rates). When these non-electric power issues are addressed, because of the pervasive role of electricity in this society, the use of electric power policy as a tool of social engineering will increase.

"Economic Malaise" Scenario (Scenario F). The generally poor state of the national economy postulated by the "Economic Malaise" Scenario is the principal determinant of important electric power policy issues in that scenario. Therefore, issues, such as the level of utility rates of return, that obviously affect electricity rates will be important. Because the scenario postulates very low electricity demand growth, power plant cancellations will continue to occur in the late 1980s and early 1990s. Ratepayers will, therefore, incur increased charges during the 1980s and early 1990s as the costs of power plants on which construction is abandoned prior to completion are amortized. Regulators can anticipate protracted ratepayer opposition to these charges.

Although electric power RD&D issues will not be a matter of much concern to the average citizen, the size of the federal government's role will be an issue of considerable importance to electric utilities, their regulators, and, when the economic recovery finally begins, industrial customers. In the face of poor economic conditions, no party will undertake much electric power RD&D. Electric utilities and their regulators will advocate transfer of most of the RD&D cost burden to the federal government. The federal government will accept only part of this burden. With needed RD&D going undone, there will be progressive deterioration of the nation's electric power technology position. This will be to the nation's detriment when the economic recovery begins. Indeed, outmoded technology will be viewed by some of the general public prior to that time as a cause of the protracted period of economic malaise.

Regulatory treatment of construction work in progress and regulatory preference for particular power generation mixes will not be issues of much importance in this scenario. They are more important in scenarios postulating significant expansion of the power system. Pollution control issues, on the other hand, are likely to remain important here. Although the scenario assumption is that no real roll back of any health safeguards occurs, the scenario does postulate some relaxation during the mid to late 1980s of environmental and health safeguards related to coal. This is consistent with the twin assumptions for that period of national re-industrialization policy and poor economic growth. However, because the consequences of coal pollution are visible and have international ramifications and because the nation's level of environmental consciousness is only slightly "elastic" to economic conditions, there will be vigorous opposition to lowering of coal-related environmental standards.

Finally, issues dealing with the social support role of electric utilities will be more numerous in the "Economic Malaise" Scenario than in other scenarios and than they have been in the past. These issues, such as lifeline rates and the "right" to electricity, will be particularly difficult to resolve within the contexts of rising electricity rates, poor national economic performance, and an increasing number of individuals in need of government social support.

Issue importance by scenario. Exhibit II.5 reflects the results of an analysis conducted by the assessment staff to determine the importance of each of sixteen policy issues in each of six futures like those described in the assessment's scenarios. As a means of focusing more clearly on potential concerns regarding each issue, the analysis first considered the segments of society that are likely to have large stakes in the outcome of each policy decision; these stakeholders are described in Chapter III, Section G. The analysis scored each issue in each scenario as to its level of importance. An issue was scored to be of high importance (H) if it will affect a large number

Exhibit II.5

Issue Importance by Scenario

Issue	INVESTOR-OWNED ELECTRIC UTILITIES	PUBLIC-OWNED UTILITIES AND COOPS	RESIDENTIAL AND COMMERCIAL SECTOR	THIRD PARTY GENERATORS	COGENERATORS	INDUSTRIAL SECTOR	FUEL SERVICES/OIL AND GAS	FUEL SERVICES/COAL	FUEL SERVICES/NUCLEAR	NEIGHBORS	REGULATORS	RELATED GOVERNMENT INTERESTS	DOMESTIC NON-GOVERNMENT INTERESTS	WORKERS AND UNIONS	CONSTRUCTION	EQUIPMENT SUPPLY COMPANIES	END USE COMPANIES	FINANCIAL INSTITUTIONS	RESEARCH INSTITUTIONS	INTERNATIONAL INTERESTS	A	B	C	D	E	F
12. Should the federal government encourage domestic nuclear fuel reprocessing?	●	●							●		●	●	●							●	M	H	H	L	L	M
13. Should the federal government encourage particular types of electricity generating facilities?	●	●		●	●		●		●		●		●						●		L	M	H	L	H	L
14. Should government encourage electricity generation by non-utility groups?	●	●		●	●						●					●			●		L	M	M	M	H	L
15. Should government encourage energy conservation beyond that motivated by market forces?	●	●	●			●		●	●	●	●	●	●				●		●		M	M	H	M	H	L
16. What should be the focus of electric power RD&D policy?	●	●				●						●	●			●	●		●	●	L	M	H	L	H	L
17. What should be the level of federal support of electric power RD&D?	●	●											●			●	●		●	●	L	M	M	L	M	H
18. What should the federal government do to control proliferation of nuclear materials?									●		●	●	●						●	●	M	H	H	M	M	M
19. What should the federal government do about the treatment of nuclear wastes?	●	●							●	●	●	●	●						●		M	H	H	L	M	L

Column groups: "Principle Affected Stakeholders" (stakeholder columns); "Issue Importance" (columns A–F).

	INVESTOR-OWNED ELECTRIC UTILITIES	PUBLIC-OWNED UTILITIES AND COOPS	RESIDENTIAL AND COMMERCIAL SECTOR	THIRD PARTY GENERATORS	COGENERATORS	INDUSTRIAL SECTOR	FUEL SERVICES/OIL AND GAS	FUEL SERVICES/COAL	FUEL SERVICES/NUCLEAR	NEIGHBORS	REGULATORS	RELATED GOVERNMENT INTERESTS	DOMESTIC NON-GOVERNMENT INTERESTS	WORKERS AND UNIONS	CONSTRUCTION	EQUIPMENT SUPPLY COMPANIES	END USE COMPANIES	FINANCIAL INSTITUTIONS	RESEARCH INSTITUTIONS	INTERNATIONAL INTERESTS	A	B	C	D	E	F
20. What should government do to control the hazards of hydrocarbon combustion in the course of electricity generation?	●	●			●		●	●	●	●	●	●	●		●				●	●	M	M	H	H	M	M
21. What should be the allocation of utility regulatory authority among the various levels of government?	●	●				●		●	●		●		●					●		●	L	M	H	M	H	L
22. To what extent should government regulation of electricity price be relaxed?	●	●	●	●		●					●		●					●			M	L	M	M	H	M
23. Should government encourage a broadening of the spectrum of activities that a utility may permissibly undertake?	●	●	●								●	●	●	●			●	●			M	L	L	M	H	M
24. What return on investment should electric utilities be allowed?	●		●			●					●		●					●			L	L	M	L	M	M
25. How should the costs of utility facilities under construction be allocated between current and future ratepayers?	●	●	●			●					●		●					●			M	M	H	L	L	L
26. How should the costs of abandoned electric power facilities be allocated?	●	●	●			●					●		●					●			M	L	H	L	H	H
27. What provisions should be made for losses resulting from major power facility accidents?	●	●							●	●	●		●	●				●		●	M	H	H	L	L	L

of stakeholders; most affected stakeholders will feel strongly that government action on the issue is called for; and the interests of the stakeholders in the issue will be largely dissimilar. An issue was scored to be of moderate importance (M) if it will cause a more limited number of stakeholders to advocate government action strongly; if affected stakeholders will largely share the same interest in the issue; or if the issue will affect numerous stakeholders but will not affect them sufficiently to mobilize strong demands for government action. Issues not falling in either of these categories were scored as being of low importance (L). None of the issues was sufficiently trivial to be considered of no importance.

BIBLIOGRAPHY

Committee on Nuclear and Alternative Energy Systems (CONAES). *Energy in Transition 1985-2000.* Report of the National Research Council. San Francisco: W. H. Freeman and Co., 1979.

Energy Information Administration (EIA). *1980 Annual Report to Congress: Forecasts,* vol. 3. U.S. Department of Energy DOE/ETA-0173 (80)/3. Washington, D.C.: U.S. Government Printing Office, 1981, Table 3.

Exxon Corporation, U.S.A. *Energy Outlook 1980-2000.* Houston: Exxon Public Affairs Department, December 1980.

Solar Energy Research Institute (SERI). *Building a Sustainable Energy Future.* Andover, Mass.: Brick House Publishing Company, 1981.

CHAPTER III

Potentials for Regret: Policy/Future Mismatches

A. OVERVIEW

In the previous chapter, six sets of assumptions of how the future of the United States might develop over the next three decades were presented. It is obvious that electric power policies of governmental bodies and private utilities will, in part, determine the course of future events. However, these six scenarios assume that non-electric factors (e.g., economic conditions, demographics, and technological change) will dominate future electric power developments. They treat electric power policy as, essentially, a dependent variable.

If the electric power policy makers had sufficient prescience to know which of these futures were going to come about, they would, of course design and implement sets of policies that would best accommodate that future. If policy makers felt confident that the "Average Future" Scenario represented the best estimate of future trends and events, they would adopt a policy set in consonance with those developments (i.e., Policy Set Alpha). If, on the other hand, they felt the future would unfold as described in "Nuclear Resurgence" Scenario, then policy makers would adopt a different set of policies better suited to that future (i.e., Policy Set Beta) (GE:III-1). Similarly, Policy Sets Gamma, Delta, Epsilon, and Zeta would be adopted for each of the other scenarios, respectively. In fact, variables 12-28 in the previous chapter's scenario comparison table represent these policy sets (see Table II.1). (That is, if one considers the items listed in rows 12-28 in the "Nuclear Resurgence" Scenario (B) column, the package of ideas presented represents the major elements of Policy Set Beta.)

Unfortunately, policy makers do not know which of the specified scenarios will best describe future developments. Therefore, it is probable that electric power policy will continue to be formulated in a fashion that is more reactive to events and pressures than it is anticipatory. To the extent that future policy formulation requires the policy maker to develop assumptions about the more distant future, it is probable that those assumptions will encompass the middle-of-the-road, average view of the future. A set of assumptions rather like that embodied in the "Average Future" Scenario seems likely to constitute the implicit set for many electric power policy decisions during the 1980s.

If the future does, indeed, develop as described in the "Average Future" Scenario, it is reasonable to expect that Policy Set Alpha will be, overall, beneficial to society. To check this assumption, a stakeholder analysis (a version of which was presented in Appendix H to the final report to the National Science Foundation) was performed on Policy Set Alpha. This helped to ensure that the set postulated policies that were reasonable in light of the external variables specified for the "Average Future" Scenario. Had many stakeholders been adversely affected by the policy set postulated for the "Average Future" Scenario, that would have indicated that the policy set was incorrectly hypothesized. The first iteration of the analysis indicated that some policy assumptions of Policy Set Alpha should be modified. This was done and again tested by stakeholder analysis. The modified set is reflected in the "Average Future" column of Chapter II's Table II.1. This is the set used as the basis for policy/future mismatch analysis reported in this chapter (GE:III-2).

Obviously, if policies were adopted based on the "Average Future" Scenario and actual future developments were more in accord with one of the other scenarios, stakeholders would be affected differently from what they would have been had a more appropriate set of policies been adopted. Although some stakeholders might actually benefit from the adoption of policies that are non-optimal for the nation as a whole, the overall societal impact would probably be disadvantageous. In Sections B-F, consideration is given to the relative impacts on the various stakeholders of adoption of Policy Set Alpha if alternative scenarios were to come about. As with the overall assessment, these sections focus on the long-term effects of relatively short-term decisions. However, because policy is dynamic and, in time, will be made compatible with exogenous societal variables, these sections pay particular attention to implications that might arise during the 1990s because of policies adopted during the 1980s.

Each of the Sections B-F is divided into four parts. In the first part, a description is given of how the nation might react to the unfolding of a future different from that assumed by the "Average Future" Scenario. In each case, it is assumed that the nation will generally follow Policy Set Alpha until the end of the 1980s. Realization that trends and events are following a different path is assumed to occur at about that time, and this part describes how the nation might react to the new situation.

The second part of each section summarizes, in general terms, the results of a detailed stakeholder analysis of the implications for society during the 1990s of policy/future mismatches. The full stakeholder analysis is Appendix H to the final NSF report. It should be borne in mind when considering the second part of each section that the impacts of interest are those that derive from the adoption of the non-optimal policy set. They are not impacts that arise simply because of external variables, which are beyond the control of policy. Thus, although investor-owned electric utilities may well be better off, overall, if the booming economy postulated by the "Mega-plant" Scenario comes about, rather than the less favorable economy postulated in the "Average Future" Scenario, investor-owned utilities would, presumably, still experience adverse impacts associated with the combination of Policy Set Alpha and the "Mega-plant" Scenario.

The third part of each section lists possible modifications which might be made in Policy Set Alpha to accommodate better the possibility that the given alternative scenario might develop. Finally, the last part of each section briefly discusses implications of these modifications.

Section G of this chapter is a summary of the stakeholder impacts detailed in Appendix H of the NSF final report and discussed in Sections B-F of this chapter. The final section of this chapter, Section H, discusses modifications to Policy Set Alpha which might make it reasonably responsive to any of the hypothesized futures.

B. POLICY SET ALPHA/"NUCLEAR RESURGENCE" SCENARIO IMPACTS

Probable national reactions. The primary differences between the "Nuclear Resurgence" Scenario (Scenario B) and the "Average Future" Scenario (Scenario A) are that the former reflects a larger gross national product, greater total energy consumption, and a greater percentage of primary energy conversion into electric power. These circumstances will result in a major increase in the demand for electricity in the "Nuclear Resurgence" Scenario.

The increased gross national product, which will drive the other listed increases, reflects the success of the nation's overall economic policies. During the 1980s, it is assumed that the nation will continue both to develop new industries based on new high-technology innovations and to revitalize the more traditional industries (e.g., steel, automobiles, aluminum, machine tools, ceramics, glass, etc.), with considerable success in both endeavors.

Because there is a substantial amount of unused capacity at the beginning of the decade, the importance of the large increase in demand for electric power will not become evident until late in the 1980s. (Otherwise, another policy set, such as Policy Set Beta, would be adopted at an early time.) Utilities will consider the early increases in demand to be temporary and will attempt to meet requirements by more effective use of existing facilities and reduction of reserve margins. These actions will generally be supported and encouraged by regulatory bodies and environmental groups. However, as the economy continues to strengthen and demand continues to grow, black outs, brown outs, and power shortages will convince all parties of the need to expand power production, transmission, and distribution facilities rapidly. Crash programs will be undertaken to construct and put into operation new facilities of all types. In addition, heretofore neglected, wide-ranging conservation programs will also be undertaken. In the short term, oil and gas facilities will be constructed. Potential cogeneration and third party power producers will face high barriers to entry because of previous policies and, therefore, will be unable to take full advantage of the changing situation. Nevertheless, many industries may be pressed to consider, if not actually to construct, their own generating facilities to ensure immediate electricity requirements for production.

The intermediate response of the utilities will be to reconstitute or to construct new conventional nuclear and coal facilities, and, because of the evident need, the public will assent to constrution programs for both types of plants. However, communities also will have become more concerned about health and environmental concerns associated with coal plants (e.g., acid rain, carbon dioxide buildup, small particulate emission, ash leaching, smog, etc.). Despite society's concern about nuclear plants, growing recognition of improving safety records, development of generally acceptable waste disposal processes, technological advances in breeder and reprocessing programs, as well as failure to overcome coal-related problems, will make this power alternative increasingly attractive. Therefore, a larger number of utilities will reassess earlier nuclear power plant decisions. Construction will be resumed on most plants on which work was suspended during the 1980s and new construction and operating permits will be secured. By the mid-1990s, nuclear fuel will be reprocessed routinely, and by 2000 a commercial-sized breeder will have been operating long enough in the United States to show its technical and economic practicality. (Because breeder research is quite limited under Policy Set Alpha, this reactor will likely be of foreign design.) By 2010, generation in nuclear power plants will almost equal that in coal plants, and there will be every indication that the proportion of nuclear power will continue to grow. Moreover, the strong growth of nuclear power will decrease the need to import power, even though the importation of power will have surged in the 1990s; by 2010 it will fall to approximately what it is today.

During the 1990s, the rush to construct needed facilities will promote turmoil for all parties involved (e.g., architectural engineering firms, construction firms, equipment supply companies, labor, etc.). Utility funding needs will be particularly critical. Not only will a large increase in facilities be required, but the crash nature of the building program will result in high unit costs. (These needs will be somewhat abated by the existence of partially completed nuclear plants.) To support these expanding costs, rates to users in the near term will be rapidly and substantially increased. Public utility commissions will be caught between the need for these increases and the public's reluctance to accept them. Rate increases will be delayed and/or greatly reduced, thus threatening the utilities' abilities to support construction. Moreover, as rate increases are approved, customers will unilaterally take steps to minimize use. Although, over time, demand will increase dramatically, the cyclical spurts and slumps in demand as prices change will cause further difficulty for utility planners. By the turn of the century, however, many new nuclear plants will be on line and electricity prices will have stabilized.

An initial shift in the environmental control responsibilities to state and local agencies will put those regulatory bodies in a position similar to that of utility commissioners. It may be assumed that the same type of pressures and interactions that exist today will exist in the 1990-2010 time frame. By the mid-1990s, power shortages will have become evident, and the crash programs to construct additional conventional facilities will overwhelm regulatory forums at the state and local levels. This crisis will lead to a re-emphasis on regulatory responsibility at the federal level.

Summary of impacts. The impact of the nation's following Policy Set Alpha during the 1980s, when in fact the "Nuclear Resurgence" Scenario unfolds, will relate primarily to the United States' failure to recognize and to plan for its rapidly expanding electric power needs. This unexpected demand for power will force utilities to bring new conventional generating facilities on line in a minimum amount of time. This will result not only in higher construction costs for these facilities, but also in the construction of oil, gas, and coal plants, despite the high fuel costs of the first two and the health and environmental problems of the third. In later years, as more nuclear plants come on line, there will be pressure to decommission the hydrocarbon-fueled plants even though their serviceable lives have not been expended. Obviously, utilities, particularly those that are investor-owned, will be faced with the problem of financing and managing construction of these new facilities in the 1990s and then recovering expended construction costs ten to fifteen years later. Moreover, because RD&D in coal-related health and environmental problems were not stressed under Policy Set Alpha, resistance to new coal plants will be high. Likewise, failure to adequately fund RD&D in nuclear-related health and safety problems will increase local resistance to this energy source as well, even though overall public acceptance is postulated. As new coal and nuclear plants are built, workers and the surrounding neighbors will be affected to a greater degree than would be the case if additional RD&D were conducted in the health, safety, and environmental areas during the 1980s. The impact of an unexpectedly high power demand will, to a large extent, be ameliorated by the existence of incomplete nuclear plants.

Many of the problems faced by the utilities will be mirrored in the problems of public utility commissions (PUCs), public service commissions (PSCs), and government environmental regulators. The need for capital to finance new construction will force many utilities to seek significant rate increases. These will be resisted by consumers who will be facing lower service standards (because of power shortages) at the same time they are asked to pay more for the service. The public utility and service commissions will have to develop policies that will generally satisfy both groups. In addition, the PUCs and PSCs will later face the problem of treating all parties fairly when fossil fuel plants built during this era are decommissioned prior to the end of their serviceable lives. At the same time, those agencies charged with promulgating and enforcing health, safety, and environmental protection regulations will find themselves torn between the nation's need for electricity and the public's resistance to new plants, particularly those utilizing coal.

Most stakeholders will be affected to some extent by the electricity shortages and price fluctuations associated with the situation described above, but industrial consumers and power equipment manufacturers will be particularly heavily affected. In case of power shortages, industrial consumers are traditionally the first to have their supply decreased or terminated. Those industrial groups that actually construct their own electrical generation facilities may not be able to fully realize their investment in the facilities before the utilities acquire sufficient capacity to meet all electric requirements at a lower cost. In addition, increased use of coal in power plants will either restrict usage in industry and/or cause higher pollution abatement costs.

For power equipment manufacturers, the early, drastically low nuclear plant construction estimates and the reluctance of the government to support nuclear RD&D will limit their ability to meet the newly developed needs for state-of-the-art nuclear plants. Hence, a major portion of the nation's nuclear equipment will be imported from foreign suppliers. A similar situation will exist for U.S. research institutions. After years of declining research support, the country's research firms will be in a poor position initially to react quickly. Thus, it is conceivable that the U.S. utilities would go elsewhere, perhaps France or Japan, for the necessary expertise to meet nuclear program requirements.

Suggested modifications to Policy Set Alpha. Modifications to Policy Set Alpha that will make it more compatible with a future similar to that described in the "Nuclear Resurgence" Scenario primarily involve means for effecting the smooth, timely, and efficient increase in nuclear power facilities to meet larger power needs than postulated in the "Average Future" Scenario. Because additional coal plants also will be needed to meet power demands, steps to minimize the adverse impacts of these plants should also be undertaken. Moreover, the increases in electricity requirements could be partially abated by expanded conservation efforts.

Although the "Nuclear Resurgence" Scenario assumes public acceptance of a much larger nuclear system than we have today, that acceptance must be based on a general belief that the presently perceived problems of safety, waste disposal, proliferation, and environmental contamination have been solved. The substantiation of this belief must be based on records of safe operation and the institution of credible programs. It is improbable that the RD&D program postulated in Policy Set Alpha will be sufficient to establish credibility for the nuclear power program. Hence, the practical availability of the nuclear power option will require an extended, continuing RD&D program throughout the 1980s. In addition, it may be desirable for the government and the utility industries to take steps to ensure better the availability of nuclear power plants in case of unexpected increases in power demand. That is to say, subsidies or other types of incentives might be given to utilities either to complete or partially complete plants in excess of near-term requirements so that they can be activated in a short time if needed. This action might not only prevent power shortages, but also might alleviate the need for new fossil fuel plants which will not be required in the longer run.

The possible need for a significant increase in coal plant generation during the 1990-2010 time frame points out the desirability of increased RD&D into methods for minimizing health and environmental risks associated with facilities of this type. This RD&D should be directed not only to reducing pollution from new plants but from old plants as well. In addition, methods should be sought to reduce industrial coal pollution. Programs to accomplish these ends are, of course, already under way, but the expansion of coal utilization envisioned in the "Nuclear Resurgence" Scenario will make amelioration of the problems even more vital.

Early identification of changes in expected electricity demand by rate-setting groups and power regulatory agencies will facilitate dealing with the associated problems. For example, efforts to complete or partially complete "stand-by" nuclear plants will probably require sympathetic rate policies by public utility commissions, and early realization of the need for a considerable number of new coal plants will permit the development of plans to minimize the impact of construction and operation. The recognition of potentially short-term shortages of electricity may encourage extended conservation programs. Overall, regulatory and rate-setting bodies should be encouraged to examine and to monitor long-term trends and to encourage utility and public action that will improve the utilities' ability to meet different power demand levels.

Implications of suggested modifications. For the most part, the modifications to Policy Set Alpha listed above involve laying a technical, educational, psychological, and facility infrastructure for expanding nuclear power generation relatively easily and quickly if the nation's electricity requirements turn out to be considerably larger than predicted in the

"Average Future" Scenario. Two major actions were discussed: increasing RD&D efforts and subsidizing completion or partial completion of nuclear plants that would not be needed to meet the "Average Future" Scenario power requirements. If the additional power predicted in the "Nuclear Resurgence" Scenario is not needed, the funding of these actions, particularly the latter, will be perceived to have been ill advised. Public respect for the policy process will lessen. Moreover, residual public opposition to an expanded nuclear power program, despite RD&D efforts, will appear to have been well founded. (It might be noted that the additional RD&D efforts to ameliorate nuclear-related problems also may be of value if the "Average Future" Scenario comes to pass.) Likewise, a smaller demand for power may reduce the incentives for RD&D in the coal area and for more extensive conservation programs. However, again, efforts in each of these areas would not appear wasted under any reasonable future.

C. POLICY SET ALPHA/"MEGA-PLANT" SCENARIO IMPACTS

Probable national reactions. As was the case for the "Nuclear Resurgence" Scenario, the basic differences between the "Mega-plant" Scenario and the "Average Future" Scenario are that the former reflects a considerably larger GNP, energy consumption, and electric power usage. Again, it is assumed that the increased GNP reflects national success in both re-vitalizing the traditional industries and establishing healthy high-technology and service industries. Here, these last industries come to represent a larger segment of the economy than in the "Nuclear Resurgence" Scenario.

As in Section B, it is assumed that the large increases in demand for electric power will not become evident until late in the 1980s. Utilities will consider early increases temporary in nature and will attempt to meet demands by more effective use of existing facilities and reduction of reserve margins. These actions generally will be supported and encouraged by regulatory bodies and environmental groups.

As the economy continues to strengthen, demand will continue to grow and black outs, brown outs, and power shortages will convince all parties of the need to expand power production, transmission, and distribution facilities rapidly. Hence, crash programs will be undertaken to construct and put into operation new facilities of all types. In addition, wide-ranging conservation programs neglected during the 1980s will be undertaken. Since oil and gas plants can be built in the shortest periods of time, many of these will be constructed. Third parties will view increased demand as an opportunity to construct new facilities, and some industries will construct their own generating facilities to ensure power supply. The bulk of new construction, however, will involve conventional coal and nuclear plants. This construction will occur despite increasing public and international alarm about health and environmental risks associated with coal emissions and nuclear waste handling. Although the public will assent to construction of plants of these types because of electric power needs, it will also press for development of technologically sophisticated power sources, such as solar satellites, that will not involve nuclear and coal's obvious risks. In the early 1990s, public fervor for development of large-capacity alternatives to coal and nuclear power plants will lead to costly and, occasionally, contradictory federal incentives to accelerate the alternatives' commercialization. The events of the "energy independence" era of the mid-1970s are analogous to those to be anticipated in the early 1990s.

The rush to construct the needed conventional facilities will again cause turmoil for all parties involved (e.g., architectural engineering firms, construction firms, equipment supply companies, labor, etc.). Utility funding needs will be particularly volatile. Not only will a large increase in facilities be required, but the crash nature of the building program will result in high unit costs. To support these expanding costs, rates to users will be increased. Public utility and public service commissions will be caught between the need for these increases and the public's reluctance to accept them. Moreover, as major rate increases are approved, customers will unilaterally take steps to minimize use. Although, over time, demand will increase dramatically, the cyclical spurts and slumps in demand as prices change will cause further difficulty for utility planners.

A shift in the environmental control responsibilities to state and local agencies will put those regulatory bodies in a situation similar to that of utility commissioners. It may be assumed that the same type of pressures and interactions that exist today will exist in the 1990-2010 time frame. Therefore, as power shortages become more evident in the 1990s, much of the regulatory responsibility that will have been transferred to the states in the 1980s will be returned to the federal government.

Summary of impacts. The shortcomings of Policy Set Alpha with regard to the actual unfolding of the "Mega-plant" Scenario lie basically in the failure of the policy set to foresee and plan for the markedly increased demand for electric power. As a consequence of this failure, utilities will be required to take emergency measures to provide the needed power. Industrial, commercial, and private users will face power shortages and failures. Regulatory bodies will be faced with strong pressures from utilities to increase electric power rates and relax health and safety standards, while the public exerts strong pressure to hold firm both rates and standards. Although the "Mega-plant" Scenario assumes a successful program to develop a new power source, a solar satellite system, the time required for such a technologically sophisticated system to provide a major fraction of the nation's future power needs will necessitate the construction of conventional facilities in the interim. Thus, utilities will be faced with the need to construct new plants under urgent conditions, with the knowledge that these plants will probably have to be decommissioned prior to the ends of their useful lives. The problems facing the utilities will include the requirement to raise large amounts of capital in a short period of time; the necessity to plan for, supervise, and monitor major construction activities; and the need for significant rate increases to fund these activities. These problems will definitely be aggravated by the developing public resistance to conventional power facilities.

Regulatory bodies will face similar problems. They must balance the public's demand for adequate power against the public's persistent concern over expanding conventional facilities. Therefore, rate-setting bodies will be caught between the requests of the utilities to fund system expansion and the public's reluctance to bear the burdens caused by what will appear to be inadequate utility and government planning. Moreover, agencies responsible for environmental regulation will be faced with pressures to accept standards degradation to allow expedient construction and operation of plants. Even those government agencies not directly involved in electric power regulation will be subject to turmoil as it is probable that the Congress and the executive branch will seek to assign responsibility for the failure to plan properly in these matters.

Although the utilities, both public and private, and regulatory bodies will be most directly affected by the upheaval described, most of the other stakeholders will also be affected. In some cases, the impacts will be positive, where certain groups, such as the oil and gas suppliers, will be required to provide a larger share of the prime energy needs than they would under a more appropriate policy set. However, in a majority of cases, the impacts will be negative because most business and private interests are poorly served by unexpected turbulence in market and economic environments. Foreign relations will probably also be strained by relaxed environmental controls; however, the unanticipated large need for electric generation, transmission, and end use equipment will present foreign suppliers with unexpectedly favorable market opportunities.

Suggested modification to Policy Set Alpha. Modifications to Policy Set Alpha that would make it more consonant with the unfolding of the "Mega-plant" Scenario fall into two general categories: those involving supply and those involving demand. In the latter case, the relaxation of efforts to promote conservation under Policy Set Alpha will limit the nation's ability to control power demand as economic conditions improve rapidly. Hence, the possibility of unexpected major increases in demand could be addressed by modification of Policy Set Alpha to put greater emphasis on conservation efforts. These efforts might include upgrading in the energy efficiency of both old and new industrial plants, commercial establishments, and private dwellings. In addition, research, development, and demonstration programs could be expanded to provide more energy-efficient products and processes and to determine effective means of improving overall energy efficiencies in both the private and industrial areas. Primary emphasis should be placed on improved industrial capabilities since this is where the bulk of new power demand is expected to occur.

On the supply side, policies are needed to increase the acceptability of coal and nuclear plants, to encourage the development of other power sources, and to promote the timely development of a technologically sophisticated system (e.g., the solar power satellites or other large power supply systems, such as fusion). The "Mega-plant" Scenario postulates that public acceptance of coal and nuclear plants is severely diminished by growing concern over health and environmental impacts of coal plant emissions and by a failure of the nation to develop satisfactory waste disposal procedures for spent nuclear fuel. Under this scenario, it is assumed that no practical environmental system can be developed in the time frames involved to completely solve either of these problems. However, it does seem reasonable that timely RD&D programs, properly funded and administered, could develop equipment and procedures that would decrease the adverse impacts of conventional electricity generation. If these impacts could be sufficiently alleviated, the coal and nuclear plants needed to meet the electric power requirements postulated in the "Mega-plant" Scenario could be constructed and operated with minimum affront to the environment. This would not only be beneficial in its own right, but might also extend the period in which plant operation is acceptable and, hence, minimize the financial problem connected with premature plant closing.

In regard to other power sources (e.g., third party generation, cogeneration, etc.), Policy Set Alpha could be modified to encourage sufficient activity to provide a knowledge and regulatory infrastructure, along with an experience and equipment supply base, on which future expanded production could be founded.

If the solar satellite system is to be developed in the time frame described in the "Mega-plant" Scenario, it is probable that RD&D levels over the next decade must be increased over those described in Policy Set Alpha. The time required to develop, test, and initiate a system of this scope will, barring unexpected breakthroughs, require long-term, consistent RD&D program support.

In regard to the problems associated with cross-border contamination, which would be exacerbated by the increased number of coal-fired plants, it appears that RD&D efforts undertaken to decrease the domestic impact of coal facilities emissions would also alleviate many potential conflicts. Other preventative actions that might be taken include continued efforts to understand the nature of coal emissions impacts, care in the selection of sites, and reasonable indemnification agreements. Groundwork should be laid to facilitate more international cooperation in utility regulation, both because of the environmental problems that become more pronounced in the "Nuclear Resurgence" Scenario and because of the multi-national impacts likely to be associated with very large capacity power generation.

Implications of suggested modifications. The most important single modification to Policy Set Alpha suggested above is a major increase in RD&D efforts. This increase should focus on the areas of coal emission control, nuclear waste disposal, very large power system development, and conservation. If the electric power demand turns out to develop at a rate equal to or lower than projected by the "Average Future" Scenario, much of this additional RD&D effort may not have been justified. In the particular case of the mega-plants, it should be borne in mind that prudence will require funding of research not only in the solar satellite area but also in other advanced project areas, such as fusion power or ocean thermal power. Although costs for research support in the 1983-1990 period will be small compared to later implementation costs, the cumulative costs even in the earlier period will be far from trivial. It should also be noted that there is no guarantee that any of the mega-systems will prove technologically, economically, or politically practical.

D. POLICY SET ALPHA/"SMALL COAL PLANTS" SCENARIO IMPACTS

Probable national reactions. The major differences between the "Small Coal Plants" Scenario and the "Average Future" Scenario are the markedly larger overall demand for energy projected in the former and the concurrent increasing reluctance of the public to accept nuclear power as a suitable energy source. Because of the decreased fraction of total energy converted to electric power in centralized stations, output from these stations is only increased slightly in the "Small Coal Plants" Scenario. The lower output from nuclear plants is more than made up by increased dependence on oil and gas plants, hydropower, increased imports, and non-utility generation.

As in previous sections, it is assumed that differences between scenarios are relatively small until the end of the 1980s. During this period, the nation will continue its efforts to develop a new technological base as well as refurbish its traditional industries. However, by the early 1990s, it will become apparent that Japan and some European Community countries will have advanced more rapidly into the so-called high-technology areas and that, for the United States, these sectors will present only limited promise. However, it will also have become apparent that the re-vitalization of the more traditional U.S. industries (e.g., steel, automobiles, aluminum, machine tools, ceramics, glass, etc.) will have proceeded to the point where the nation is once again a world leader. Hence, there will have been and will continue to be a resurgence in facility construction and production in these areas. Because these industries utilize most of their energy in primary forms and because oil prices escalate very moderately, electric power demand will increase only slightly. Moreover, because of the energy intensities of these industries, total energy growth will more closely track gross national product growth.

The "Small Coal Plants" Scenario assumes that throughout the 1980s public concern over nuclear power safety, long-term environmental effects, and possible nuclear proliferation continues to grow. This concern will result in utilities continuing to postpone new nuclear construction. Nuclear facilities will also face increasingly restrictive operating rules and, periodically, shutdowns. Thus, although electric power demand growth will be moderate in the 1990-2010 period, local and short-term shortages will develop. This will lead many industrial concerns to construct their own electricity-generating facilities or to contract with third party producers for needed power. On the other hand, many utilities will react with stopgap measures that increase the use of relatively low cost gas and oil and, where practical, will import electric power from Canada.

However, for long-term needs, the utilities will rely principally on coal. Although total coal-generated electric power will not be much greater than in the "Average Future" Scenario, increased use of coal by the industrial sector will magnify and alter coal emissions problems. The nature of coal transport, storage, and burning problems will also be altered as utilities increasingly utilize relatively small (200-400 megawatts) plants to accommodate slow demand growth and better meet the needs of a more dispersed populace. Most of the new small plants will take advantage of the more sophisticated fluidized bed technologies, first atmospheric and, later, pressurized. The adoption of these technologies, as well as development of reasonably effective means of controlling emissions from existing coal plants, will help to minimize resistance to the wide-spread use of the coal resource for electricity generation.

Summary of impacts. The impact on the United States of following Policy Set Alpha through the 1980s and finding in the early 1990s that conditions were developing more in accord with the "Small Coal Plants" Scenario would appear to be low or moderate for most of the stakeholders considered. Because the overall growth of utility-generated power is approximately the same in both scenarios, the only major differences in utility programs as the "Small Coal Plants" Scenario unfolds would be the decreased role of nuclear power and the shift to smaller, geographically dispersed coal plants. Since Policy Set Alpha depended, in part, on moderate growth of nuclear power, it is probable that funds will have been expended for nuclear plant planning and construction that will prove inappropriate. Reduced load utilization, cancellation of plant construction in progress, and increased operating costs will add to the utilities' expenses. Rate-regulating bodies will be faced with the problem of fairly allocating these costs between customers, stockholders, and the general public.

The hardest hit group will, of course, be those organizations responsible for servicing the nuclear fuel cycle. During the 1980s under Policy Set Alpha, they will have been encouraged to develop fuel supply and service facilities which will ultimately not be used. In particular, all fuel-reprocessing activities will have to be abandoned. This will, of course, materially affect both domestic government agencies involved in reprocessing and foreign nations that depended on U.S. reprocessing of their spent fuel. Initially, the repercussions may be much more political than economic.

Another problem that may be associated with the unexpected evolution of this scenario is coal pollution. Although the scenario projects no greater use of coal in power production than does the "Average Future" Scenario, it does assume wide-spread dependence on coal (and other hydrocarbons) in the industrial sector. Therefore, total emissions problems, as well as other coal related problems, will be greater. This could cause increased restrictions on industrial use of coal as an energy source.

Suggested modification to Policy Set Alpha. There are two basic modifications that could be made of Policy Set Alpha to make it more appropriate for the "Small Coal Plants" Scenario. First, it is desirable to diminish the early commitment to nuclear power until long-term public reaction is better ascertained. Second, additional RD&D dedicated to better understanding of coal- and other hydrocarbon-related health and environmental problems and to the development of equipment and techniques to mitigate these problems is also advisable. Taken together, these modifications imply that the policy set adopted should examine as early as practical the possibility that expansion of nuclear power will not be politically feasible and that fossil fuels, principally coal, will have to bear a much larger portion of energy supply than envisioned. Obviously, these modifications imply a postponement of reprocessing activities and even more limited support of nuclear breeder reactor research. Moreover, additional encouragement also should be given to conservation programs and to non-utility generation. Simultaneously, groundwork could be laid for increased importation of power.

Implications of suggested modifications. The policy of discouraging nuclear power expansion during the 1980s will prove imprudent if demand for electric power increases materially, if environmental and health problems associated with coal prove insurmountable, and if nuclear power production proves to be relatively benign. Discouragement of nuclear power will result in decreased RD&D in this area, and if this resource must be utilized, plant safety and efficiency may not be as well developed as they might have been. Likewise, delay in establishing a reprocessing program may cause fuel users to turn to other nations for reprocessing. This may lead not only to loss of income for the nation but may also decrease U.S. control over weapons-grade nuclear material.

Expanded RD&D in coal emissions and other hydrocarbon-related problems may result in a misallocation of RD&D resources. (However, it is hard to believe these activities would not be justified in the long run.)

E. POLICY SET ALPHA/"POST-INDUSTRIAL" SCENARIO IMPACTS

Probable national reactions. Of all the scenarios considered, the one most divergent from the "Average Future" Scenario is the "Post-industrial" Scenario. Although the gross national product projected for the "Post-industrial" Scenario is 26 percent higher than that projected for the "Average Future" Scenario, the total energy consumption is 20 percent less. This reflects a much lower BTU/real dollar GNP ratio than projected in the "Average Future" Scenario and one that is less than one-half the 1980 ratio. This low ratio is assumed to be made possible by the vigorous expansion of the service sector of the economy; by the development of new high-technology industries, which produce high value-low energy products (i.e., integrated circuits, bio-engineered specialty chemicals, practical results based on advanced industrial RD&D, etc.); by strongly encouraged conservation and efficiency improvement programs; and by significant increases in the importation of energy-intensive products (i.e., steel, cement, glass, and automobiles). There is also expansion in the use of imported electricity in areas geographically near foreign power sources. Despite the much lower total energy consumption predicted in the "Average Future" Scenario, the two scenarios assume nearly equal electric power generation. However, the "Post-industrial" Scenario assumes that almost a fifth of the power will be provided by non-utility generation, with over half of this amount being produced by small power producers and residential or neighborhood solar facilities.

As in the "Small Coal Plants" Scenario, it is assumed that during the 1980s the nation will continue its efforts to develop new technologies, as well as refurbish its traditional industries. However, in this scenario, it will become increasingly apparent during the 1980s that the nation will not be able to compete efficiently with foreign producers in the more mature industries. Although the government will seek for several years to protect these industries by trade restrictions, indirect subsidies, and favorable regulatory practices, by the end of the decade, federal policy will shift toward emphasis on development in the high-technology industries at the expense of the more traditional ones. Recognizing the thrust of the policy change, both the government and business will expend major efforts to retrain displaced people. The government will also help affected corporations to find viable roles in the economic transition. However, major economic dislocations will occur and many employees and companies will not be able to make the necessary conversions. Considerable social and economic dislocation will result, particularly in the short term.

The movement from concentrated urban development to suburban and exurban environments will continue throughout the remainder of the twentieth century. Moreover, the new and emerging industries will accompany the population movement and will site facilities in a dispersed manner, too. This diffused society is feasible because of technological development in the electric power industry, coupled with advances in small solar photovoltaic power systems and strong policy support for integration of non-utility generated power into the overall grid. Concurrent development in power systems integration will make the systems more compatible. Utilities will face requirements for considerably more wheeling and load management practices in the 1990-2010 time frame than they have previously. The energy service functions of utilities will expand significantly.

However, central station plants will continue to produce the bulk of the nation's electricity needs. The development of alternate power supply systems, however, will be pervasive enough to affect the utilities' traditional markets and to change forever many operating practices. In addition, the apparent feasibility of alternate electric energy sources will increase public opposition to coal and nuclear plants. This opposition will lead to abandonment of nuclear fuel reprocessing programs.

Since the utilities will have based expansion plans on the "Average Future" Scenario, the unfolding of the "Post-industrial" Scenario will leave them with excess and/or unwanted capacity. Since this capacity will normally be included in the base on which electricity rates are calculated, the price of central station power will be relatively high, presenting a problem to consumers and increasing the relative attractiveness of solar systems. Since, in general, these higher prices will more often fall on the sectors of the society most distressed by the industrial shift previously discussed, government subsidies of some type will probably be enacted.

Summary of impacts. The impacts of the "Post-industrial" Scenario's evolving after the nation's having followed Policy Set Alpha during the 1980s will fall most heavily on the utilities and the government bodies charged with their regulation. Based on the expected evolution of the "Average Future" Scenario, utilities will have built central station generating plants that will not be needed and will have constructed transmission facilities that may not be compatible with the new dispersion of industry and people. Recovering the costs of these capital investments will become a major point of contention for the utilities, regulatory bodies, and customers. More affluent customers, particularly businesses, will be likely to have the capital resources necessary for investment in solar power. As these customers drop off the utility grid, the burden associated with capital recovery of grid facilities will be less widely distributed. In effect, if the regulatory agencies allow increased rates to offset the utilities' "mistakes," those who can least afford to pay will be required to do so.

The utilities also may not have developed wheeling and load management techniques and equipment of the type that will be needed in the new system configuration. In general, failure to anticipate and to plan for a new power supply system that involves an extensive non-utility generation element will cause both operational and economic hardship. Since most of the new technology and solar power development is postulated to occur in the South and West, utilities will be particularly strongly impacted in these areas.

The distress of the utilities will be transmitted directly and quickly to the regulatory agencies. Again, assuming the unfolding of the "Average Future" Scenario, these agencies will have done only minimal planning and reorganization to prepare for the new electric power system envisioned in this scenario. They will be faced with a myriad of problems involving the sale of power to utilities by dispersed generators, the management of large systems with many generation nodes, the wheeling of power from non-utility generators, and so on. They will also be faced with the question of fair allocation of the costs of excess capacity. Rate increases will be required to cover at least part of the costs of unneeded capacity. In addition, those agencies responsible for environmental, health, and safety standards will be under increased pressure to strengthen restrictions on coal and nuclear plant operation, since solar generation will appear to be a feasible option.

The effect of the situation outlined above on coal and nuclear fuel services will also be adverse, as these groups will have invested in facilities to provide more fuel than will be needed. In fact, there will be varied impacts on most stakeholders in this society because of the turbulence associated with changing federal policies. However, the only other sizeable impact will fall to companies involved in power supply and electric end use equipment. In the former case, the failure of policy makers to foresee the coming change in equipment needs will lead to excess capacity to produce traditional power equipment, and to insufficient capacity to produce equipment needed in the new situation. In the latter case, failure to envision new societal directions will result in insufficient emphasis on developing new efficient and versatile end use products.

Suggested modification to Policy Set Alpha. The modifications that are needed to make Policy Set Alpha more compatible with the actual unfolding of the "Post-industrial" Scenario center on changes that either permit earlier identification of the developing trends or that postpone irreversible decisions until adequate evidence is in hand. Since the total amount of power generated in the two scenarios is not very different, analysis should be focused on the greatly increased role of non-utility generation in the "Post-industrial" Scenario. Specific policies might include expanded research into photovoltaic systems to assess their potential better, more active analysis of coal emissions and radioactive waste problems, discouragement of construction of coal or nuclear plants until the need for additional facilities is clearly demonstrated, regulations to allow utilities to invest in small-scale generating facilities, and additional encouragement of third party generators and cogenerators. This last action would promote development of non-utility generation techniques and equipment and lead to experience in the integration of such parties into the overall power grid.

In general, these policies would prevent premature construction of central station capacity, which might prove unneeded; prepare for the possibility that non-utility generation might grow more rapidly than commonly predicted; and lay the groundwork for smooth adjustments to new utility, non-utility generator, and consumer relationships. (It would, of course, be desirable for the nation to realize at an earlier date the emergence of the "Post-industrial" Scenario and to take a more proactive stance in the fields of trade relationships, industry subsidies, worker retraining, immigration control, and so on. However, policy decisons of these types are beyond the scope of this study.)

In order to improve the ability of regulatory bodies to appreciate and to prepare for the special problems that will accompany the shift from Policy Set Alpha to one more in line with the "Post-industrial" Scenario, reorganization and reorientation of such bodies would appear in order. (To a large extent, acceptance and encouragement of non-utility generation reflects a trend toward deregulation of the industry.) Early modification of the structures and operating procedures of the regulating bodies might not only better prepare them for the electric system changes that are indicated, but might also provide an earlier realization of the nature, extent, and timing of such changes.

Finally, the growing importance of electric power to society makes it highly desirable that electricity end use equipment be as efficient, versatile, and economical as possible. Hence, it would appear that more attention to RD&D in this area would be very appropriate.

As was indicated in the discussion of the "Mega-plant" and "Small Coal Plants" Scenarios, complete abandonment of the breeder reactor and fuel reprocessing programs would avoid unprofitable expenditure of resources. More careful analysis of possible problems for nuclear power might provide at an earlier time the information needed for a rational decision.

Implications of suggested modifications. The principal adverse implication of the modifications suggested above is that the discouragement of power plant and supporting facilities construction in the 1980s might lead to major shortages if the need for power unexpectedly increased and solar photovoltaic generation proved impractical. Moreover, anticipatory policies by the government to promote photovoltaics may result in inappropriate programmatic decisions by all parties concerned. Possible implications of modification involving complete abandonment of breeder reactors and fuel reprocessing programs are discussed in previous sections.

F. POLICY SET ALPHA/"ECONOMIC MALAISE" SCENARIO IMPACTS

Probable national reactions. The "Economic Malaise" Scenario presents a much less optimistic view of the nation's future than does the "Average

Future'' Scenario. This scenario postulates a lower GNP, a lower total energy consumption, and a lower fraction of primary energy converted to electric power. This scenario also outlines a more tumultuous future than any of the others. As with the other scenarios, it is assumed that during the 1980s the nation will continue to support both the development of new high-technology industries and the re-vitalization of the traditional ones. However, in the ''Economic Malaise'' Scenario, by the early 1990s it will become evident that the nation cannot compete efficiently with foreign concerns in the mature industries. Hence, the nation will greatly diminish its support of traditional industries in order to stress development of ''new wave'' technologies.

Because the nation has split its support efforts in the 1980s while a number of other nations concentrated their support on high-technology areas, the United States will, however, find that it is in a much weaker position in terms of technology, trade relations, and support systems. Although there will be periods of economic advance as well as decline during the 1990s, the result will be that overall economic growth through the turn of the century will be very low. As the first decade of the twenty-first century progresses, the United States' position in the new technology areas will improve. It is also postulated that by 2010 the nation will be on the verge of a major economic upswing, despite the relatively low level of the economy at that time.

The results of the economic turbulence associated with this scenario will be felt by all of the stakeholders considered in this study. As before, it is assumed that Policy Set Alpha will be followed throughout the 1980s. As demand increases fail to materialize at even the levels postulated by the ''Average Future'' Scenario, some utilities will be burdened with excess capacity. Increases in capitalized costs attributable to unnecessary capacity will become a point of contention between the utilities, their customers, and the rate-setting commissions. Because of the general economic malaise, rate increases will have significant impacts on the consumers. Utility commissions, caught in a political vise, will attempt to follow some middle ground which will satisfy no one, and will, in actuality, compound already tense relationships. As rates rise, consumers will react strongly with both conservation measures and direct political action. Utilities will face business operating environments much worse than those of the late 1970s. The financial integrity of many utility companies will be threatened and some will not survive the turbulence.

During the 1990s, it is postulated that little more construction will take place in the electric power industry. Research and management efforts will be focused on means of operating plants at the lowest possible costs and on means of extending the operating lives of existing equipment. Because of the prolonged period of low construction, the electric power support infrastructure (equipment suppliers and their personnel, construction firms and sub-contractors, etc.) will deteriorate. Some industrial consumers, driven by the rather high cost of electric power and deteriorating quality of utility service, will develop internal generating capacity. Throughout the nation, major efforts will also be made by this sector to conserve power.

The general economic hardship throughout the country will create more popular sentiment for the relaxation of current regulations on health, safety, and environmental protection in hopes that this will have a positive effect on the economy. However, significant public opposition to any relaxation of these regulations will remain. In some areas, supporters of an environmental relaxation movement will be successful. The country's overall commitment to environmental protection, however, will not be much eroded.

When the postulated economic upturn finally begins, the electric utility industry, the regulatory agencies, and the electric power support structures will have developed very conservative attitudes. Having experienced severe difficulties as a result of 1980s overbuilding, they will be reluctant to support the aggressive building programs that will be needed to support economic expansion. This bodes ill for the quality of power supply in the early years of next century's second decade.

Summary of impacts. The major impacts of the nation's following Policy Set Alpha during the 1980s if the "Economic Malaise" Scenario comes about are caused by the construction of more generation, transmission, and distribution facilities than will be justified by actual demand. Investor-owned utilities will be directly affected, as they will face the problem of paying for the excess facilities at a time when usage levels are nearly flat. They will have difficulty convincing public utility commissions to raise rates in the face of general economic troubles. Power consumers, believing that the utilities have shown poor management judgment, will not be sympathetic to the industry's problem; and governments, federal and state, may increase their control of the utilities. The large public utilities will also be adversely impacted because of overcapacity. Municipal utilities and rural electric cooperatives may, in fact, be helped by the excess capacity as they, for the most part, will be buyers in a buyers' market.

Obviously, rate-setting agencies will be affected as they will be called on to determine how the costs of overbuilding will be apportioned in a period in which both suppliers and consumers are sensitive to costs. Since at least part of the costs of excess construction will be passed on to consumers, these groups, particularly in the industrial sector, also will be impacted. Failure to support end use equipment RD&D during the 1980s will exacerbate their problems. Fuel services groups, particularly nuclear fuel supplies, will be adversely impacted because they will have costly facilities which will not be utilized.

All groups will be affected to some extent by the turmoil that will result from the nation's changing electric power policy. Although the industry infrastructure would have faced major financial problems during the 1990-2000 decade even if more appropriate policies had been followed during the 1980s, a more gradual buildup by these groups during the 1980s might have made their survival at a low level during the 1990s less difficult.

In the international arena, some nations which will have depended on the United States for support in the breeder and nuclear fuel reprocessing areas will be hurt by the outright cancellation of those programs. However, the weakening of the electric and end use equipment supply companies will provide foreign suppliers increased opportunities for sales when recovery begins.

Suggested modifications to Policy Set Alpha. Since the major problems that would result from following Policy Set Alpha during the 1980s if the "Economic Malaise" Scenario actually comes about involve the construction of unneeded power facilities, modifications should center on policies that will result in programs that are responsive to demand changes. That is to say, construction programs should be developed that would postpone actual construction as long as possible to prevent overbuilding, but that would lay the groundwork for rapid construction if a high growth scenario should come about. Such programs would require detailed planning and, in many cases, the early undertaking of preliminary construction activities (land preparation, meeting permit requirements, ordering of selected long-lead items, etc.). Since, in general terms, smaller plants can be completed more rapidly than larger ones, RD&D might be undertaken that would concentrate on methods for making smaller plants more cost efficient.

Because postponing plant construction will probably decrease power reserves, RD&D should be undertaken to increase reliability, simplify maintenance, and reduce down time on existing plants and transmission systems.

Many of the modifications listed above will call for utility expenditures for facilities that may not be used for a considerable period of time; perhaps, never. Since these costs will be incurred primarily to ensure power availability while preventing unneeded construction, it seems reasonable that they be largely borne by the consumers. Therefore, public utility commissions should develop effective mechanisms for encouraging responsible action in this area by the utilities.

If the "Economic Malaise" Scenario were to develop, cost control would be very important to industrial organizations in the country. Given the importance of electricity to these companies, it would appear that government, institutional, and utility support of major RD&D projects to improve end use equipment efficiency and flexibility would be highly advantageous.

Finally, recognition of the possibility of a long term of limited economic growth in the United States should be accompanied by policies that will assist in the survival of the electric power infrastructure during that period.

Implications of suggested modifications. The major implication of the policies listed above to limit construction during the 1980s is the vulnerability of the system to unexpectedly high demands for power. Although the policy modifications suggested should provide sufficient flexibility to meet the power needs postulated in the "Average Future" Scenario, the utility system and, indeed, the economy as a whole might be put in serious jeopardy if one of the higher electricity demand scenarios were to develop. Moreover, construction programs that encourage early completion of preliminary tasks may not be economically or technologically optimal. In like manner, RD&D projects undertaken to meet the specific needs of the "Economic Malaise" Scenario may not be appropriate if any other scenario comes about.

G. SUMMARY OF STAKEHOLDER IMPACTS

Table III.1 summarizes the stakeholder analyses on which the preceding five sections were based. The table cells show the impact on each of 20 stakeholders (rows) in the event future electric power policy is as specified by Policy Set Alpha, but future events are more as specified by one of the alternative scenarios (columns). The degree of impact is scored as none (0), low (L), medium (M), or high (H). Favorable impacts are denoted with a "+," unfavorable impacts with a "–." In some cases, favorable and unfavorable impacts were very nearly balanced or subject to serious definitional ambiguity. In these cases, only the degree of impact was scored.

Following the table are definitions of the stakeholder categories.

H. AVENUES TO VERY FLEXIBLE ELECTRIC POWER POLICIES

In the previous sections, consideration has been given to modifications that might be made to Policy Set Alpha to make it more commensurate with the needs of the future, were a future other than the "Average Future" Scenario to come about. In actuality, of course, it is possible that any one of the described scenarios might unfold in any one of an infinite number of possible variations. Hence, the real question is whether a policy set developed to meet what appears to be the most likely future can be modified in such a way that it can, at an acceptable cost, accommodate a myriad of other possible futures reasonably well. This is the question that is addressed in this section.

In examining the modifications suggested in Sections B-F, two factors are common to all: a need for RD&D programs beyond those postulated in Policy Set Alpha and a need for regulatory actions that permit and encourage both power producers and consumers to take actions that will contribute to their common good. Although different types of RD&D programs were suggested for the different possible futures, to be successful each program must be characterized by careful planning, detailed coordination, and reasonable continuity. When the programs are considered in concert, it is evident that the magnitude of an overall program with sufficient scope to provide the desired policy flexibility will be substantially beyond that specified in Policy Set Alpha. However, it is evident that the cost of even a large, long-term RD&D program will be low in comparison to the costs involved in over- or underbuilding, in investing in inappropriate types of facilities, or in ill-advised changing of the nation's overall power system.

Determination of the precise size and nature of the suggested enhanced RD&D program will, obviously, require a great deal of analysis and planning. However, certain aspects of the overall program would appear evident. The program should be broad in nature in order to give decision makers as much latitude as possible. Wide participation in the program should be encouraged, with individual utilities, the Electric Power Research Institute, universities, private research organizations, federal laboratories, and other relevant groups contributing to the overall program. The program should be reasonably well coordinated. Support for the program should be organized in such a way as to ensure continuity over a reasonably long period. Activities undertaken under the program should not be restricted to technical projects, but should include analyses of the overall goals being sought by society, the preferences of the people as to how those goals be met, specific concerns of the populace, and means for alleviating those concerns. Special attention also should be given to development of new technologies which might provide vast new capabilities to the electric power industry; examples include robotics, information manipulation and communication, materials development, space travel, and biotechnology. Policy makers involved with the nation's power systems have, for a long time, sought to develop and implement effective RD&D programs. What is suggested here is that future programs be planned to provide flexibility in the decision process eight to ten years in the future.

The second factor that appeared in each of the modification sections was the appropriate role for regulatory bodies, those involved in both rate determination (normally, public utility or public service commissions) and in operational procedures. The ability of both utilities and power consumers to make changes in ongoing and planned programs rapidly and efficiently is often restricted by these agencies. Increased responsiveness of these regulatory bodies to changing electric power realities may require alterations in organization, personnel, and incentives. Again, determination of the exact nature of these changes will require study and analysis. However, in general, they should be such as to give proper recognition to long-term impacts and the need for flexibility to meet shifting load patterns, system structure, technological capabilities, and specific consumer requirements.

Two technological advances that may alter the nature of the nation's electric power system and that may require innovative approaches to government regulation are the development of mega-plants, such as solar space

Table III.1
Summary of Stakeholder Analyses

Stakeholder	Scenario B	C	D	E	F
Investor Owned	M−	H−	M−	H−	M−
Public, Municipal, and Rural Electric Cooperatives	L−	M−	L−	M−	L+
Residential and Commercial Sector	L−	L−	L−	M−	L−
Third Party Generators	O	M+/−	L−	L+	O
Cogenerators	L−	L−	M−	L+	O
Industrial Sector	M−	M−	M−	L−	M−
Fuel Suppliers/Oil and Gas	L+	M+/−	L−	O	O
Fuel Suppliers/Coal	L+	L+	L−	L−	L−
Fuel Suppliers/Nuclear	LH−	M+/−	H−	M−	M−
Neighbors	M−	M−	M−	L−	L+/−
Regulators	M	M	M	H	M
Related Government Interests	L−	H−	M−	M	L−
Domestic Non-government Interests	M−	H−	M+/−	M	M−
Workers and Unions	L−	M	L−	L+/−	L−
Construction	L−	L−	L−	L+	L+
Equipment Supply Companies	M−	L−	M−	M−	L+/−
End Use Companies	L−	L−	O	M−	H−
Financial Institutions	L−	M−	O	H−	M−
Research Institutions	M−	M−	L−	M+	L−
International Interests	L+/−	M−	L−	L+/−	M+

Stakeholder Definitions

Investor-owned electric utilities. This group includes both the management and the shareholders of investor-owned electric utilities.

Public, municipal, and rural electric cooperatives. This group includes the management of federally chartered electric utilities, such as the Tennessee Valley Authority; municipally owned utilities; and rural electric cooperatives.

Residential and commercial sector. This group includes all non-industrial purchasers of electricity. It includes, therefore, all individual urban rural ratepayers and various commercial establishments, such as retail stores, office buildings, and small agricultural enterprises.

Third party generators. This group includes non-utility generators of electricity (also referred to as small power producers, SPP) that generate electricity primarily for resale.

Cogenerators. This group includes industrial, municipal, and, potentially, residential and commercial power consumers that generate electricity incidentally to other energy production. Cogenerators differ from small power producers in that, for the latter, electricity generation is a business objective in and of itself.

Industrial sector. This group includes industrial purchasers of utility electricity.

Fuel suppliers/oil and gas. This group includes the producers, refiners, transporters, and distributors of oil and natural gas.

Table III.1 (*continued*)

Fuel suppliers/coal. This group includes companies that mine, prepare, and transport coal.

Fuel suppliers/nuclear. This group includes all companies that extract or prepare nuclear fuel for consumption by electric utilities. It also includes companies that assume responsibility for spent nuclear fuel after it leaves storage at utility power plants.

Neighbors. This group includes individuals and institutions that compete for the fuel, water, land, and other resources that are "consumed" in electric power generation. It includes those individuals and institutions that are in close geographical proximity to electric utility generation, fuel supply, or waste disposal activities.

Regulators. This group includes regulators at all levels of government who have significant and direct responsibility for the financial regulation of electric utilities. It includes, therefore, such groups as the Federal Energy Regulatory Commission, the Securities and Exchange Commission, and state public utility commissions.

Related government interests. This group includes all government agencies that are not covered under "Regulators," and that have either direct or indirect responsibility for electricity generation by electric utilities, small power producers, or electricity cogenerators. Probably, the most significant members of this group are the Environmental Protection Agency, the Occupational Safety and Health Administration, and the Department of Energy. It also includes such government agencies as the Office of Management and Budget, federal and state congressional committees on energy and natural resources, and agencies, such as the Department of Defense and the Department of Transportation, that have some peripheral interest in electricity systems.

Domestic non-government interests. This group includes all the non-government special issue advocates. It includes, therefore, environmental conservation groups (e.g., the Sierra Club), professional and industrial lobbying groups (e.g., the Slurry Transport Association), sectarian organizations, and state and municipal associations.

Workers and unions. This group includes the non-management employees of electric utilities and the unions representing those employees. It also includes employees and unions that, though not directly affiliated with electric utilities, are particularly dependent upon electric utilities; for example, members of the Chemical and Atomic Workers Union and the United Mine Workers are covered in this stakeholder group.

Construction. This group includes management and shareholders of major architectural engineering and construction firms, such as Bechtel and Brown and Root.

Equipment supply companies. This group includes the manufacturers, such as General Electric and Westinghouse, who supply equipment for power generation, transmission, and distribution. These companies supply equipment to both utilities and non-utilities.

End use companies. This group includes all companies that manufacture electricity end use equipment. Its principal constituents are those companies that supply machinery to industry.

Financial institutions. This group includes banks, savings and loan associations, investor rating services, and pension fund managers.

Research institutions. This group includes all organizations that do extensive research, both technological and sociological, that bears on electric power. Therefore, it includes the Electric Power Research Institute and the National Laboratories. It includes university research programs and the research programs of electricity equipment suppliers.

International interests. This group includes foreign governments, foreign citizens, and foreign industries. In particular, it includes the Canadian and Mexican governments and foreign manufacturers of electricity generation or end use equipment.

satellites, and the development of practical, small local power-generating facilities, such as photovoltaic generators. In the former case, it is envisioned that such plants will have enormous capacities, large construction costs, and low fuel and operating costs. In addition, it is presumed that the bulk of RD&D costs will have been borne by the federal government. Thus, in many respects, these plants will resemble the huge hydroelectric plants built during the 1930s. However, their locations will have greater national and international significance. Thus, new principles and procedures may be necessary for allocating the benefits that would accrue from these plants in an economically and politically acceptable manner.

The emergence of small-scale power production facilities presents even more interesting possibilities for change in the electric power system of the nation. Although home generating units are a possibility, local community generating facilities seem more likely. If such facilities could be built and operated economically, they would obviously affect the overall power supply system. It is probable that such facilities would be interconnected with a central station power system similar to the one in existence today. Therefore, the nation would have, in effect, a dual power system: a centrally owned and controlled central station system and a coexistent decentralized system. Many of the present regulations and rate-setting procedures would require major modifications to accommodate these changes. In fact, the existence of a realistic alternative to central station power may end the need for any rate-setting authorities. Certainly, the mere potential for small-scale power production militates for explicit examination of current regulation-mandated cost subsidies among customer classes. Contingency plans are needed to direct the gradual elimination of these subsidies in the event technological change renders them untenable.

In addition to expanded RD&D efforts and modified regulatory activities, several other alterations to Policy Set Alpha seem appropriate. It would appear, for example, that enhanced conservation programs are desirable, regardless of which scenario develops. These programs should be directed not only to preventing waste, but also to improving efficiency, particularly in industrial operations. Likewise, efforts to understand the causal linkages of environmental pollution and to design programs for abatement would seem desirable regardless of which scenario should unfold. Special programs to encourage cogeneration and third party generation will probably add to the nation's ability to respond to rapid economic and technical change. Moreover, close coordination of domestic programs with those of other nations undoubtedly would prove desirable.

In summary, it would appear that reasonably modest changes can be made to Policy Set Alpha that will enhance the capabilities of decision makers in the electric power field to react in an effective, efficient, and timely manner to diverse realities.

Possible modifications are suggested and described above. In general, these suggestions will involve higher costs in terms of human resources, facilities, and programs than a policy set planned for a single, determinable future. However, given the uncertainty unavoidably coupled with the future, the additional flexibility afforded to future choices of action would seem to justify the additional cost.

CHAPTER IV

Principles for Electric Power Policy

A. OVERVIEW

The purpose of this assessment is to identify and analyze electric power policy issues of potential importance to the United States during the next three decades (GE:IV-1).* These issues will arise both from an increased dependence by the nation on electric power and from potential changes in the way electric power is generated, distributed, and utilized.

Because the extent of future dependence on electric power and the various ways in which electric power is generated, distributed, and utilized are highly uncertain, this assessment considers policy within the contexts of six scenarios. Each scenario makes explicit assumptions about key elements of uncertainty. The scenarios and the means by which they were developed are described in some detail in Chapter II. The importance of the scenarios, for current purposes, is that each provides a context within which to analyze future electric power policy. The six, together, allow analysis of a broad spectrum of electric power issues and policies at a meaningful level of detail.

This chapter attempts to organize future electric power issues, problems, and opportunities in a way that will highlight their importance, their implications, and, particularly, their interrelationships. By design, no list of recommendations on specific courses of action is given. An attempt is made, however, to clarify options, identify significant policy variables, and highlight major decisions with which public policy makers will be faced in the future. Although the assessment makes projections for a period of approximately 30 years, this chapter places primary emphasis on decisions that will be made over the next 8 to 10 years. In essence, it seeks to examine

*Dr. Enk has a number of general comments concerning this chapter. They are reproduced in Part Four.

the long-term implications of electric power policies initiated and implemented during this shorter period.

The substance of this chapter is presented in two parts. The first part presents two electric power principles that do not embody new wisdom but, rather, make explicit realities that are essential to an understanding of electric power policy (SM:IV-1). The second part of the chapter presents seven additional principles that assessment analyses show should guide fair, effective, and efficient electric power policies. Discussion of each of the nine principles begins with a statement of the rationale for its validity. Then, each discussion addresses the significance of the principle for electric power policy makers. Each concludes with a partial listing of policy questions associated with the principle. Associated questions that are particularly global ("overarching") are segregated from the more narrowly focused ("operational") questions.

In summarizing the nine principles, the most important observation is of the need for electric power policies that are sufficiently adaptable to meet a wide variety of social, technological, economic, and environmental developments. To say the future is uncertain is to state the obvious. However, there are strong indications that changes bearing on electric power may be inordinately dramatic, rapid, and surprising over the next three decades. Therefore, it will be unusually difficult during this period to strike a reasonable policy balance between efficiency, equity, and risk. It is the purpose of this chapter to lend some guidance to the striking of that balance.

B. UNDERLYING PRINCIPLES

1. Electric Power Policies Are Inexorably and Intricately Interrelated with Most Other Public Policies

Rationale. This assessment concentrates on policy issues directly related to electric power generation, distribution, and use. (GE:IV-2.) The electric power system, however, is often affected as much or even more by less directly "electric" policies (e.g., those involving finance, labor, anti-trust, and defense). Further, there are few policy decisions that are so specifically targeted to electric power that they do not also affect other elements of society.

In the first case, consider, for example, the impact that immigration policies, seemingly unrelated to electric power decisions, can have on electric utilities: the size and profile of immigration cohorts affect the structure of the nation's economy and thereby affect electricity consumption; the origins and languages of immigrants may affect utility billing procedures; and national efforts to control immigration may affect individuals' expectations of "reasonable" civil liberties, thereby helping expand or limit the legitimate authority of electric power institutions. In the second case, consider the ripple effects of utility commission treatment of certain funds used for power plant construction (SB:IV-1). (This topic is discussed more thoroughly in Chapter V.) The extents to which public utility commissions direct electric utilities to treat as current "income" potential earnings on money that is, in fact, invested in the construction of power plants apparently affects the value of electric utility stocks and bonds. Thus, utility commissions' policies on this obscure electric power issue affect the wealth of a great many individual and institutional investors.

The higher order impacts of these policies subsequently emerge in the electric power arena when individuals and institutions discount the value of securities of electric utilities for which this "income" is a significant source of revenue. To offset this, utilities must offer investors the inducement of higher returns on their investments. This, in turn, marginally increases the cost of doing business as an electric utility and thereby marginally increases the price that the consumer must pay for electricity.

It has for some time been the message of the electric power establishment that electricity is one of the important engines of U.S. economic growth (Starr and Searle, 1982). There are certainly credible arguments for the proposition that electricity is of large and growing importance to the nation's economy. (These arguments are summarized in the discussions of Principles 3 and 4.) As the linkages between electricity and the economy become better understood, the interdependent nature of policies that are nominally either "electric" or otherwise will become more readily apparent.

Significance to policy decisions. Although electric power policies and other public policies are interrelated, it is probable that the latter most often drive the former. Chapters IX and X discuss how public interest in low energy costs, environmental protection, and energy independence successively caused changes in electric power research and development efforts as well as in facility construction. Similarly, each of the scenarios described in Chapter II is driven primarily by economic and environmental variables. This dominance of electric power policies by exogenous factors means that those people charged with formulating and executing electric power policies must be alert to possible developments across a wide spectrum of events. As discussed in Chapter III, the optimal policy set for one future might be quite inappropriate for a different future. Therefore, people devising electric power policies should employ formal methods to project possible societal changes, to devise contingency policies for a wide range of changes, and to monitor continuously for the emergence of important changes.

Although changes in society at large will, generally, have a greater impact on electric power than will electric power changes have on society at large, decision makers in the electric power arena should place a premium on understanding the ways in which electric power policies reinforce or retard more basic trends in society (SM:IV-2). These policies are, of course, the ones most directly in the decision makers' control and the impacts of these policies on basic trends can be significant. A review of potential interactions associated with the "Post-industrial" Scenario illustrates this. This scenario postulates a continuation of the demographic trend, discussed in Chapter VIII, to geographic dispersion of the population. It also assumes economic trends that favor growth of service and information industries. Under these circumstances, it becomes increasingly important that electricity supply in rural areas be reliable and of high quality (e.g., of uniform voltage and frequency). In the absence of formal consideration of demographic and economic trends, electric power policy might fail to lay the groundwork for supplying high-quality power to rural areas and thereby retard evolution of demographic and economic trends. Conversely, the laying of groundwork to support those trends would promote smooth evolution to a future like the one postulated by the "Post-industrial" Scenario.

This reciprocal relationship between electric power policy and larger societal trends is also highlighted by analysis of details in the "Mega-plant" Scenario. In that scenario, the robust high-technology economy provides both the financial resources and the technological resources that make possible the deployment of solar power satellites. However, the prospects of large non-military payloads, the solar power satellites themselves, also serve to stimulate the evolution of the space delivery systems and the high-technology industries that contribute to the robust economy. The prospects of assured energy supplies, in spite of postulated petroleum scarcity and limitations on the use of coal and nuclear fuel, likewise would serve to spur economic momentum.

This potential for symbiotic relationships argues for additional investment, as discussed in Chapter IX, in more extensive and formal efforts to transfer information and ideas between the electric power industry and other scientific and engineering research agencies.

Associated Questions: Overarching

1. What "non-electric" factors will most affect electric power policy decisions in the future?
2. By what chain of interactions do these non-electric factors become relevant to electric power policy decisions? In what respects are the factors themselves likely to change over the next three decades?
3. By what process can the effects of electric power policies on non-electric aspects of society be foreseen? What can be done to minimize the undesirable effects of electric power policies throughout society?
4. By what theories can one judge whether it is appropriate to promote particular broad societal objectives through electric power policy?
5. How can the processes by which electric power policy makers and other policy makers interact be improved?

Associated Questions: Operational

1. How will the prices of labor, raw materials, and primary energy sources affect the cost of electric power?
2. How will the price of electricity affect other prices in society? What effect will this have on the overall inflation rate?
3. How will world conditions affect the choice of electricity-generating sources?
4. In what ways will the predicted shortages of engineers and other technical people affect the operations of electric power plants? Will other technical advances needed to keep electric power plants operating with increasing effectiveness and efficiency be slowed by these shortages?
5. Will shortening of the work week cause new uses for electric power, particularly in the recreational areas?
6. If civil disobedience directed at power system components increases, how should law enforcement and utility security activities and information be coordinated?
7. To what extent does future U.S. economic structure, hence, electricity demand, depend on economic growth in less-developed countries?
8. In what ways will attempts to control "acid rain" affect electric utilities?

2. Electric Power Policies Typically Have Long-Term Implications and Are Difficult to Modify

Rationale. Major electric power decisions have historically involved complex technological questions, large expenditures of funds, and implications experienced over decades. Once initiated, electric power programs have often developed self-reinforcing inertia. They have had long-term effects on capital markets, building and equipment contractors, labor forces, and potential power users. Program modifications or terminations typically have been costly and have caused wide-spread distress. The recent experience of the Washington Public Power Supply System (WPPSS) (see the accompanying case history) illustrates this point, as do the high electricity prices many utility customers currently pay because policies in the 1960s favored heavy oil and gas generation mixes (GE:IV-3). Much of the discussion in Chapter V is concerned with future electricity price implications arising from power plant cost escalation during the 1970s.

The long-term implications of policy decisions have neither been, nor will they be, only monetary. Consider, for example, nuclear power policies adopted during the 1950s. These encouraged development of light water reactors for civilian electricity production. They appeared sound at the time because they joined civil electric power and naval submarine reactor programs to the same, and most advanced, technology development cycle. However, these policies were also instrumental in shaping the nuclear proliferation and waste debate that continues to this day. Many experts now believe that heavy water or molten salt reactors would have been more appropriate for civilian programs (Bupp, 1978; Rochlin, 1979).

WPPSS Case History

The Washington Public Power Supply System (WPPSS) epitomizes the problems that can plague an overly ambitious nuclear power program. Roots of the WPPSS misfortune began in 1968, when authorities concluded that an energy shortage was inevitable in the Northwest and proposed to supplement cheap hydroelectric power with electricity from new nuclear and coal-fired generating plants. By 1974, five nuclear power plants were under construction or being planned and 115 utilities in eight states had been recruited for the supply system network. The federal government's Bonneville Power Administration (BPA) guaranteed financing for two of the facilities and most of a third.

Unfortunately, the energy shortage forecast proved wrong and the WPPSS nuclear power program became a political, legal, and financial fiasco. Regulatory delays, construction snags, management problems, and soaring interest rates pushed the cost of the project from $6.67 billion projected when work began in 1972 to nearly $24 billion in 1981. This disastrous record has earned the supply network the derisive nickname of "WHOOPS."

What has been the result of the fifteen years of funding? Of the five plants, two have been canceled and two have been postponed for three to five years. The fifth "should" be completed by February 1984—seven years behind schedule and at a cost of six times the original estimate. The two canceled plants left $2.25 billion in WPPSS bonds in default. Also, electric rates for the WPPSS customers have skyrocketed. BPA, for example, has raised its wholesale electric rates three times, a compounded 478 percent increase. Finally, WPPSS customers are organizing, marching, packing legislative hearings, and withholding payment of rate increases, making WPPSS the most politically sensitive issue of the region.

Note: For an in-depth policy history of WPPSS, see Darryll Olsen, "The Washington Public Power Supply System: The Story so Far," *Public Utilities Fortnightly* (June 10, 1982), pp. 15-26; and "The Fallout from 'WHOOPS'—A Default Looms, Casting a Pall Over the Entire Municipal Market," *Business Week* (July 11, 1983), pp. 80-87.

The potential environmental and health consequences of hydrocarbon and nuclear wastes offer the most striking examples of still prospective impacts associated with electric power policies. For instance, the Nuclear Waste Disposal Act, signed into law in January 1983, allows four-and-a-half years for selection of one permanent burial site and four more years for the selection of a second. The first sited facility would not be operational until the late 1990s. In the event agreement cannot be reached on permanent burial sites in these time frames, the act calls for policy on "monitored retrievable storage" on a 100-year "temporary" basis.

The length of the technology cycle from research through development and demonstration to commercial use also ensures that policy decisions made in the 1980s will affect posterity. This is reflected in the responses to several questions included in this assessment's expert opinion survey (Appendix B). For example, respondents to Item 57 believed, on average, that it will take 20 years for the super-conducting electricity generator to progress from the small capacity demonstration phase to its first large capacity commercial application. These experts projected (Item 58) about the same time span before there will be any meaningful penetration of the intracity personal transportation market by predominantly electric vehicles. This is despite the fact that the demonstration phase for electrically powered vehicles is already under way in this country. Of course, commercialization of terrestrial solar photovoltaic technology, if this comes to pass this century, will have been made possible by favorable RD&D policies during the 1970s.

Significance for policy decisions. The long-term inertia that frequently characterizes electric power policies makes changes in those policies difficult; it also creates barriers to experimentation with alternative options. It is unfortunate, therefore, that the social and technological uncertainties of, at least, the next three decades indicate that flexible electric power policies (e.g., policies that can be modified in light of future experience and that do not foreclose future options) will be even more important in the 1980s than in the past (SM:IV-3).

In general, flexibility is enhanced by adopting programs and projects that are diverse in nature and incremental in application. Normally, stress on flexibility leads to policies that will be viewed as non-optimal (hence, "unnecessarily" costly) from the hindsight perspective of the future. For example, if a future like that postulated in the "Average Future" Scenario (i.e., about 2 percent annual growth in electric power demand and nuclear and coal processes that are, at least, minimally acceptable in terms of environmental and health concerns) comes to pass, then funds devoted to developing and introducing new power technologies will appear to have been misspent. There is, then, a potential real cost associated with flexibility.

This cost must be balanced against the cost of relying on a single view of the future. If the demand for electricity were to expand significantly, if public resistance to nuclear and/or coal power plants were to intensify, or if diversified power production were to prove technologically feasible and socially popular (all conditions reflected in the scenarios in Chapter II), then failure to consider these possibilities at an early date will impose costly constraints on future options. In an era as uncertain as that immediately ahead, the greater risk surely lies in adopting and acting solely on a single view of the future.

Part of the cost of flexibility will be a long-term commitment to research, development, and demonstration programs. As mentioned, the technology development cycle is normally a lengthy one. To some extent, the costs of policies that address technology development can be lessened if the policies allow postponement of decisions involving major expenditures of time and money. Nonetheless, the reality usually is that research at some level must begin many years prior to a technology's commercialization. For example, in considering the "Mega-plant" Scenario, it is obvious that solar power satellites simply will not be a power supply option in the year 2010 unless that technology receives considerable research support during the 1980s. Of course, the planning of expensive research projects such as those bearing on solar power satellites can itself be a lengthy process and should mostly precede the research being planned.

Associated Questions: Overarching

1. What are specific techniques that could add flexibility to the electric power decision-making process?
2. By what processes can electric power decision makers be more quickly advised of the need for policy modifications or redirection?
3. By what processes can decision makers more effectively and equitably balance short-term costs and benefits of electric power decisions with long-term costs and benefits?
4. By what processes should the costs of electric power policy flexibility be distributed across society?

Associated Questions: Operational

1. How can the lead times associated with building electric power facilities be decreased?
2. To what extent will the construction of smaller generating plants permit flexibility in electric facility planning?
3. To what extent can improved engineering in initial power plant design simplify future power plant modifications?
4. To what extent does manufacturing investment in electric power efficiency improvements depend on stability in the electric power industry?
5. To what extent do decisions about construction of coal generation facilities depend on stability in electric power supply system and the electric power policies of government?
6. To what extent do decisions regarding construction of mass transit facilities depend on stability in the electric power supply and electric power pricing policies?
7. To what extent does public reaction to relatively short-term problems motivate changes that have long-term implications for the electric utility industry?
8. What are the potential long-term consequences of the carbon dioxide buildup? How much impact can U.S. policy have on this?
9. How could improved understanding of the technology innovation process improve long-range electric power decisions?

C. SYNTHESIZED PRINCIPLES

3. The Importance of Electric Power to U.S. Society Will Continue to Grow for the Foreseeable Future

Rationale. Electric power is today so widely available and used that many people have come to take its benefits for granted. This was not so in earlier eras. When Tennessee Valley Authority and Rural Electrification Administration projects brought electricity to areas of rural and remote America, people gained explicit awareness of the benefits of electric power. Electric lighting, refrigeration and, later, television helped make life both less perilous and more diverse. Because electric power was highly transportable, it facilitated the establishment of industry at points remote from primary energy resources and thereby helped open to individuals the practical opportunity to migrate to less densely populated areas. The air-conditioned cities of the South and Southwest, of course, owe much of their appeal to the availability of electricity.

Exhibits IV.1 and IV.2 provide some quantitative indication of electricity's increased importance to U.S. society during the past 30 years. Exhibit IV.1 shows that approximately 17.5 percent of all energy consumed

Exhibit IV.1
Energy for Electricity Production, 1950–1980

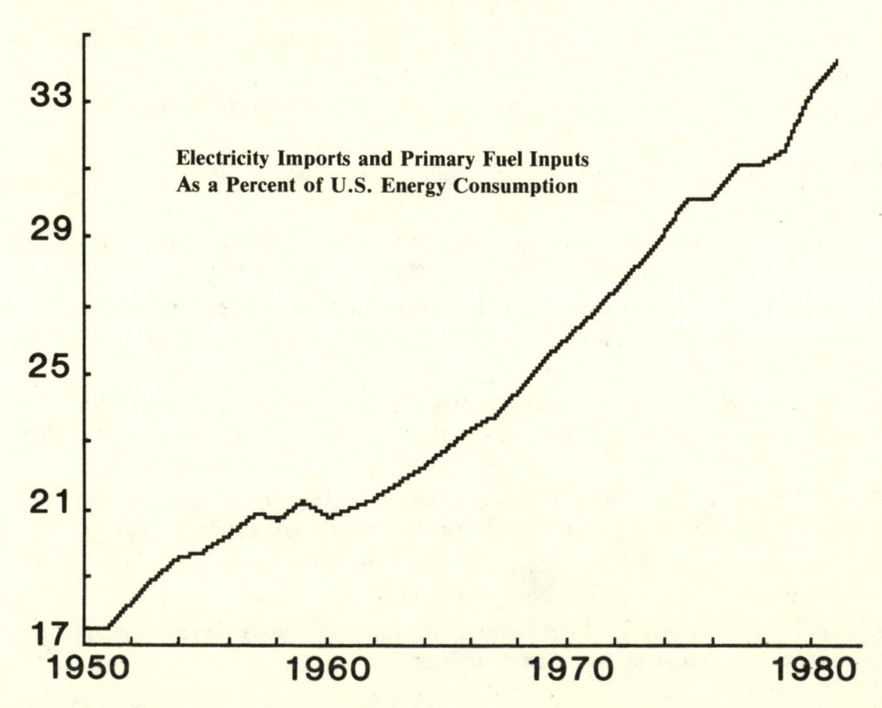

Exhibit IV.2
U.S. Energy Consumption and Real GNP, 1950–1980

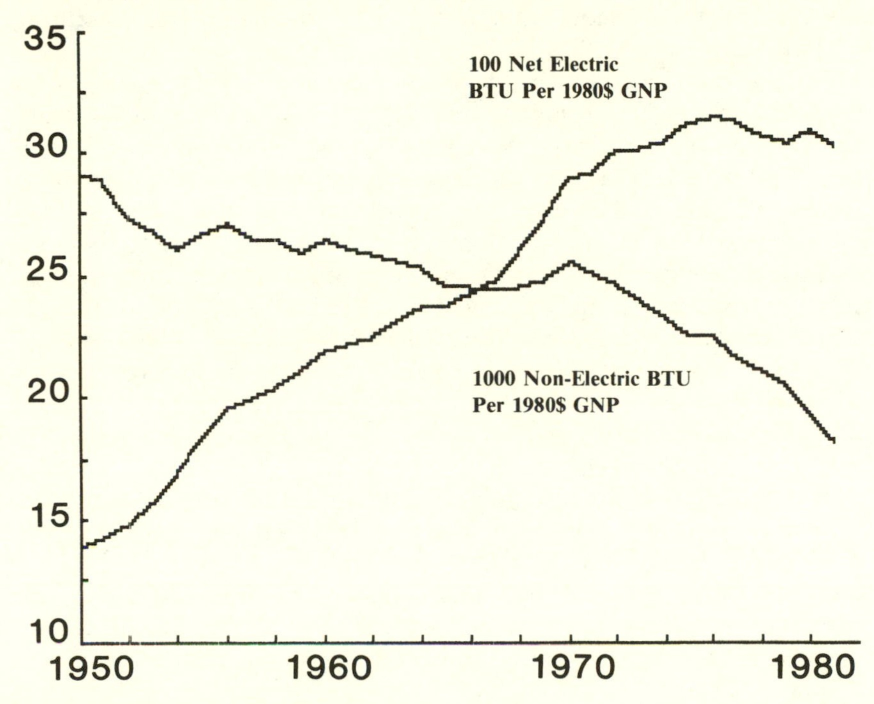

Sources: BEA, 1982; EIA, 1982.

in this country in 1950 was consumed for the production of electricity. This fraction grew steadily over the past 30 years and, as of 1981, had approximately doubled to stand at 34 percent. Exhibit IV.2 provides some insight to this phenomenon. It shows that the ratio of electrical energy inputs to U.S. gross national product more than doubled during the past 30 years, while the ratio of non-electric energy inputs to gross national product declined by nearly 35 percent. This suggests, though it does not prove, that electricity has been becoming increasingly important to the money-making activities of this country during the same period of time that the importance of other energy forms has declined. (Exhibit IV.3, which accompanies Principle 4, shows the relative importance during the past three decades of electricity and other energy forms to American industry.)

The next 30 years will likely witness a continued flow of benefits for individuals from electricity. Electricity, of course, will make possible the widely predicted computer/information economy. There is strong evidence that industry will increasingly turn to robot manufacturing processes that will relieve humans of tedious and dangerous work. Electric power is essential for these processes. Electricity will also make very distributed manufacturing processes economically feasible. A number of emerging electricity-based industrial processes, such as smelting, drying, and curing processes, are much less sensitive to scale economies than are conventional processes and are, therefore, more compatible with distributed manufacturing (MITRE Corporation, 1982; EPRI, 1982).

Significance for policy decisions. Prior to the mid-1970s, it was an axiom of electric power policy formulation that the reliable supply of high-quality, reasonably priced electricity was essential to the well-being of the nation. Since the Arab oil embargo it has become progressively more respectable to question this belief. However, the analyses of this assessment strongly support the continued and growing national importance of electricity. The potential cost of electric power irregularities and shortfalls are real, large, and to be avoided if at all practical. For example, the analyses of Chapter III indicate that policies promulgated during the 1980s in anticipation of lower levels of electricity demand growth in the 1990s than actually materialize will lead the country to take stopgap actions in the early 1990s. Those will be expensive in terms of power costs, environmental degradation, and undesirable health and safety risks.

Though there is little substantiated, quantitative support for the theory that constrained electricity supply also constrains economic growth, correlation (reflected in Exhibit IV.2) lends considerable credibility to this theory. Further, the limited quantitative support that does exist comes from the efforts of highly respected economists working outside, though partially funded by, the electric power establishment (Hudson, Jorgenson, and O'Connor, 1980).

Each of the scenarios postulated by this assessment encompasses an expanded role for electricity, relative to other energy forms, in the future. Such an expanded role is common to most models of the future (Schurr et al., 1979:195; EIA, 1982). For example, robotics are an integral part of most postulated futures that envision a re-vitalization of this nation's industrial base. High-quality, reasonably priced electricity supply obviously promotes the probability that these futures will come about. Futures that postulate a post-industrial America deriving economic vitality from information services and high-technology, cottage industry manufacturing, likewise, depend on the ready availability of high-quality, low-cost electric power.

On balance, the foremost maxim for those who formulate electric power policy should be that a reliable supply of high-quality, reasonably priced electricity is essential to the well-being of the nation.

Associated Questions: Overarching

1. What, specifically, are the potential economic and non-economic losses associated with electric power supply shortages? How undesirable (large) are the losses at various levels and types of shortages?

2. Which groups within society are most affected by electric power shortages? In what ways will they be affected?
3. What balance should the nation strike between resource commitment and attainment of an acceptably high probability of adequate electric power supply? By what process should this balance be struck?
4. Over which groups and in what proportions should the costs of attainment of an acceptably high probability of adequate electric power supply be distributed? By what process should this determination be made?
5. By what process can more meaningful correlations be drawn between the adequacy of electric power supply and the economic, physical, and psychological well-being of Americans?

Associated Questions: Operational

1. How can a reasonable balance between the cost of ensuring adequate power supply and the expense of building overcapacity be achieved?
2. What measures are available for evaluating the relationship between the quality of life and the availability of electric power?
3. To what extent can electric power substitute for end uses of other types of energy sources that might become more expensive and scarce?
4. To what extent can electric power support the development of new, efficient mass transit?
5. To what extent can electric cars contribute to pollution abatement?
6. To what extent can electric power assist the country in competing in world trade?
7. What additional requirements will advances in end use technologies, such as computers and robotics, place on electricity quality?

(GE:IV-4.)

4. Projections This Decade of the Long-Term Size and Distribution of Electric Power Demand Will Be Increasingly Uncertain

Rationale. Efforts to project future electricity demand growth conventionally begin with projections or assumptions of future economic growth. From 1950 through 1973, electricity demand growth averaged 7 percent annually. This was about twice the average annual rate of growth in the nation's real gross national product. Since 1973, average annual electricity demand growth has been only slightly greater than average economic growth (see Exhibit II.4 in Chapter II and Exhibit V.4 in Chapter V).

Although no one is confident about the future correlation between the two, one common assumption is that electricity demand growth and economic growth will in the future be approximately the same. This assumption is probably reflected in the mean responses to the real economic growth and electricity demand growth questions in this assessment's expert opinion survey (Appendix B, Items 20 and 46). The mean response to each of these questions translates to a compound annual growth rate of approximately 2.1 percent.

Unfortunately, even if one were confident of a particular long-term real gross national product/electricity demand linkage, one could not confidently project future electricity demand because there is wide-spread uncertainty concerning future economic growth rates. Economists who participated in this assessment's expert opinion survey provided 30-year estimates that varied from 1.15 percent per year to 3.56 percent per year (Appendix B, Item 20).

Actually, there are two major reasons why long-term electricity demand growth would be unclear during the 1980s, even were there a consensus about long-term economic growth. The first of these turns on the matter of electricity price. This had been essentially constant (declining in real terms) for three decades prior to 1971. Then the real price of electricity began to increase. This trend continued throughout the 1970s and into the 1980s. The analyses in Chapter V indicate that real prices are unlikely to fall again for any meaningful period this century. The mean estimates of respondents to this assessment's opinion survey also bear this out; they predict residential and industrial electricity price escalations of, respectively, 1.3 percent and 2 percent in real terms annually through the year 2000 (Appendix B, Item 50). These mean estimates are more conservative than many (*Electrical World*, 1982).

Electricity demand is sensitive to price, especially in the longer term, because the long term allows time for turnover of energy using capital stock, such as appliances and machinery. The degree of sensitivity, however, apparently varies among customer classes and is, in general, poorly specified (Berndt, Cowing, and Wood, 1983:5.7). It follows, then, that the demand reduction influence of uncertain price increases is also uncertain.

The second major reason why future electricity demand growth would be unclear even if future economic growth were known is that structural change may be under way in the U.S. economy. If so, the linkage between economic growth and electricity demand growth will be much altered.

Since Daniel Bell coined the term in the early 1970s, there has been a general awareness that the United States is moving in the direction of a "post-industrial" economy (Bell, 1973). Unfortunately, this awareness is of the general nature of future society. It is not sufficiently defined to allow realistic analyses of future energy consumption patterns. In general, an "industrial" society is characterized by various forms of materials fabrication (i.e., by manufacturing and heavy construction). A "post-industrial" society is characterized by the offering of services, such as the supply of utilities, financial advice, recreation, and information (Bell, 1979). In fact, one would expect a number of fabrication activities to coexist with service activities in a society that is nominally post-industrial. Consider, by way of analogy, the continued existence and importance to this country of pre-industrial agricultural activities. Furthermore, one would expect that services apart from transportation and recreation would continue to be electricity intensive. However, it is the mix of economic activities that must be understood to make meaningful projections of future energy consumption patterns. Thus, the future linkage between electricity demand and economic growth will remain unknown until it is possible to determine whether post-industrial America will accommodate services and much-expanded industrialization, at one extreme, or at the other extreme, very little beside services.

As is indicated by the material accompanying Exhibit IV.3, there are persuasive arguments to support the proposition that industrial activities that remain in this country will be more electricity intensive than they have been in the past.

Finally, there are technological and social factors that obscure future electricity demand projections. Most electricity end use technologies are already much more efficient than were oil and gas end use technologies during the 1970s. Therefore, one would expect much less dramatic electricity demand impacts associated with more efficient end use equipment and processes. New electricity technologies, nonetheless, will be more efficient than their predecessors, and this will tend to lower demand growth (RB:IV-1). Implementation of time-of-use electricity pricing is also widely expected. However, its implementation time frame is a matter of conjecture, as are its effects on load profiles and total electricity consumption. Population migration patterns are difficult to predict and likewise strongly affect regional electricity demand (RB:IV-2) (SM:IV-4).

Significance to policy decisions. If future demand for electricity is uncertain and if it is important that power demands be met at any reasonable level then it follows that appropriate policies must run the risk of developing capacity that may not be needed. This "insurance" can only be had at some cost. To the extent that flexibility to supply future demand is explicitly acknowledged as a criterion for the formulation of electric power policy, costs incurred to meet this criterion will not only be prudent, but will obviously be so to knowledgeable observers.

In Chapter III, several actions are discussed that might be taken to lessen the cost of maintaining flexibility to meet a range of potential power demands. One important set of possible actions involves programs to understand better the linkages between economic growth and electricity demand. Especially important is a clear concept of how different national economic structures will affect the demand for electricity. The results of such programs will define more clearly the range of future demand that is reasonable. Also, definition of how demand could vary between geographic regions will be important.

Policies that encourage experimentation with load management techniques and time-of-use rate schedules should also bear useful results. They

Exhibit IV.3
Industrial Electricity Consumption, 1950–1982

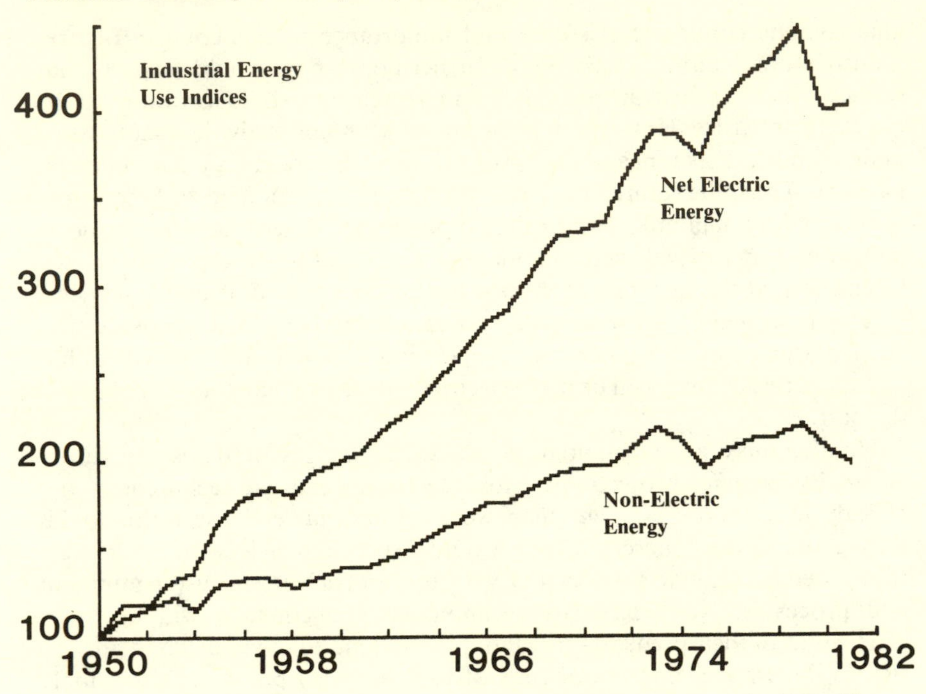

Sources: EIA, 1982; MITRE Corporation, 1982; EPRI, 1982; Schurr et al., 1979; Schurr, 1982.

Note: Industrial use of electricity has grown more rapidly during the past 30 years than has industrial use of non-electric energy. The following arguments support the proposition that this trend will continue:

1. *Price advantage*. Electricity price is likely to escalate less than the prices of its industrial substitutes, oil and gas, because electricity can be generated from other primary fuels, such as coal, for which prices will escalate less than those for oil and gas. This argument assumes that the capital amortization element of electricity prices does not escalate enough to offset fuel cost advantages or that industry energy conversion plants experience similar capital cost escalation.

2. *Materials savings*. Resource scarcity places a growing premium on heat treating, forging and drying technologies which lessen materials waste (e.g., through irregular shrinkage, scaling, and discoloration). Electrical processes can do this. The relatively high energy costs of electric processes are increasingly offset by the materials savings made possible by the precision and uniformity with which electrical energy can be applied.

3. *Cleanliness*. Electrical energy is very clean at the point of end use. This can enhance product quality and lessen work place health hazards.

4. *Robotics and synergy with robotics*. Robot automation of industrial processes is widely anticipated during the next 20 years; these robots will be electrically fueled. There may also be synergistic factors (e.g., low bulk purchase electricity rates and electrically knowledgeable work forces) which militate in favor of electrifying other industrial processes.

5. *Total factor productivity*. There is an emerging body of productivity research that suggests electrification of industrial processes enhances productivity. The research is still largely anecdotal but points to reasons such as improved work flow, lower loss per unit of equipment downtime, and lessened material waste.

will provide hard data on the demand-reduction potential of these innovations. This, again, could cut the monetary costs of flexibility. Experimentation would help to define the probable effectiveness of tools that might be employed in the event future electric power demands are greater than anticipated.

In the development of Chapter III, when consideration was given to society's probable reaction to a belated realization that future power needs exceeded those postulated in the "Average Future" Scenario, it was apparent that cogenerators and small power producers would be helpful in meeting electricity demand. However, it is not likely that these will be around unless their interim existence is encouraged by regulations during the 1980s. Thus, regulations that currently offer varying degrees of encouragement to small power producers and cogenerators would seem to be prudent, even though the short-term economic justification for encouraging power supply from these resources during the 1980s may be questioned.

Associated Questions: Overarching

1. By what process can this nation's future economic growth be projected?
2. What will be the relationship between economic growth and electric power demand over the next three decades?
3. To what extent will future improvements in end use efficiency affect electric power demand?
4. To what extent will future differentials between the price of electric power and the price of other energy forms affect electric power demands?

Associated Questions: Operational

1. At what rates will the economy in the United States grow over the next three decades?
2. How will the price elasticity of electric power change over the next three decades?
3. To what extent will increased conservation measures in both industry and consumer markets change demand for electric power?
4. How will the demand pattern for electric power be affected by an increasingly affluent society?
5. How will electric power demand patterns be changed by new rate structures, such as time-of-use pricing?
6. How will the increasing size of the older population of the United States change demand patterns?
7. How will changes in household size (probably decreases) affect electricity power usage?
8. How will changes in exurban, suburban, and urban population patterns affect electric power use?

9. To what extent will the increasing percentage of women in the labor force change power patterns?
10. How will the change in the nature of the work week affect electric power patterns?
11. How will U.S. movement to a post-industrial or high-technology society affect electricity demand?
12. Does electric power policy need to plan for the advent of electric automobiles?
13. What will be the regional electricity demand growth patterns in this country during the next 30 years?
(SB:IV-2.)

5. The Range of Practical Electric Power Generation Sources May Expand Dramatically Over the Next Three Decades and the Relative Attractiveness of Traditional Sources May Change Significantly

Rationale. Most industry authorities predict primary dependence on coal technology for new electricity generation capacity for the remainder of this century. Some authorities predict renewed dependence in the late 1980s to early 1990s on nuclear power plants. However, there are reasons to believe that public resistance to plants of either type could increase in the future. If so, this could coincide with technical developments that make other generation sources more attractive than they have been in the past. Although it is hard to conceive of major changes in the nation's electricity generation mix over the next decade, there could be important shifts in later years that should be anticipated during the 1980s.

Presently, the generation source most subject to public resistance is nuclear power and, therefore, its future role is most fraught with uncertainty. There remains the finite chance that another Three Mile Island, or something worse, could occur. Even in the absence of obvious failures, the strength of current opposition to nuclear power creates doubt about its future use (Rosa, 1983). Considering coal generation technologies, it is possible that public awareness of environmental and health problems (e.g., acid rain, carbon dioxide buildup, and carcinogenic particulates) associated with power production from hydrocarbons will increase during the 1980s (see Chapter VI). These environmental and health problems may prove to be inconsequential or may prove to be remediable through technical modifications. On the other hand, they could foster public opposition to coal analogous to that developed to nuclear energy in the 1970s.

It is also possible that unconventional power generation technologies will prove commercially viable before the end of this century. The large research investments made in unconventional electric power technologies in the 1960s and 1970s should generate technological return. Magnetohydrodynamic power, perhaps even fusion power, could prove commercially feasible by the end of this century (RB:IV-3). Power from solar satellites or ocean thermal engines are other possibilities. It is widely assumed that electricity from solar energy will not prove commercially feasible, at least on a large scale, this century; however, many people also believe that continuing solar technology progress will prove these assumptions wrong. Although rooftop solar generation is the archetype for small-scale power generation, there are other generation alternatives, such as fuel cells, that could quickly become feasible at capacities much smaller than those of today's coal and nuclear units.

The impacts of social movements on future generation options are very difficult to anticipate. This is particularly so when the movements are international in scope. Increasingly, it appears that generation options are affected by such movements. The acceptability of nuclear power will continue to be influenced by international power plant safety and international political movements. World reaction to the environmental consequences of carbon dioxide buildup will have a direct impact on utility reliance on hydrocarbon fueled generating facilities; carbon dioxide, of course, is generated worldwide and the debate on its buildup will be of international scale.

Significance to policy decisions. The significance of this principle for electric power policy largely parallels the significance of the preceding principle, which dealt with demand uncertainty. Because one cannot be sure in the 1980s of the acceptability of various power generation technologies early next century, policy makers should seek to avoid disproportionate commitment to any one technology and should maintain future options by formal support of a broad spectrum of generation technologies. Most projections for future power generation, and all scenarios considered in this assessment, postulate a dominant role for decades to come for coal-based technologies. Serious problems of power generation from coal would cause very serious problems in meeting future electricity demand, so special attention should be devoted to coal and coal technology research. Were it to prove that projected expansion of coal usage for power generation presents unacceptable problems, nuclear generation is the most likely candidate to fill the supply gap. Therefore, continued efforts to improve nuclear fission reactors and waste-handling technologies seem warranted, regardless of wide-spread current disenchantment with nuclear power.

The possibility that conventional generation units might become inappropriate by the late 1990s undermines, with reason, the confidence with which long-term capital investment decisions can be made during the 1980s. Consider, for example, the plight of those utilities that invested heavily in the 1960s in oil- and gas-fired plants. Except in special circumstances (e.g., long-term, low-cost gas supply contracts), the market value of those plants decreased more rapidly in the 1970s than the rates at which utility companies were allowed to depreciate them. The difference between market value and book value was a capital loss that, by and large, was borne by utility shareholders as their shares traded for lower prices in the stock markets (Berndt, Cowing, and Wood, 1983:5.2).

Now the capital costs of oil and gas capacity are low compared to capital costs of nuclear or coal capacity. Were political or technological forces to render coal or nuclear plants prematurely obsolete, the aggregate loss would be very large. These losses could be too large to be absorbed solely by utility shareholders and, therefore, might accrue to ratepayers and, in extreme cases, taxpayers. It would behoove state and federal regulators to review their policies on depreciation and loss sharing while the need is still speculative and deliberate consideration is possible. The continuous reassessments given nuclear power prospects are reason enough to honor this type of contingency planning.

Analyses conducted within the context of this assessment's two scenarios that contemplate radically unconventional power generation technologies, the "Mega-plant" Scenario and the "Post-industrial" Scenario, point to problems in attracting capital for the construction of conventional power-generating facilities. These facilities will be needed during the transition to the new technologies, but they will have their economically useful lives foreshortened by the new technologies. This is a potentially important policy problem, though it is not as immediate as others.

Associated Questions: Overarching

1. By what process can appropriate criteria for future generation mixes be determined? What are reasonable criteria?
2. How should the development process for generation mix criteria and the criteria themselves change with the passage of time?
3. For each present and potential generation technology, what are the most important unknowns affecting its future viability?
4. What are the appropriate roles of state and federal government in encouraging investments in particular power generation options?
5. What can be done to influence the evolution of future generation mixes that are not favored by financial analysis alone? (SB:IV-3).

Associated Questions: Operational

1. How well does the public understand health, safety, and resource problems associated with renewable energy sources, such as biomass, solar, geothermal, and wind?
2. To what extent can electric power assist in the developing of liquid fuels, such as hydrogen and hydrozene, for use in transportation?
3. How does the availability of synthetic liquid and gas fuels affect generation choices in the electric power industry?

4. Will the capital costs of coal-based generation escalate in the future, as the capital costs of nuclear-based generation have in the past?
5. At what real rates will oil prices escalate in the 1980s and early 1990s?
6. What should be done with solid wastes from pre-combustion coal cleaning?
7. How can nitrogen oxide and sulfur oxide atmospheric transport/transformation phenomena be accurately modelled?
8. Are long-term cogenerator supply contracts inconsistent with the flexibility cogenerators might provide?
9. Is it desirable to encourage cogeneration and small power production as insurance against *future* need? If so, are current incentives sufficient?
10. How can harmonic content, circuit protection, and so on be assured if cogenerators and small power producers proliferate?
11. How should the "marginal unit" of power production be defined for purposes of calculating the "avoided cost" of that unit?
12. What are the most promising contingency plans for generating power if reliance on coal plants proves misplaced? (What are time frames to develop these alternatives?)
13. What grid control/management problems are peculiar to large numbers of dispersed generating sources?
14. Are there useful correlations between the demographic characteristics of a society and the degrees of acceptability that society accords various power generation options?

6. The Roles, Structures, and Procedures of Electric Utilities Could Change Significantly in the Relatively Near Future

Rationale. The image of the privately owned electric utility as a conservative, highly regulated, monopolistic entity has developed steadily over the years since the passage of the Public Utilities Holding Company Act in 1935. There are indications that changes in this image are taking place and will accelerate in coming years. The Public Utility Regulatory Policy Act of 1978 mandated that private utilities buy power from other producers on a marginal-cost basis and that they wheel power for other electricity producers. The National Energy Conservation Policy Act, also of 1978, requires that they promote loans for customer conservation activities. These requirements not only define new roles for private utilities, but also encourage the expansion of third party and cogenerator power production. Thus, the roles of utilities may be shifting from the ones traditionally attributed to them to those of more broadly based energy service corporations.

Arguments favoring such a transformation are persuasive. For one, it is consistent with the interests of financial institutions, utilities, and energy end users that utilities serve as financial conduits and energy advisers when electricity demand growth is uncertain and energy prices are high. For another, an expanded common carrier role offers utilities risk reduction through product diversification and an entry into the field of power brokerage services should need for those services arise. Two of the scenarios considered by this assessment, the "Post-industrial" and the "Small Coal Plants" Scenarios, describe futures in which ultimate consumers will likely need the services of intermediate power brokers.

By the turn of the century, the development of new types of generating technologies could also change the basic structure of the electric utility industry. For example, very large capacity units, such as fusion power plants or solar power satellites, could result in new relationships between utilities and the federal government. The latter might play a major role in financing these large units. It is also possible that solar satellites might be owned by a semi-public organization, such as the Communication Satellite Corporation (COMSAT), with the utilities providing only transmission and marketing functions.

At the other end of the spectrum, the emergence of practical solar photovoltaic power generation could beget major new electric power institutions. Although it is possible that solar-based power production could be most efficiently realized via central stations operated by utilities, it is also possible that the technology would allow generation on a very small scale, perhaps even on a home-by-home basis. Local generation by community corporations or large businesses would also seem a possibility. Hence, it could be that a two-tiered supply system would develop. The traditional central station utility system might exist even as new, locally owned and locally controlled systems come into existence.

In considering changes to the electric utility system, however, it must be kept in mind that there is tremendous inertia in the present system (RB:IV-4). The technologically useful life of power plants is long: 30 to 40 years. Any scenario that acknowledges the social benefit of economic efficiency has to postulate that at least two-thirds of the electric plant capacity on line in 1980 will be on line in 2010. Further, the electric utility industry has tremendous economic leverage, with approximately $250 billion invested in capital assets (in 1980 after depreciation) (Edison Electric Institute, 1981; Lovins and Lovins, 1982). A vast infrastructure, which in some way touches almost every middle or upper income American, has developed to support this industry. Under these cirumstances, institutional restructuring will likely be more incremental than revolutionary.

Significance to policy decisions. Those responsible for electric power policy should remember that institutional change can originate within the industry, largely in the absence of regulatory direction, or it can occur in response to regulatory direction. In the former case, regulators need to reach an early accord as to what serves the public interest in areas where regulation is not currently a concern. For example, several of this assessment's scenarios, particularly the "Small Coal Plant" Scenario and the

"Post-industrial" Scenario, contemplate growing acceptance of utilities as energy service corporations; this expanded role places regulators in the position of having to decide whether it is in the best interests of society to have utilities or financial institutions make energy conservation loans to consumers. In like manner, regulators may be required to determine whether it best serves the public interest for utility power brokerage services to be offered on a franchise or competitive basis.

Where electric power institutions are induced to change because of regulatory initiatives, it is important that regulators be aware that these initiatives may lead the institutions into areas in which they lack necessary resources, such as specialized management expertise. The effect of the inducement, therefore, may not be exactly as intended by the regulator. Also, note that movement of a regulated utility into new business areas may create conflicts between conventional electric utility regulators and other regulators whose jurisdictions encompass or could encompass the utility's new activity.

One of the major ways in which the electric power institutional structure might be different in the future than it is today is by the addition to that structure of small power producers. It is common to think of a small power producer as a risk-taking entrepreneur carving a profitable niche in power supply by dint of regulatory inducement, management skill, and innovative technology. The prospect should be borne in mind, however, that the small power producer could also be a giant energy company or even an electric utility.

Associated Questions: Overarching

1. To what extent should utility structural change be induced by regulatory policy, and to what extent should this change be a matter of utility discretion? By what process can this question be answered?
2. What are appropriate new commercial roles for electric utilities? By what process should the appropriateness of potential new roles be determined?
3. If extensive physical centralization or decentralization of power generation comes about, what new types of electric power institutions might result? What would be the relationships among these new institutions and conventional electric utilities?
4. What new problems might result from movement of electric utilities into business areas where they have not historically operated?

Associated Questions: Operational

1. Will changes in the public's demand for reliability affect the organizational and operational rules of electric power groups?
2. Will the need for large amounts of capital change the ways in which electric power utilities interact with financial institutions?

3. What would be the electric power impact of major movement of U.S. capital to other countries?
4. To what extent would foreign investment in electric utilities affect those utilities?
5. If the regional nature of the electric power industry increases, will there be a tendency toward mergers in the industry?
6. Will government insistence that utilities provide minimum power to all customers change the relationship between the electric utilities and the government?
7. Is the role within the industry of publicly owned utilities likely to increase?
8. Are there special maintenance problems of importance to small power producers and cogenerators?
9. Does diversification adversely affect regulated activities through intangibles, such as loss of skilled management?
10. Is "goodwill" enjoyed by a regulated utility a shareholder asset only? If not, does an unregulated utility subsidiary receive goodwill for which it owes ratepayers?
11. Will presence of good returns in unregulated businesses cause regulators to depress regulated rate of return?

7. The Means, Extents, and Purposes of Electric Utility Regulation Could Change Significantly in the Relatively Near Future

Rationale. Some of the societal, technological, and institutional transformations discussed in the previous four principles will surely become evident during the next three decades. Some will be evident during the 1980s. As these changes occur, agencies responsible for regulating electric utilities will likewise be transformed to accommodate new realities.

Consider the element of regulatory jurisdiction. Electric utility regulation was once exclusively a state function. Initially, electric utilities were local concerns. Their power was produced largely from local fuels and sold to local customers under terms sanctioned by regulators no further removed than the state capital. Electric utilities began shedding the trappings of "local institutions" in the early 1900s with the advent of public utility holding companies. Since then, and particularly since World War II, the interconnection of utility grids and the formation of utility power pools have elevated utility interaction and coordination to regional and supra-regional levels. Utility regulation, particularly non-price regulation, has become correspondingly less local. Especially since the advent in the 1960s of the environmental movement and the Arab oil embargo of 1973, the jurisdictions of, respectively, the Environmental Protection Agency and the Federal Energy Regulatory Commission (previously, the Federal Power Commission) have expanded and brought a more national perspective to electric utility regulation. The global nature of potential coal and nuclear waste problems suggests that internationalization of some aspects of utility regulation is not unlikely. Further, attention focused during the 1970s on resource scarcity, both real and potential, has added legitimacy to the theory that the earth's natural resources are to some extent the property of all nations. The debate over seabed mining, for example, reflected this. Along these lines, this assessment's consideration of regulation under the conditions of, particularly, the "Mega-plant" Scenario point to the potential for greater international regulation of utilities.

Of potentially more importance than shifts in the geographic jurisdictions of utility regulatory bodies are shifts in the scope and process of regulation. Consider first the scope of price regulation.

The business of electricity supply has historically been a monopoly business. The reasons for this are discussed in some detail in Chapter V. Basically, this circumstance arose because the unit cost of generating and delivering power was observed to decrease as the amount of power generated and delivered increased. Competition among power suppliers within one market would have deprived society of the opportunity to realize scale economies fully, so competition within a single market was generally forbidden. Utility price regulation developed as a surrogate for competition. Regulation allocated the benefits of declining power costs between the owners (shareholders) of the generation and delivery system, on the one hand, and the electricity consumer, on the other.

When the costs of electricity supply began to rise in the 1970s, the regulatory system, with varying degrees of efficiency, translated these cost increases into end user price increases (see Exhibits V.1 and V.2 in Chapter V). Thus, during the 1970s, utility price regulators came to allocate not benefits but costs. Though it is by no means obvious that theoretical justification for electricity price regulation no longer exists, the fact of rising electricity prices has focused debate on the price-setting process. This debate arose during the time when policies concerning other regulated business activities, such as trucking, airlines, and interstate communications, increasingly favored competition as a more efficient means than regulation of suppressing price. This debate also arose within the context of the highly integrated power transmission and distribution grid and multiple utility power pools that have developed in this country since World War II. These have effectively created a national market for each electricity generator. Under these circumstances, previously suspect suggestions that some form of electricity price deregulation might be in order have recently gained greater credibility (Joskow and Schmalensee, 1982; Bohn et al., 1982; Niskanen and Zych, 1982; Killian and Trout, 1982).

It is probable that the circumstances that last decade focused debate on electricity price deregulation will gain momentum in the coming decade. As discussed briefly under Principle 4 and as discussed at some length in Chapter V, this assessment has found no reason to believe that electricity prices will again decline for any meaningful period during this century; even should variable (fuel) power generation costs decline, fixed costs (plant depreciation) will increase.

Furthermore, many people believe there are technical, economic, or political reasons why the generation of electric power by small units is preferable to the generation of electric power by large units. If this opinion becomes more prevalent, it is probable that small power producers and industrial cogenerators will begin to penetrate the electricity supply market, at least as long as they are allowed to do so on an unregulated basis. If they do begin to penetrate the electric power market, the likelihood of price deregulation of electricity generation will grow more probable. The small power producer phenomenon will presumably not occur unless there is a competitive advantage in deregulated small power production, and, if there is such an advantage, it is difficult to believe that regulations would be maintained that locked the utility industry into competitive obsolescence. This is the conclusion to which this assessment came in its consideration of policies that would be consistent with the technological and societal assumptions of the "Post-industrial" Scenario and, to a lesser extent, the "Small Coal Plant" Scenario.

Rather rapid change in the process of utility regulation also seems likely to continue. In part because of rising electricity prices, electric power policy formulation throughout the 1970s was characterized by an unprecedented level of factiousness. Especially in the setting of rates and siting of power plants, countervailing considerations were frequently debated at adversarial extremes. The policy formulation process grew more dilatory as hearings took on the trappings of media events. Certainly, electric power policy formulation ceased to be principally in the arena of experts working quietly within bureaucracies; it entered the public political domain. An examination of already visible electric power policy issues (nuclear waste disposal, acid rain prevention, implementation of lifeline rates, etc.) for the coming decade strongly suggests that they will be addressed in the same highly political fashion.

Significance to policy decisions. Given the impact of regulations on electric utility structure and operations, it is obvious that this principle and the previous one are closely related. Together, they indicate that the institutions most affected by electric power policies are changing, that the substance about which policies are formulated is changing, and that the mechanisms for implementation of policies are also changing. Under these circumstances, with three important variables in flux, policy makers must be particularly alert to the possibility that the effects of their actions may be quite different from those intended. Thus, it will probably be desirable that regulatory changes be made in an incremental manner and that carefully monitored pilot programs be undertaken.

One area of particular importance in this regard is that of price deregulation. The concept of price deregulation is relatively new in modern electric power policy considerations, and the potentials for miscalculations are many. Nonetheless, material in Chapter V and consideration of scenario policies that could be consistent with cogeneration and small power production suggest that some forms of price deregulation are likely in the reasonably near future. Therefore, encouragement of controlled experimentation in the application of this concept would seem well directed.

In considering varying degrees of price deregulation, it is important to remember that the choice is not between imperfect regulation, on the one hand, and some type of perfect, free market pricing mechanism, on the other. Rather, the choice is between pricing mechanisms that are imperfect. Further, it is important to consider that price deregulation at one stage of the electricity supply system (e.g., the deregulation of bulk power sales) will probably call for modification of regulations pertaining to other, regulated stages of the power supply system.

This assessment's analyses of utility regulation within the contexts of the "Mega-Plant" Scenario, the "Small Coal Plant" Scenario, and the "Post-industrial" Scenario raise the possibility that the state role in utility regulation may not be as dominant in the future as it is today. Technological or demographic trends associated with those scenarios point to the wisdom of increased regulatory responsibility at, respectively, the international, regional, and local levels. Regulatory infrastructures are commonly not well developed at these levels, and public policies promoting and supporting development in these areas would appear well advised (RB:IV-5).

Associated Questions: Overarching

1. By what techniques can the higher order implications of potential utility regulatory actions be anticipated?
2. By what methods can criteria be developed to determine the most rational geographical jurisdictions for various electric power system configurations and activities? What are these criteria?
3. By what means could jurisdictions based on these criteria be implemented?
4. By what procedures should regulatory test programs be evaluated for their degrees of efficiency and equity?
5. What should be the regulatory approach to determining the means and degree of electricity price deregulation?

Associated Questions: Operational

1. Will increased dispersion of power generation sources cause more interest in local regulation of power facilities or utilities?
2. How can the additional cost of pollution control equipment be allocated between stockholders, ratepayers, and the general public (SB:IV-4)?

3. How might power plant operator training and performance standards be made international in the event that becomes desirable?
4. How might international agreement on carbon dioxide, acid rain, particulates, and nuclear wastes be reached and policed?
5. How can federal and state regulation be coordinated to reduce regulatory duplication without subordinating states' roles?
6. What are the consequences of allowing utilities or large corporations to be small power producers?
7. How much autonomy from regulation can small power producers and cogenerators be allowed, consistent with high electric service reliability?
8. What will be the fate of cogenerators and small power producers, whose existence has been encouraged (subsidized), in the event of electric power price deregulation?
9. How can regional regulatory cooperation be enhanced?
10. If publicly owned power systems become more numerous or if their power sales expand much more rapidly than power sales of privately owned utilities, does this importantly affect PUCs' ability to set power policy?
11. Are traditional utility capital depreciation and loss allocation policies still sound?
12. Is power generation price deregulation inconsistent with the planning environment required for successful large-scale power projects?

8. It Will Be Increasingly Difficult for Electric Power Policy to Strike Acceptable Balances among Considerations of Efficiency, Equity, and Risk

Rationale. Efficiency (optimum allocation of resources), equity (fairness), and risk (exposure to loss or peril) are fundamental human concerns of obvious importance to the future of electric power policy. The achievement of an acceptable electric power policy balance among these concerns has always been difficult. However, public and private value systems changing within the contexts of an increasingly diverse and complex society will make the striking of this balance even more difficult in the future. The 1980s promise to be a decade during which risk concerns, particularly environmental and health risk concerns, will remain high, while rising power costs will argue against further compromise of efficiency and equity concerns.

The historical influences on 1980s electricity regulatory policies derive from at least the early 1900s. At that time, utility regulation stressed engineering and economic efficiency (Garfield and Lovejoy, 1964). During the depression of the 1930s, society came to expect electric power policy to give explicit consideration to matters beyond efficiency. Equity was the area of general concern. This general concern is reflected in federal regulatory legislation of the period: the Social Security Act (1933), the Federal Communications Act (1934), the Public Utilities Holding Company Act (1935),

Civil Aeronautics Act (1938), and the Natural Gas Act (1938). Creation of the Tennessee Valley Authority (TVA) in 1933 and its history since then vividly show the extent to which the public believed equity should be a consideration in the formulation of electric power policy. The TVA had, or at least took, as its mandate the improvement of material well-being throughout much of the southeastern United States. The nation supported millions of dollars of federal TVA expenditures to bring this underdeveloped region to closer parity with the rest of the country.

During the late 1960s and 1970s, society began to expect electric power policy to give explicit consideration to environmental and health risk. Again, the fact of public concern with these issues in general is reflected by the federal legislation of the period: the National Environmental Policy Act of 1969, the Federal Clean Air Act Amendments of 1970 and 1977, the Federal Water Pollution Control Act Amendments of 1972, the Resource Conservation and Recovery Act of 1976, and the Toxic Substances Control Act of 1976. This assessment uncovered no support for the proposition that the 1970s' environmental movement will prove to be short lived. Indeed, the mean response to this assessment's opinion survey predicts a steady increase during the next 30 years in citizen support of environmental and conservation efforts (Appendix B, Item 32). This conclusion is consistent with the conclusions of numerous other inquiries into public environmental and conservation values (see, for example, Continental Group, 1982; Milbrath, 1981) (GE:IV-5).

The rise in public concerns about environmental and health risks greatly affected electric power institutions during the 1970s. For example, a study of all 100 megawatt and greater coal and nuclear plants completed between January 1972 and January 1979 found that capital costs per kilowatt of capacity for coal plants increased 68 percent over the period and those for nuclear plants increased 142 percent. These increases were in addition to general cost inflation in the steam electric power plant construction industry and, in each case, were hypothesized to have arisen "primarily because of efforts to prevent total accident and environmental risks from expanding in proportion to the growth of either sector" (Komanoff, 1981:2). The impact of environmental and health risk concerns on electric power could be even greater in the next two decades because it is during this period that the activist, health-conscious "baby boom" generation will both gain political dominance in the country and become aware of its own mortality. Nuclear wastes, carbon dioxide buildup, acid rain, and small particulate carcinogens may be particularly worrisome to this generation as it ages.

Efficiency, equity, and risk considerations can rarely be simultaneously maximized. There is, therefore, great potential for strife and contention surrounding attempts to find acceptable balances of the three during the 1980s (RB:IV-6). With electricity costs rising, there will be opposition, for example, to operating coal-fired boilers at less efficient rates; that, however, seems to be the most feasible means of controlling nitrous oxide pollution. Retrofitting older coal-fired plants with costly new pollution control devices will be sought by some groups and opposed by others. Lifeline rates for electric power may be equitable, but they are inconsistent with economic efficiency in some situations. The burial of high-level nuclear wastes in the lowest risk geological formations may, nonetheless, seem inequitable to the citizens and their representatives in states where those formations are found.

Most potential points of policy contention are related to electricity generation as opposed to the availability of electricity once it is generated. With important exceptions (e.g., nuclear proliferation), the negative aspects of power generation are not experienced equally within society and frequently discriminate as to location and time.

Issues that promote divisiveness based on geography are principally those involving the coal and nuclear fuel cycles. The interests of high sulphur, deep-mined coal producers in the Midwest are not, in many cases, those of low sulphur, surface-mined coal producers of the West. What is perceived as equitable by coal producers in Wyoming and Colorado is, in many cases, not perceived as equitable by electric power users in Texas and Oklahoma. To the extent that electric power is generated at sites remote from the sites at which it is consumed, the environmental, health, and social burdens borne at the generation site are not obviously equivalent to the financial burdens of the remote power consumers.

There are also numerous potential intergenerational conflicts associated with electric power production. The most obvious of these involve nuclear waste processing and storage. At a more pedestrian level, regulatory treatment of the costs of construction work in progress will continue to cause price conflicts between current ratepayers and future ratepayers. There may also be serious long-term conflicts related to worker health. For example, fugitive emissions from hydrocarbon fuels could cause chronic health problems among workers exposed to them (Ramsey, 1979; MITRE Corporation, 1980) (i.e., a problem analogous to asbestosis could materialize).

Opportunities also continue to exist for contention between electricity ratepayers and utility shareholders. Fuel price adjustment clauses were a highly visible source of such divisions during the 1970s and early 1980s. Depending on one's perspective, these clauses either unwisely reduced utilities' incentives to minimize fuel costs or wisely insulated utilities' revenues and capital assets from unpredictable surges in fuel prices. If fuel prices rise predictably in the future, there will be less need for fuel price adjustment clauses; if fuel prices rise unpredictably or fall (perhaps prompting fuel price credits), adjustment clauses will continue to divide ratepayer and shareholders. A more probable source of future divisions is the allocation of losses from what are seen in hindsight to have been poor capital investment decisions by utilities.

Finally, old opportunities will remain and new ones will be opening for conflict among classes of customers. Historically, the commercial customer has not had the policy influence of either the large industrial customer or the politically important residential customer. This customer has been "underrepresented" in the determination of equitable treatment for customer classes. As the importance of the service sector to the economy grows, this imbalance is likely to change and, in changing, aggravate friction among customer classes.

It seems obvious that the process of utility regulation and policy formulation will undergo some adaptation in an effort to strike an acceptable balance among competing efficiency, equity, and risk concerns. At the least, it appears that the process must come to value more explicitly not only that which is measurable, but also that which, though intangible, is valued.

Significance to policy decisions. The most significant policy implication of this principle is that it postulates an electricity regulatory process that will, in the future, be even more political than it is today. With this politicalization of the regulatory process, one would expect that the role of the electric power specialist will further diminish and that it will be increasingly difficult to maintain a long-term view of electric power issues. Analysis within the contexts of each of the scenarios postulated for this assessment indicates that a long-term perspective often cannot be maintained in a politicized regulatory process. This was particularly evident when considering the "Mega-plant" Scenario (to which a visionary regulatory perspective is essential) and the "Economic Malaise" Scenario (in which financial distress raises imperative short-term problems).

There are steps that can be taken to modify the regulatory process to improve its performance in a political environment. First, processes that give formal recognition to the importance of equity and risk as well as efficiency will lead to a balance of these concerns that will be and, as importantly, will appear to be consensual. Greater use of citizen advisory boards and, perhaps, the electoral process are examples of modifications that would have this effect (SM:IV-5). The creation of formal posts for ombudsmen or guardians *ad litem* to represent explicitly the interests of future stakeholders in electric power decisions would also further this objective (KH:IV-1). In fact, it will be particularly important for policy makers to formalize mechanisms for understanding and giving weight to future impacts of policies formulated in current political environments.

Associated Questions: Overarching

1. By what process can important but, conventionally, unquantifiable electric power concerns be integrated with quantifiable concerns in the electric power policy-making process?

2. By what body or bodies and by what process can international equity regarding electric power policy be assured?
3. By what process should standards be developed to allocate across society the economic costs of reasonable electricity supply decisions that ultimately prove unwise? What should these standards be?
4. In what ways can political processes be made more effective in determining the balance of efficiency, equity, and risk?
5. To what extent is it the responsibility of government to ensure adequate expression of the concerns of electric power stakeholders with limited resources?

Associated Questions: Operational

1. In what ways can electric power policy decisions be made more sensitive to intergenerational equities?
2. To what extent should the government subsidize the development of particular new energy sources?
3. How can the government protect the rights of citizens' groups without unduly affecting the times required to build electric power facilities?
4. To what extent should the people located in the immediate vicinity of power facilities be allowed to impose specific health and safety demands?
5. How could the effect of electric power plants on other nations be adequately considered in the political choice process?
6. What special environmental and health protection problems are posed by small power producers and cogeneration?
7. Can the nuclear waste storage issue be "resolved" through the electoral process (e.g., referenda)?
8. Under what conditions, if any, should the federal government, hence, taxpayers, assume cost burdens for incomplete or economically obsolete power plants?
9. Does increased politicalization of the regulatory process imply increased election of PUC commissioners? If so, are adequate campaign and election standards in place?
10. Would "expensing" of utility environmental safeguarding/cleanup costs promote environmental quality at an acceptable price (KH:IV-2)?
11. How might electric power community impact insurance be structured?
12. How "elastic" is environmental concern to economic hardship?
13. By what theories can policy makers quantify tangible and intangible benefits of technology and policy alternatives?

9. Broadly Based, Long-Term Research, Development, and Demonstration Programs Are the Single Most Efficient Means of Permitting Flexibility in Electric Power Policies and Decisions

Rationale. Properly conceived and executed RD&D programs contribute in a number of ways to efficient and equitable electric power planning, operations, and utilization. RD&D gives earlier and better definition of future technological options and provides analyses of opportunities and problems these new options may present. The spectrum of technological choices it can provide will allow policy makers greater latitude in their decisions; it can create policy opportunities where none would otherwise exist. In addition, RD&D can identify and evaluate trends in demographics, economics, and human value systems (GE:IV-6); it can help anticipate critical decision points and assist in analyzing the trade-offs characteristically involved in policy decisions. In short, appropriate RD&D programs can provide both new technological and social science capabilities and analyses of the best uses of those capabilities.

The history of electric power over the last century has been characterized by continuing technological advance (see Chapters IX and X). In some cases, change has been dramatic and revolutionary, as in the switch from direct to alternating current and the introduction of nuclear-powered steam generation. In other cases, change has been incremental, as in the continuing increases in generation capacity and reliability. There are a number of reasons to believe that technological change will continue in the next three decades at a rate at least equal to, and probably greater than, that of the past. The environmental movement, the oil embargo, and the drive to increase industrial productivity spurred greatly enlarged government support of many RD&D programs of direct interest to electric power technology. With the formation of, first, the Energy Research and Development Administration and, later, the Department of Energy, programs were started and expanded in, among other areas, solar energy, nuclear power, fusion power, conservation, load management, ocean thermal energy, low head hydropower, and geothermal energy. Similar projects were also supported by the electric power industry, both through individual utilities and the newly formed Electric Power Research Institute. The results of many of these programs were not immediately available for practical utilization. The payoffs for a number of the RD&D efforts, however, should begin within the next few years.

Some recent theories on innovation patterns support the anticipation of major technological breakthroughs in the near- to mid-term future (Mensch, 1979; Graham and Senge, 1980). These theories postulate that the introduction of basic technological innovations does not take place at a uniform rate but, rather, in dramatic, periodic bursts. According to these theories, it appears that the application of new technologies is about to enter a "burst" period. Recent developments in electronics, materials, biotechnology, and space utilization would seem to confirm this contention.

Each of the scenarios described in Chapter II assumed the nation would build on past RD&D discoveries and continue to conduct RD&D at some level. However, each assumed different overall programs. RD&D programs appropriate for each scenario are discussed in Chapter IX and a program designed to serve a spectrum of possible futures is outlined in Section H of

Chapter III. As indicated in Chapter III, technological RD&D can accomplish two ends. Product and process RD&D can, at any point in time, provide technological choices for the present. Exploratory RD&D can define what is possible for the future, as well as estimate necessary support levels, reasonable development times, relative costs and benefits, and associated problems and opportunities.

Given the uncertainties highlighted by the preceding principles, a wide range of technological alternatives would be of major value to future electric power decision makers. This observation is borne out by discussions in the previous two chapters, where different technologies were required to meet the demands of various postulated futures. One factor was not stressed in those chapters, however: the value of RD&D in defining technological capabilities and special technological and non-technological support requirements. For example, all of the scenarios, but, particularly, the three assuming high energy use, postulate heavy dependence on coal, both for industrial and power generation purposes. Experience could prove this impossible. Late discovery of this reality would result in calamity. However, expanded research into the nature, mechanisms, and seriousness of the risks of using coal might reveal the need for developing other energy sources in time to minimize the trauma. It is also possible that means for fairly and economically alleviating the deleterious impacts might be developed.

An expanded range of available technologies can also permit more latitude in policy implementation. To illustrate this point, compare the "Average Future" Scenario with the "Mega-plant" Scenario. One of the major differences between the two is the approximately 50 percent greater use in the latter of coal for electricity generation. Policy makers would, obviously, regret decisions made in anticipation of the latter scenario, should the "Average Future" Scenario more nearly reflect the ultimate reality. Since small coal plants can, in general, be built in less time than large coal plants, construction of small plants can be delayed until power needs and prospects of alternate power source development are better clarified. RD&D that results in smaller power plants that are more economically competitive with larger ones might make such delays economically feasible.

Recent years have seen progress by a number of foreign countries in advanced electricity technologies. The Japanese have arrived at the leading edge of thermal nuclear power technologies and are now essentially self-sufficient in this area. France is well ahead of the United States in the breeder reactors and many nations lead this country in high-voltage transmission capabilities. In the end use area, Japan now dominates many consumer electronics areas and is planning the construction of an electric railroad in this country. Effective RD&D programs will be necessary to forestall further erosion of the nation's electricity technology position (SB:IV-5). Note that the support of electricity technology RD&D is not only a matter of economic importance. If, in the absence of policy support of nuclear fission improvement innovations, the U.S. nuclear power industry loses the ability to compete successfully with foreign industries, U.S. influence over the course of nuclear proliferation will also wane.

In this assessment, most of the discussion of RD&D has concentrated on technology programs. However, throughout the assessment the need for research in the behavioral sciences was apparent. (Examination of the Appendices B, D, and E to the NSF report—see also Appendix B to this book—will bear this statement out.) The span of relevant areas is so broad that it is difficult even to list them. Examples include analyses of the relationships between electric power and the satisfaction of human needs, the economic and social impacts of intermittent power outages, and the psychological value of non-dependence on central station power. The importance of such research is well illustrated by today's nuclear power situation. At present, one of the biggest single deterrents to nuclear power expansion (or, even, continuation) is the lack of a publicly acceptable means for disposal of radioactive wastes. Considerable effort is being expended to develop disposal techniques. However, it is legitimate to question whether the general public will accept these techniques, regardless of their technological adequacy. Research into the exact nature of public concerns and the means required to address those concerns might be extremely valuable in guiding technical RD&D.

Large-scale and extensive RD&D programs entail significant costs. Not all programs will prove successful or even needed. However, given the costs of failing to give policy makers sufficient flexibility to deal with an uncertain future, carefully planned and managed RD&D programs represent a significant bargain for the nation (RB:IV-7) (SM:IV-6).

Significance to policy decisions. Because of the close association between electric power and the economic and psychological well-being of the citizenry, federal and state governments have sought to influence electric power decisions by laws, regulations, and subsidies almost from the beginning of public electric service. Although direct government involvement in the electric power field has waxed and waned through the decades, the overall trend throughout the century has been increased government involvement. Given the increased societal importance of electricity argued in this assessment, it seems probable that, despite current governmental emphasis on deregulation, the role of the federal and state governments in influencing electric power programs and decisions will grow over the next decade. As the influence of government policies increases, one would expect that the nature, size, and pace of both public and private electric power RD&D programs will be altered. Hence, it is imperative that decision makers in federal and state governments have clear understandings of the full scope of electric power RD&D policies and act only after consideration of those policies.

In formulating or modifying electric power RD&D policies, decision makers must first define the objectives of those policies. In Chapter IX, a typology of six types of objectives for electric power RD&D programs is presented. These objectives are: improving present systems; permitting increases in conventional capacity; providing new unconventional capacity; supporting fundamental science and engineering; expanding technology transfer programs; and improving end use technologies. Other typologies, of course, could be offered (KH:IV-3). However, it is important to note that the emphasis in this organization is on the "objectives" of electric power RD&D and not on specific technologies. The implication of this emphasis is that policy makers, when allotting resources to RD&D projects and programs, should first decide how much support to allocate to meeting each objective. Once this has been determined, further allocation to specific technologies can be better undertaken. This procedure would likely result in more balanced, better understood programs that are less subject to shifting public moods.

For example, in examining the proper levels of support for new large-scale, high-technology systems, policy makers might first determine an appropriate support level for all such systems. When this is accomplished, decisions about supporting multiple-system research, about the timing of research, and about allocation of funding between systems will be more easily made. A rough analogy is presented by Defense Department funding procedures. In past years, the military services were funded directly, and each made its own allocation of funds for specific missions (e.g., strategic deterrence, conventional warfare, and unconventional warfare). Present procedures involve, first, determination of a total allocation for each basic mission. Funds are then allocated to individual services to support each mission. Each service has roles in support of more than one mission; the matching of funding with missions, however, provides guidance for force composition and support.

Once the objectives of RD&D policies have been defined and an appropriate level of support for each determined, the means of providing that support must be chosen. As discussed in Chapter IX, over the last century, a custom of RD&D support and performance has emerged with fairly well defined roles for the federal government, the government laboratories, the equipment manufacturers, the individual utilities, and the Electric Power Research Institute. As the overall electric power environment has changed, these roles have, as a consequence, been altered. Likewise, as public policy changes in the future, the roles will continue to be transformed. However, it is neither desirable nor necessary that these changes be unanticipated. It is, in fact, a major purpose of electric power policy to decide appropriate roles for the major participants in RD&D efforts. Policy can, then, encourage appropriate participation by the application of policy levers such as rate structure changes, tax incentives, regulatory restrictions, and direct subsidies.

By following an anticipatory RD&D policy, the present generation can fulfill its part of an unwritten contract with future generations. Despite earnest efforts to the contrary, the current generation, like those before it, will take actions and implement policies that burden the future. However, through a commitment to aggressive research, this generation can afford posterity additional information, insight, and experience with which to address unknown future problems; by its RD&D commitment, this generation can also help to assure future generations expanded opportunities to control their own destinies (GE:IV-7).

In considering electric power RD&D policies, a final irony emerges. Most experts in the electric power field stress the necessity for federal RD&D programs that are clearly defined, generally stable, and supported at levels that are predictable and reasonably consistent. The continuing shifts in goals, emphasis, and funding levels that have characterized federal electric power programs in the last decade are seen as inefficient, costly, and counterproductive. It appears that RD&D policies that are steadfast, keenly focused, and predictable will be necessary if the nation is to confront successfully a future that will be mercurial, diverse, and surprising.

Associated Questions: Overarching

1. By what groups and by what processes should national electric power research, development, and demonstration objectives be determined?
2. By what process should the nation's total resource commitment to electric power RD&D be determined?
3. What portion of the nation's resource commitment to electric power RD&D should be allocated to each RD&D objective?
4. What portion of the nation's resource commitment to electric power RD&D should be borne by which segments of society?
5. Which segments of society can best pursue each of the electric power RD&D objectives?
6. By what process should the benefits of electric power RD&D be distributed among the members of society?

Associated Questions: Operational

1. How can multiple centers of RD&D initiative be balanced with establishment of national objectives?
2. To what extents should the government support improved efficiency in end use products in order to promote conservation?
3. To what stage of the innovation process should the government support research, development, and demonstration?
4. To what extent should parallel development paths be followed in longer-term RD&D projects?

5. How can an effective program for research in non-technological areas be developed?
6. How can pro-active technology transfer from frontier technological areas, such as aerospace, computer science, robotics, and biotechnology, to electric power sciences be promoted?
7. Should the government directly support the training of engineers and technicians in electric power supply, distribution, and use?
8. What should be the roles of each of the major participants in research and development funding and actual conduct of research?
9. What should be the patent policy for developments made under government sponsorship?
10. To what extent is research and development within the electric power industry assisted and injured by the tradition of free sharing of information on developments?
11. How can long-term health research on generation options be promoted?
12. What are feasible actions that might compress the commercialization cycle for new power technologies?
13. What actions can PUCs take to encourage utility RD&D investments?
14. How can one compare RD&D costs and regret costs arising from failure to undertake RD&D?

D. SUMMARY

The first eight principles discussed above present the policy maker with a very taxing paradox. The first two indicate that electric power decisions will tend to have long-term impacts, be difficult to modify, and have societal effects far beyond the electric power area. Principle 3 stresses that the importance of wise electric power policy decisions is increasing, while Principles 4-7 point up the inordinate uncertainties that will characterize the next decade. Finally, Principle 8 describes how the options practically available to the decision maker are becoming increasingly fewer.

In short, the policy maker is called upon to make fair, long-term decisions that have sufficient flexibility to accommodate unexpected changes without unreasonable disturbance or cost. To accomplish this end, it will be necessary for policy makers to project as accurately as possible future social, economic, political, and technological developments; to analyze with perception the interrelationships among these developments and between them and electric power policies; to assure the availability at reasonable costs of alternate technological choices; to contrive means for deferring irreversible decisions without undue costs and risks; and to frame means for changing policies, if circumstances require, with minimal social trauma. Fortunately, Principle 9 offers hope that these ends may be met.

In this assessment, an attempt has been made to lay groundwork for addressing each of these tasks. In fact, one of the early precepts of the assessment was that flexible policy is the most intelligent tool with which to meet future uncertainties. The alternate scenario approach utilized in this assessment should help to frame such policies by describing a spectrum of feasible future trends, events, and developments; identifying decisions that will be required in the future; improving communications among participants; encouraging innovative new approaches to problems; and identifying factors of special importance that should be analyzed and monitored.

One further point stands out in reviewing the results of this assessment. Overall, the strongest consensus among assessment participants was that well-conceived and executed RD&D programs, in both technological and non-technological areas, can provide the most efficient means of assuring the fairness and flexibility that will be essential to effective electric power policies.

BIBLIOGRAPHY

Bell, Daniel. *The Coming of Post-industrial Society*. New York: Basic Books, 1973.

Bell, Daniel. "The Information Society." In *The Computer Age: A Twenty-Year View*. Michael L. Dertonzos, editor. Cambridge, Mass.: MIT Press, 1979, p. 163.

Berg, Sanford. "Federal Policies on Utilities' Research and Development Expenses." *Public Utilities Fortnightly* (October 11, 1979), 18-24.

Berndt, Ernst R.; Thomas G. Cowing; and David O. Wood. *Recent Developments in U.S. Energy and Electricity Markets*. Cambridge, Mass.: MIT Energy Laboratory, 1983.

Bohn, Roger E.; Richard D. Tabors; Bennett W. Golub; and Fred C. Schweppe. "Deregulating the Electric Utility Industry." In *Electric Power Regulation, Deregulation and Structural Reform*. Massachusetts Institute of Technology Energy Laboratory Symposium. Cambridge, Mass.: MIT Energy Laboratory, October 1982, Chapter VIII.

Bupp, Irvin C., and Jean-Claude Derian. *Light Water, How the Nuclear Dream Dissolved*. New York: Basic Books, 1978.

Bureau of Economic Analysis (BEA). *Survey of Current Business*. Washington, D.C.: U.S. Department of Commerce, October 1982, p. 43.

The Continental Group, Inc. *The Continental Group Report: Toward Responsible Growth*. Stamford, Conn.: The Continental Group, Inc., 1982.

Edison Electric Institute. *Statistical Yearbook of the Electric Utility Industry*. Washington, D.C.: Edison Electric Institute, 1981.

Electric Power Research Institute (EPRI). "Electrotechnology: Sparking Productivity in Industry." *EPRI Journal* (June 1982), 8.

Energy Information Administration (EIA). *1981 Annual Report to Congress*, vol. 2. U.S. Department of Energy. Washington, D.C.: U.S. Government Printing Office, 1982.

Garfield, Paul and W. F. Lovejoy. *Public Utility Economics*. Englewood Cliffs, N.J.: Prentice Hall, Inc., 1964.

Graham, Alan K., and Peter M. Senge. "A Long-Wave Hypothesis of Innovation." *Technological Forecasting and Social Change* (1980), 283-311.

Hudson, Edward A.; Dale W. Jorgenson; and David C. O'Connor. "The Impact of Restrictions on Electric Generating Capacity." In *Advances in the Economics of Energy and Resources*. Greenwich, Conn.: JAI Press, 1980, pp. 111-157.

Joskow, Paul L., and Richard Schmalensee. *Deregulation of Electric Power: A Framework for Analysis*. Massachusetts Institute of Technology Energy Laboratory for the U.S. Department of Energy. Springfield, Va.: National Technical Information Service, 1982.

Killian, Linda R., and Robert R. Trout. "Alternatives for Utility Deregulation." *Public Utilities Fortnightly* (September 16, 1982), 34-39.

Komanoff, Charles. *Power Plant Cost Escalation*. New York: Van Nostrand Reinhold, 1981.

Lovins, Amory B., and Hunter L. Lovins. "Electric Utilities: Key to Capitalizing the Energy Transition." Presentation before the 1982 Mitchell Energy Conference, Houston.

MacAvoy, Paul. "Economic Prescriptions for Developing the Regulated Industries." *Review of Social Economy* (March 1971), 20-30.

Mensch, Gerhard. *Stalemate in Technology*. Translated from *Das Technologische Patt* (1975). Cambridge, Mass.: Ballinger Publishing Co., 1979.

Milbrath, L. W. "Environmental Values and Beliefs of the General Public and Leaders in the United States, England, and Germany." In *Environmental Policy Formation: The Impact of Values, Ideology, and Standards*. Dean Mann, editor. Lexington, Mass.: Lexington Books, 1981.

The MITRE Corporation. *Health Effects of Coal Technologies: Research Needs*. MTR-79W15 902. McLean, Va.: The MITRE Corporation, September 1980.

The MITRE Corporation. *Presentations: Electrotechnologies in Industry*. McLean, Va.: The MITRE Corporation, 1982.

Niskanen, William A., and Benjamin Zych. Letter to the Editor. *Harvard Business Review* (September-October 1982), 180, 185.

Ramsey, William. *Unpaid Costs of Electrical Energy*. Published for Resources for the Future. Baltimore: Johns Hopkins University Press, 1979, pp. 108-18.

Rochlin, Gene I. *Plutonium, Power and Politics*. Berkeley and Los Angeles: University of California Press, 1979.

Rosa, Eugene A.; Marvin E. Olsen; and Don A. Dillman. "Public Viewpoint toward National Energy Policy Strategies: Polarization or Compromise?" In *Nuclear Power and the Public*. William R. Freudenburg and Eugene A. Rosa, editors. Boulder, Colo.: Westview Press, 1983.

Schurr, Sam H. "Energy Abundance and Economic Progress." Prepared for Georgia Institute of Technology course on "Societal Cost of Energy Alternatives: Methodologies and Results," April 1982.

Schurr, Sam H., Joel Darmstadter, Harry Perry, William Ramsay, and Milton Russell. *Energy in America's Future*. Washington, D.C.: Resources for the Future, 1979, p. 18.

Starr, Chauncey, and Milton F. Searle. "U.S. Generation Capacity Requirements for Economic Growth, 1990-2000." *Public Utilities Fortnightly* (April 29, 1982), pp. 17-24.

"33rd Annual Electrical Industry Forecast." *Electrical World* (September 1982), 82.

Part Three

BACKGROUND

The chapters in this part were developed during the middle stage of the assessment as inputs to the analyses of Part Two. Five of the chapters present background information on subject matter areas pertinent to the future of electric power. The sixth chapter is a historical perspective on the development of the nation's electric power system. These are survey chapters and do not break new theoretical ground. They are reproduced here so that the lay reader may have the benefit of an expanded knowledge base when considering policy problems potentially associated with future electrification. The chapters contain copious references to materials from which the reader can obtain more detailed information.

Events of the past decade created a number of financial stresses for electric power institutions and consumers. These stresses and the responses to these stresses will have effects that continue in and beyond the present decade. Chapter V discusses these financial stresses, both historic and prospective.

The 1960s and 1970s witnessed an awakening of environmental and health consciousness in American society. This awakening had profound impacts on electric power institutions. Chapter VI reviews the factual bases for environmental and health concerns related to electric power supply and discusses the ways in which those concerns might change in the future.

The nation's electric power supply system historically has been very reliable. Future electric power policy should obviously strive to maintain high reliability in electric power supply. Chapter VII reviews the historic sources of threats to the electric power system's reliability and considers the extents to which these threats might be increased or decreased under various assumptions about the nature of future electric power supply.

Many technological and societal variables will be influenced by and will influence the course of electrification in this country. The probable influence of electrification on demographic variables is a particularly important policy consideration because of the strong influence these variables, in turn, have on the character of U.S. society. Chapter VIII reviews the state of knowledge about influences of electrification on demographic variables.

National electric power research, development, and demonstration policy very strongly and directly influences the nature of the nation's electric power supply system. Chapter IX reviews the history of electric power RD&D and elaborates upon the interrelationships between this RD&D and the nation's electric supply system. It also considers RD&D programs that would be particularly appropriate under various assumptions about the future needs of the nation's electric power supply system.

If one knows the past, it is possible to learn many lessons of value for the future. Furthermore, the near-term future is largely determined by the momentum of events originating in the past; because of its technological interdependencies and capital intensity, this is particularly true for the nation's electric power system. Chapter X presents a very readable summary of the history of electrification in this country.

CHAPTER V

Finance

A. OVERVIEW

Financial concerns related to increased electrification are, at the most fundamental level, really quite obvious. How much will increased electrification cost and which segments of society will bear which portions of the cost? Unfortunately, once one proceeds to develop answers to these questions, very little about the future financing of electrification is obvious.

The future cost of supplying electric power will most likely be higher than it is today. The actual level of cost, however, can not be predicted confidently because it is influenced by unknowns, such as future fuel prices, future electricity demand, and future technologies for generating electricity. The inertia of events currently under way does lend some help to prediction of future costs. If utility regulatory bodies continue to allocate the cost burdens of electrification, one would expect that utility ratepayers, as opposed to utility shareholders, will bear the bulk of these costs (AK:V-1). This is consistent with the principle, not always realized in practice, under which utility regulation sets ratepayer fees high enough to allow shareholders reimbursement for utility costs and a reasonable return on their equity holdings. It is possible, however, that advances in technology could effectively take cost allocation out of the hands of utility regulators. Changed economic conditions could also lead to this result via price "deregulation." Changed political pressures could lead to regulation that, at least in the short run, redistributed the cost burden from ratepayers to shareholders.

This chapter attempts to shed light on the determinants of future costs and future costs allocation. It does this principally by clarifying the historical behavior of a number of the more important determinants. More specifically, the chapter demonstrates that real electricity costs declined for a long period prior to the 1970s. During the 1970s, this real cost decline reversed. The chapter then hypothesizes that this cost reversal can be explained largely in terms of (1) inefficient allocation of capital resources (i.e., underutilized generating facilities); (2) at least temporary exhaustion of technical improvements to generating plant efficiency; (3) increase in real power plant fuel costs; (4) increase in the real per kilowatt capital cost of new generating facilities; and (5) allocation of capital to construction that is abandoned before yielding a return. The chapter then discusses ways in which the change of electric power supply from a declining cost activity to an increasing cost activity has recently affected utility regulation and utilities as institutions. With this background, the chapter finally turns to a discussion of future financial concerns within the contexts of this assessment's six scenarios.

B. HISTORIC TRENDS

Electricity price. Exhibit V.1* reflects the history of the real price of electricity purchased by both residential and industrial customers. Electricity

Exhibit V.1
Real Electricity Prices, 1935–1980

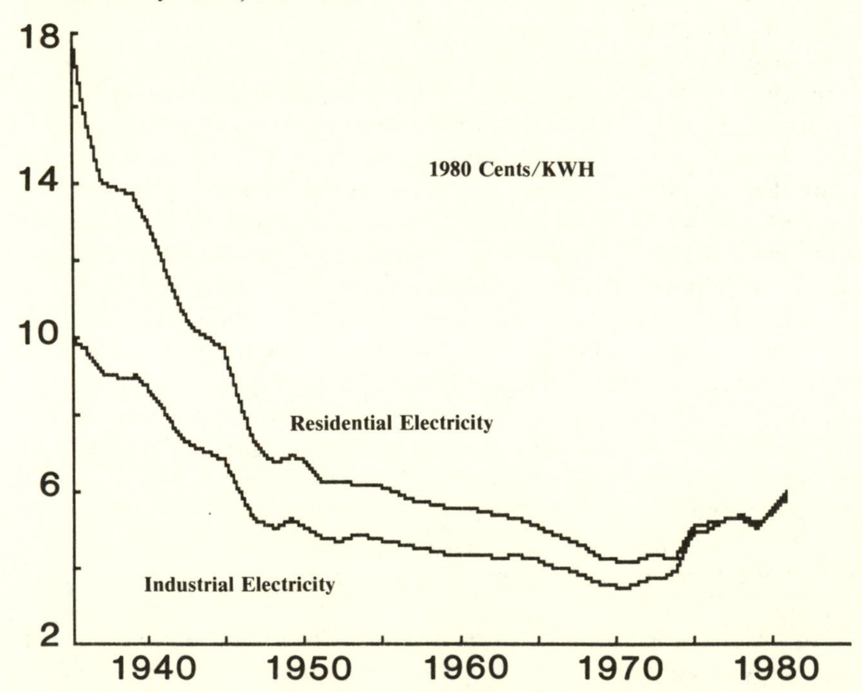

Sources: Bureau of the Census, 1975, S-115, 117; EIA, 1982, Table 72.

*Data sheets from which the exhibits in this chapter were derived are to be found in Appendix D.

Exhibit V.2
Comparison of General Inflation and Electricity Price Inflation, 1935–1980 (1970 = 100)

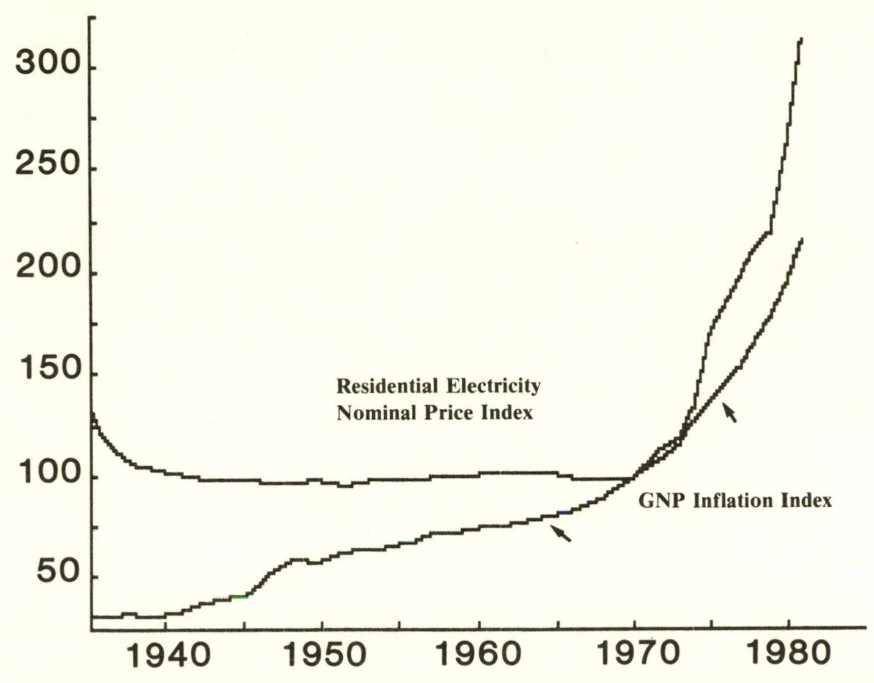

Sources: Prices: Bureau of the Census, 1975, S-115, 117; EIA, 1982, Table 72. GNP: Joint Economic Committee, 1980, Table 3; Bureau of Economic Analysis, 1982, p. 43.

prices for each customer group declined from the 1930s through the 1960s and rose throughout the decade of the 1970s. Exhibit V.2 reflects what appears to have been happening to the price of electricity from the perspective of the moderate (500 kilowatt hours per month) residential consumer. For 30 years prior to 1970, the price of a kilowatt hour of electricity appeared to change very little (i.e., the nominal dollar price of a unit of electricity did not increase each year). During this same period, the price of goods and services appeared to increase more than 200 percent (i.e., the nominal dollar price of a unit of general goods and services was three times as great in 1970 as in 1940). During the most recent decade, however, while inflation caused general prices to double, the nominal price of residential electricity nearly tripled.

Utility utilization of generating plant. Exhibit V.3 reflects a 50-year trend in utility utilization of generating facilities. It shows the number of kilowatt hours of electricity actually generated per year for each kilowatt of generating capacity installed at year-end. Plant utilization improved during the 1930s and 1940s, remained approximately constant during the 1950s and 1960s, and declined noticeably during the 1970s. This inefficient use of

Exhibit V.3
Utility Plant Utilization, 1930–1980

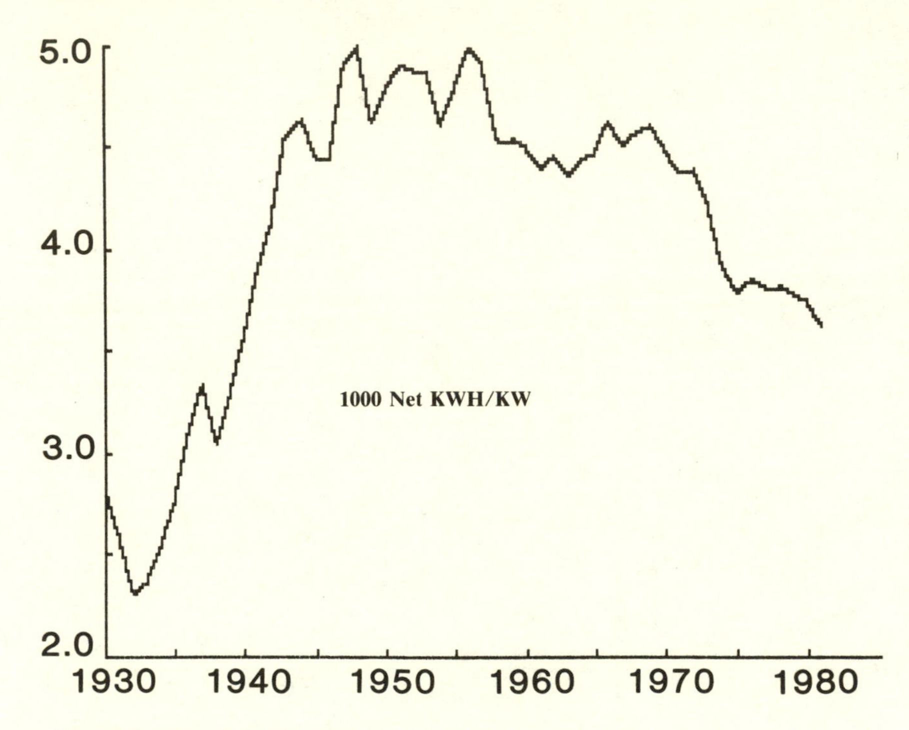

Sources: Bureau of the Census, 1975, S-36; EIA, 1982, Table 64.

capital contributed to the 1970s electricity price rise because utility regulation allows a uniform return on capital, whether or not that capital is efficiently generating revenue.

The principal reason for idle capacity during the 1970s is reflected in Exhibit V.4 (KH:V-1). The growth in demand for electric power slackened unexpectedly in the early 1970s and left utilities with excess capacity that had been or was being built in anticipation of continued 7 percent annual demand growth. Furthermore, oil price increases raised the fuel costs of some plants to levels higher than the combined fuel and capital costs of alternative generation technologies; prudent management called for "underutilization" of high fuel cost capacity (SB:V-1).

Conversion efficiency of fossil fuel power plants. Since 1950, fossil fuel (coal, oil, and gas) power plants have supplied approximately 75 percent of the nation's electricity. Exhibit V.5 reflects the history of energy conversion rates at utility fossil fuel power plants. For 40 years prior to 1969, the average conversion efficiency of fossil fuel power plants improved, although at a declining rate. During the 1970s, this improvement halted

Exhibit V.4
Total Electricity Supply, 1935–1980

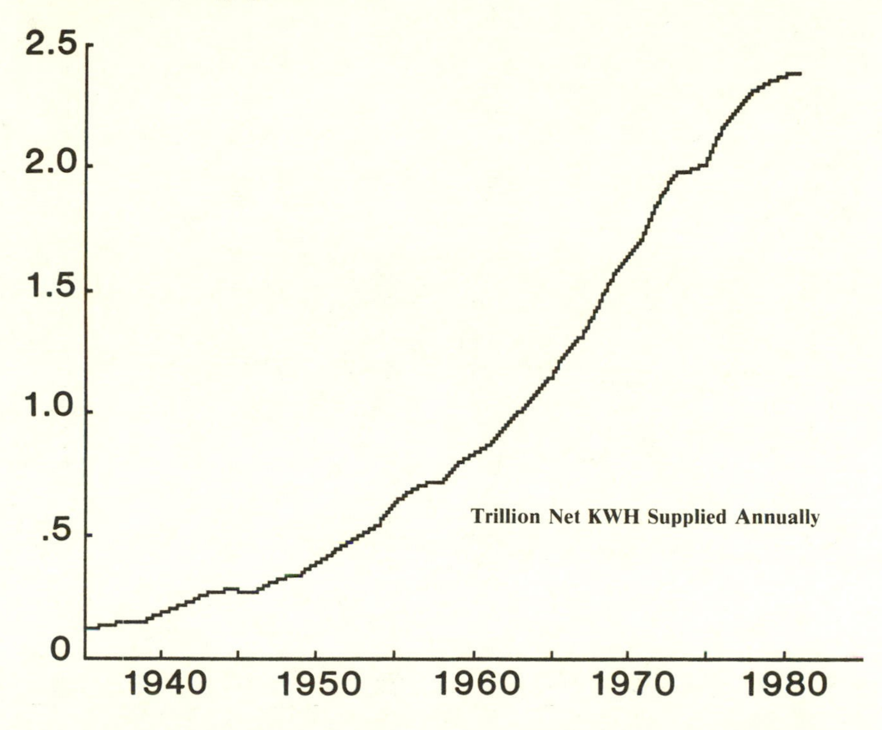

Sources: Bureau of the Census, 1975, S-36, 40; EIA, 1982, Table 64; EIA, 1981b, Table 67.

and, indeed, deteriorated slightly. The decline in efficiency during the 1970s is generally attributed to environmental protection measures installed at fossil fuel power plants. These have had a major negative impact on the efficiency of fuel conversion, but it is not clear how much of the lessened efficiency improvement can be explained solely by environmental protection measures. Regardless of its causes, the halted improvement of energy conversion rates during the 1970s removed one of the factors which had previously helped to produce lower electricity prices.

Fossil fuel costs. Exhibit V.6 reflects the history of fossil fuel energy costs to electric utilities. These costs actually declined in real terms through the 1960s. Then, during the same period of time when energy conversion rates were either flat or deteriorating, the cost of fuel tripled; nominal (not inflation-adjusted) fuel prices during the 1970s increased more than five-fold. This contributed greatly to the dramatic rise in electricity prices during the 1970s. It was, indeed, the principal factor causing those price rises (EIA, 1981a).

Generating plant costs. Utility regulators and executives acknowledge

Exhibit V.5
Utility Fuel Conversion Rate, 1930–1980

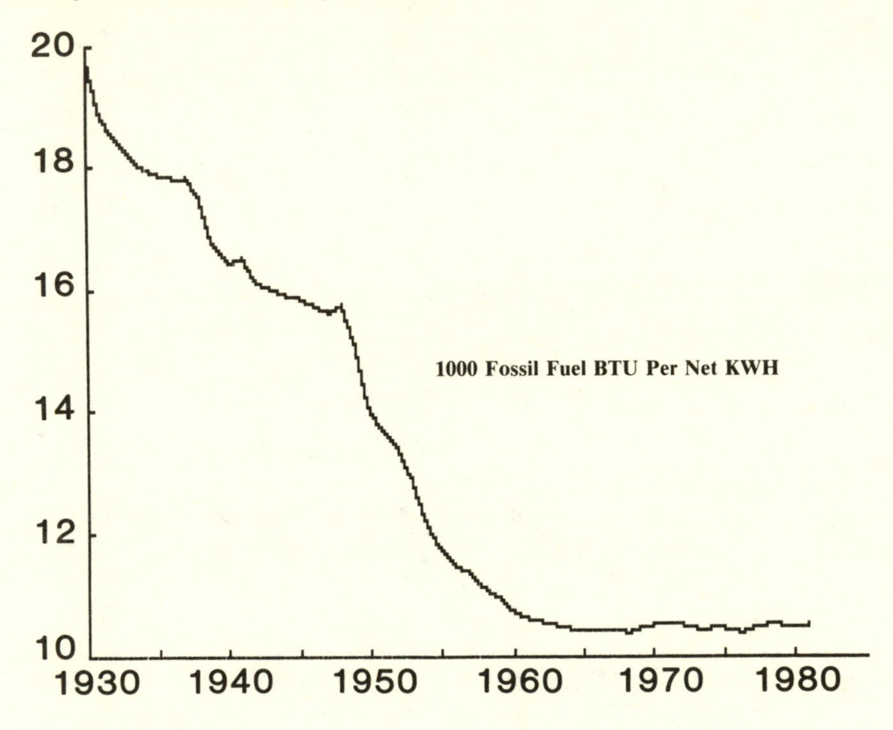

Sources: Bureau of the Census, 1975, S-107; Edison Electric Institute, various years.

that capital costs of both coal and nuclear generating plants increased dramatically during the 1970s. Statistics marshalled to clarify this increase are frequently at variance with one another, though they do support the fact of capital cost increases. There are many reasons for these variances, the principal two being that (1) there is no standard convention concerning the treatment of interest costs incurred during construction and (2) almost all capital cost statistics are reported in "mixed current dollars." "Mixed current dollars" reflect the sum of nominal dollar costs incurred during each of the years of plant construction. One comprehensive study, which adjusted reported capital costs to a constant dollar basis and included real interest costs of construction, concluded that the per kilowatt capital costs of a "standard" nuclear plant increased 142 percent between 1971 and 1978; the same study found that the per kilowatt capital costs for "standard" coal-fired plants increased 68 percent during the period (Komanoff, 1981:20).

One element of plant capital costs is the cost per unit of power plant

Exhibit V.6
Real Utility Fossil Fuel Cost, 1954–1980

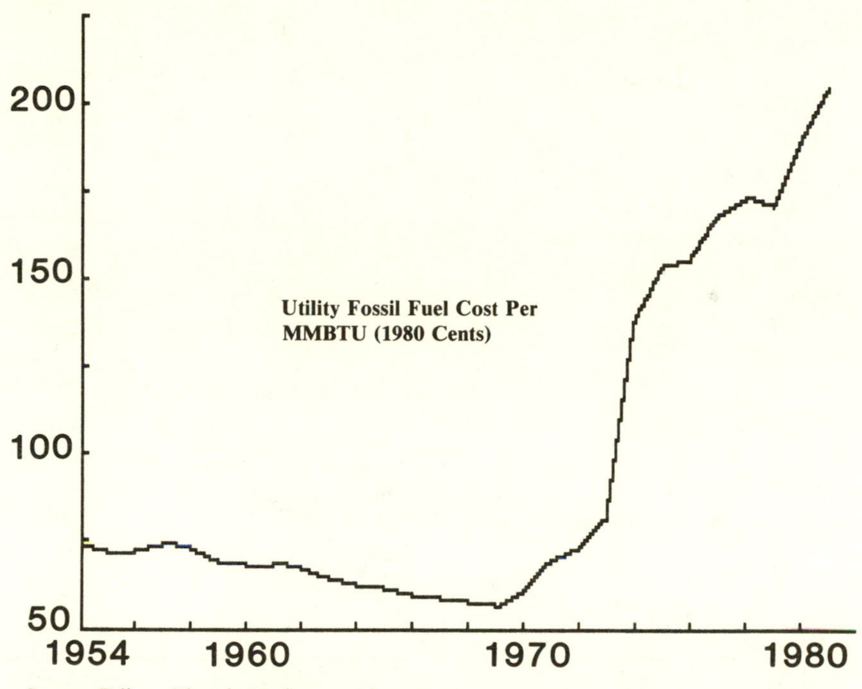

Source: Edison Electric Institute, various years.

construction materials and labor. Exhibit V.7 reflects the historic cost trend for this element (as measured by the Handy-Whitman Index of Public Utility Construction Costs). Although it demonstrates persistent escalation (i.e., cost increases above the rate of inflation as measured by the Gross National Product Implicit Price Deflator), the escalation has been substantially less than the overall escalation in power plant capital costs.

Factors beside the cost per unit of materials and labor contributed to power plant cost escalation during the 1970s. Important among these factors was a marked increase in the time elapsed between the start of site preparation and the actual generation of electricity at a new plant. One comprehensive study found that among comparable nuclear units this time averaged less than five years for plants coming on line in 1970 and averaged more than eleven years for plants coming on line in 1982 (Budwani, 1982:38). Construction duration for coal plants did not escalate as rapidly as construction duration for nuclear plants during the 1970s. The Komanoff study of coal-fired power plants completed between January 1, 1972, and December 31, 1977, found that the construction duration for a hypothetical

Exhibit V.7
Power Plant Construction Cost Inflation: Materials and Labor, 1954–1980 (1972 = 100)

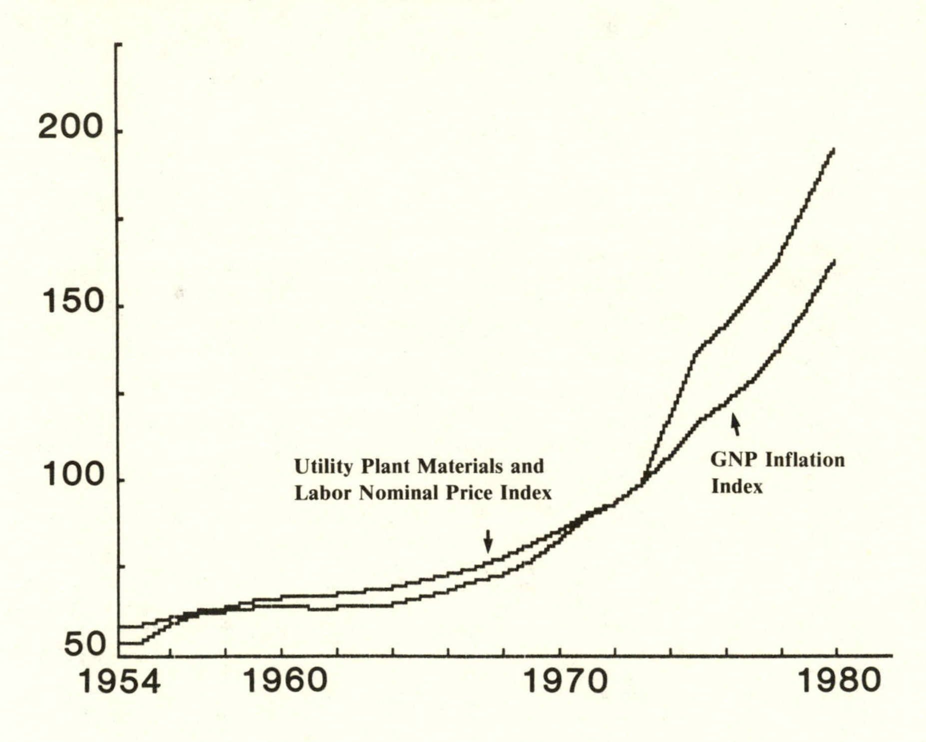

Sources: GNP: Bureau of Economic Analysis, 1982, p. 43. Handy-Whitman Index: Bureau of the Census, 1975, N-131; Bureau of the Census, 1979 and 1981, Price, Wage Scale, and Cost Indexes for Construction.

"standard" plant increased approximately 16 percent during this period (Komanoff, 1981:229). The Budwani study found an average 17 percent difference in construction duration between coal plants completed in the 1970s and those due for completion in the mid-1980s (Budwani, 1982:43); this study found wide variances in coal plant completion times during the 1970s.

The increase in the construction duration for both nuclear and coal plants in the 1970s increased the accumulation of real interest charges on the borrowed money with which the plants were built. Utility real interest charges (the difference between the nominal rate of interest charged on borrowed money and the rates of inflation during periods for which the money is borrowed) have historically been in the range of 2.5 to 3 percent annually. However, real interest charges customarily increase as the lenders' perceptions of the riskiness of projects for which money is borrowed increase. To the extent then that increased construction duration influenced lenders to perceive higher risks in coal and, particularly, in nuclear power plant projects, increased construction durations also increased the real interest rate charged by lenders (KH:V-2). (Anticipation of high future rates of general inflation also raises real interest rates, should future inflation be less than anticipated.)

The overriding reason for capital cost escalation during the 1970s, however, was that substantially more materials and labor were required to install a kilowatt of capacity at the end of the decade than had been the case at the beginning. These additional costs for coal-fired plants were incurred almost exclusively to reduce environmental pollution, principally from particulates, sulfur dioxide, and nitrogen oxide (Komanoff, 1981:24). Of the 68 percent real coal plant cost escalation that the Komanoff study documented, half could be attributed to the cost of sulfur dioxide scrubbers alone.

The reasons for 1970s capital cost escalations for nuclear plants are not as easily traced as those for coal-fired plants. The Atomic Industrial Forum estimates, however, that Atomic Energy Commission and Nuclear Regulatory Commission design and construction standards for nuclear reactor systems and equipment approximately doubled the material and labor required to bring on line a kilowatt of nuclear capacity during the 1970s (Komanoff, 1981:25). The Komanoff study found that the number of design and construction guidelines for nuclear power plants increased from 21 in 1971 to 143 by the end of 1978. This sixfold increase in design and construction guidelines occurred as plants were under construction and, therefore, led to very inefficient construction practices, such as the modification or removal of work already completed. Komanoff ultimately concluded that the capital cost increases for nuclear power plants could be attributed primarily to efforts to prevent total accident and environmental risks from expanding in proportion to growth of the nuclear power sector.

Cancelled and suspended plants. The final significant element that contributed to electricity price increases during the 1970s was the amortization of the costs of nuclear power plants on which construction was abandoned prior to completion or on which construction was suspended on an interim (e.g., five years) basis prior to completion. Though there is some debate as to what constitutes a power plant on which construction has been "suspended" and what constitutes a power plant which has been "cancelled," it appears that approximately 100 nuclear plants have been cancelled since 1972 (Raloff, 1983:12); more than 30 were cancelled in the three years following the 1979 incident at Three Mile Island (Budwani, 1982:37; Raloff, 1983:12).

The costs of power plants abandoned prior to completion are customarily amortized over a period of years (e.g., ten years) and added on a pro rata basis to the price of each kilowatt hour of electricity sold. These cancelled plant costs can be huge. For example, Washington Public Power Supply System's well-publicized decision to cancel two of its nuclear power plants resulted in a loss of at least 2.3 billion mixed nominal dollars (*Wall Street Journal*, February 22, 1983:6). Costs attributable to power plant construction suspension, as opposed to cancellation, are less clearly calculable and less significant. However, these costs are also passed along to ratepayers as part of the price of each kilowatt hour of electricity purchased.

C. 1970s RESPONSES TO CHANGED HISTORICAL TRENDS

Prior to the 1970s, electricity supply had been a declining cost business. The protocol for interaction among all the electric power institutions (e.g., consumer groups, regulators, electric utilities, and equipment suppliers) developed consistent with the declining cost environment. When the environment shifted to one of rising costs, the institutional arrangements among the parties involved in electric power supply and consumption also shifted, but in rather slow and frequently inefficient fashions.

Utility return on investment. When the supply of electric power was a declining cost activity, there was no adverse impact to electric utility financial condition associated with slow regulatory response to changes in cost factors. Consequently, there was no mechanism in place in the early 1970s to provide utilities quickly with additional revenues to cover escalating fuel costs. Utilities, therefore, diverted funds from internal sources to pay for fuel. Then, they were forced to borrow money, incurring new interest expenses, to replenish internal funds and pay shareholder dividends. Attempts to raise capital through the sale of stock were often only qualified successes because stock frequently sold at prices below company book values. Exhibit V.8 sheds light on this latter circumstance. Inflation-adjusted return on shareholder equity investment was declining at the same time the riskiness of the investment was increasing.

Utilities went to regulatory commissions and requested that fuel purchase costs be passed through automatically to ratepayers. They also requested increased rates to cover increased operating expenses, such as debt service.

As utility rates and fuel adjustment charges went up, demand for electricity began to slacken. There also came to be more ratepayer resistance to all regulatory actions that had the potential to increase electricity prices.

The cumulative result of all of this was that utilities went from positions of moderate financial strength to positions in which their debt burdens were too high, the ratings given their new debt issues were too low, their average returns on equity investment were too low and there was public and regulatory resistance to increased price for their electricity product, for which demand growth was in any event slackening. (Thompson, 1982, and Navarro, 1982, present comprehensive reviews of utility financial trends during the 1980s.)

In the early 1980s, however, actions were taken by utility regulatory commissions and the federal government that were calculated to raise rates of

Exhibit V.8
Utility Real Return on Equity, 1950–1980

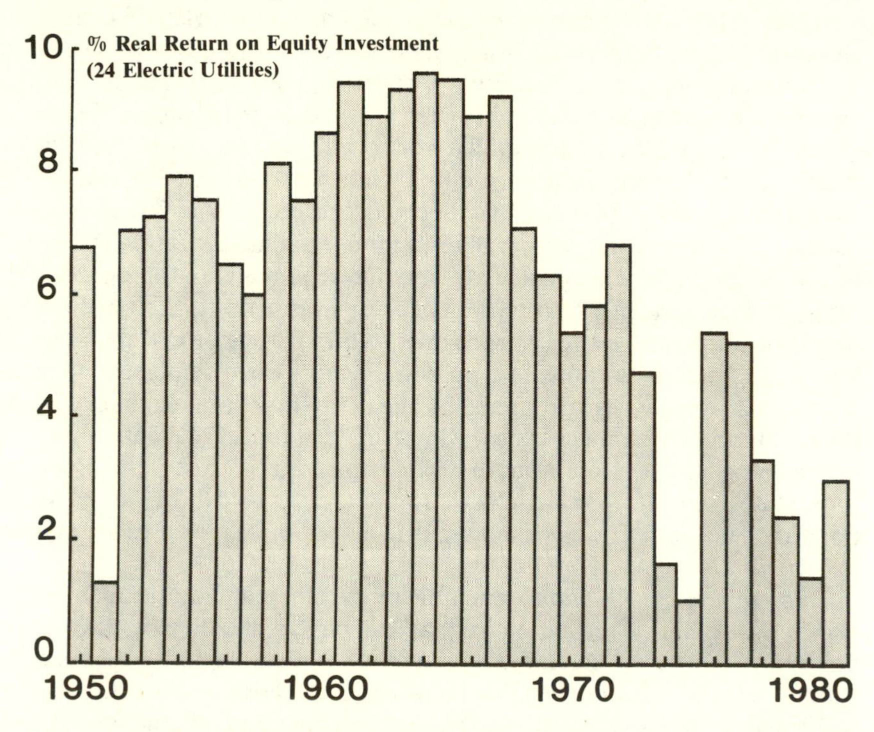

Sources: Nominal ROE: Moody's Investor Services, 1982, a13 and a14; GNP Inflation: Bureau of Economic Analysis, 1982, p. 43.

return and otherwise make ownership of electric utility stock more attractive. Initial indications suggest these efforts will be successful, at least, for the industry as a whole (*Fortune*, 1982:109). In 1981, a year in which general inflation as measured by the GNP Implicit Price Deflator rose at approximately 9.5 percent, the electric utility industry average return on common stock was 12.6 percent. In 1982, when real inflation ran at approximately 6 percent, electric utility average return on common stocks averaged approximately 13.4 percent. 1981 changes in federal tax law also worked to the benefit of electric utilities. Congress passed an exemption to individuals from federal income taxation on the first $750 per year in electric utility dividends reinvestment under specified conditions. This exemption is scheduled to run through 1985.

Institutional response to "regulatory lag." With the advent of escalating fuel prices, utilities sought and in many cases obtained authorization to automatically include in the price of their product a fuel adjustment surcharge. This surcharge was generally paid by customers prior to its review by regulatory authorities. This was a deviation from conventional regulatory practice, which calls for pre-implementation justification of rate increases; it was deviation that arose to circumvent the delay of regulatory hearings. This mechanism for distributing increased fuel costs was criticized by consumer advocates because it operated to reduce electric utility incentive to purchase fuel at the lowest possible price (Baumol, 1982).

Prior to the 1970s, electric utility price regulation was based on historic "test years." This may have been a financial help to utilities because future costs were generally lower than had been historic costs. When costs began to escalate, however, utilities generally sought to establish a "future test year" as the gauge by which to calculate revenue requirements for establishing rates. These efforts met with success in some jurisdictions, though determination of future test year costs continues to turn largely on extrapolation of historic cost trends. Another and potentially more easily implemented regulatory modification, also prompted by the desire to reduce regulatory lag, is the trend to interim rate hearings, which are held more frequently than full rate hearings and explore a more limited spectrum of cost items. Finally, some states now allow "emergency" rate increases for which full justifications are developed after the increase is in effect; these increases are subject to refund.

Treatment of construction costs. One of the oldest and most fundamental principles of utility regulation is that ratepayers should not pay shareholders a return on capital equipment until that capital equipment is actually used and useful to supply the utility service. The escalating costs of construction and the protracted periods of construction that developed during the 1970s, however, brought this principle under attack.

Escalating construction costs and protracted plant completion schedules operated to keep large amounts of utility company money from earning a return. This money was committed to the construction of power plants and would not begin to earn returns until plants were used. During the period of construction, conventional regulatory practice is to credit the utility with an accounting income flow that corresponds to the foregone opportunity cost of money tied up in construction. No revenue is realized from this accounting income flow. The credit, commonly referred to as allowance for funds used during construction (AFUDC), becomes a real, income-producing asset only when construction is finished and the plant is used. Then, compounded AFUDC is capitalized and added to the rate base (Brigham, 1981:36.11).

The fact that AFUDC is not an income-producing asset has apparently led financial markets to downgrade the value of securities issued by utilities for which AFUDC constitutes a significant portion of all income (Thompson, 1982:23). If one does not adhere rigorously to the principle of "used and useful," then some of the adverse financial impacts of AFUDC can be avoided by allowing electric utilities to include in their rate bases construction work in progress (CWIP) on plants that have not yet come on line. This transfers from future ratepayers to current ratepayers part of the burden for financing construction of plants that supply power only to future ratepayers. However, it also serves to convert what would otherwise be paper assets into real, income-producing assets and thereby improves the perceived financial condition of the electric utility. This benefits current and future ratepayers because it lowers the overall cost of supplying electric power by lowering the cost of money to utilities.

As of 1980, 35 states and the District of Columbia allowed utilities to include within their rate bases at least some portion of CWIP. About half of these jurisdictions allowed CWIP to be included in rate bases without conditions. In the other half, however, CWIP was admitted to rate bases only conditionally, as for pollution control equipment or only if interest paid on borrowed money were not capitalized (Congressional Research Service, 1982:15). The Federal Energy Regulatory Commission allows utilities to include in their wholesale customers' rates up to 50 percent of CWIP and all pollution control costs (*Wall Street Journal*, March 11, 1983:40).

Because rate base treatment of CWIP is contrary to the long-standing principle that ratepayers should only pay for equipment that is used and useful, it represents an important example of institutional adaptation to changed conditions. It is not clear whether the overall cost to consumers of a new generating facility is increased or lowered by rate base treatment of CWIP. As a gross generalization, the inclusion of CWIP operates to move nearer in time ratepayer payments for new plants. Rate base treatment of CWIP means ratepayers do not have the use of their money for the period preceding plant completion. However, they also avoid paying much of the imputed return on investment that compounds prior to capitalization under AFUDC practice. Furthermore, rate base treatment tends to smooth electricity price increases because the rate base grows as construction progresses; price signals to consumers are more predictable and, perhaps, more clear.

Conservation, alternate sources of supply, and diversification. As both fuel and plant construction costs rose during the 1970s, utilities and utility regulators became increasingly interested in mechanisms for meeting electric power demand without additional utility generation. This interest became particularly intense toward the end of the decade and federal regulation reflected the growing consensus that neither cost trends nor slackening of customer demand were short-term phenomena.

One example of utility interest in alternative sources of electricity supply is the increase in importation of electric power from Canada during the 1970s. Between 1970 and 1980 U.S. consumption of utility-generated electricity increased approximately 50 percent; during the same period, electricity imports increased more than 300 percent. (The relative importance of imported electricity to overall electricity supply remains low despite the large increase in imports; imports accounted for only 1 percent of total utility electricity supply in 1980 [EIA, 1982:149].)

A rather detailed study by the California Energy Commission reflects the extent to which utility reliance upon alternative sources of supply came to be the regulatory expectation in the late 1970s. The Energy Commission's "current trends" case for year 2000 electricity supply in California foresees an increase to 42 percent from 30 percent for generation from renewable and alternative sources (i.e., hydropower, geothermal power, electricity imports, cogeneration, wind, solar, fuel cells, and biomass) (California Energy Commission, 1981:175).

Cogeneration, the coordinated production of electricity and heat for process or space conditioning applications, became a favored source of power generation under federal policies during the 1970s. Cogeneration most commonly occurs within industries that have large process heat requirements. It can make electricity available for sale by electric utilities without utilities directly incurring either fuel or plant construction costs. Under the Public Utility Regulatory Policy Act of 1978, utilities are required to buy electricity from cogenerators and other non-utility generators of electricity at the cost that the utility has avoided by non-generation. Also under the Public Utility Regulatory Policy Act, utilities are required to "wheel" (transmit) through their transmission and distribution grids power from cogenerators and other small power producers. Whether these federal policy provisions will serve to encourage non-utility generation of electricity depends principally on how the "avoided costs" of utility generation are calculated and on the nature of fees that are charged for electricity wheeling.

Utility efforts, late in the decade, to avoid construction of additional generating capacity also led utilities and utility regulators to adopt a more liberal definition of the appropriate businesses of an "electric utility." In particular, electric utilities increasingly took on business activities beyond the generation and supply of electricity. Some utilities (with prompting from federal law) began to transform themselves into energy service corporations. These make available to the public expanded energy services such as audits and consulting. In some cases, utilities became sources of financing for customers' energy conservation efforts, such as the installation of insulation, fluorescent lighting, or other equipment that improved end use efficiency (California Energy Commission, 1981:108; Bryson, 1981:10).

Electricity price regulation questions. Rising electricity prices in the 1970s led authorities both in and out of the electric power industry to re-examine the manner by which electricity prices were determined. More particularly, these authorities called into question justification for continued electricity price regulation. Utility price regulation has conventionally been justified as a mechanism by which to allocate the benefits of declining costs made possible by scale economies. As costs rose during the 1970s, utility price regulation began instead to allocate liabilities. This change in regulatory function and the suspicion raised by rising costs that scale economies had been exhausted brought justification for electricity price regulation under reexamination.

The supply of electric power has always been considered a natural monopoly enterprise. A natural monopoly enterprise is one in which all of the products or services provided by the enterprise can be provided at a lower total cost by one firm than by more than one firm. If there is only one product or service produced by the enterprise, then a natural monopoly exists if each new unit of goods or services can be produced at less cost than any preceding unit. It also exists if there is a sufficient "threshold" of production for which this is true, even though production of units beyond this threshold might occur at constant costs or even, under special circumstances, at increased costs. An enterprise that produces multiple products may qualify as a natural monopoly enterprise even if there are no scale economies as to some of the products in the product mix; it may be that there are multi-product production economies (economies of scope) which allow the product mix to be produced at less cost by one firm than would be the case were there more than one firm. Similarly, enterprises in which the final output of goods or services is accomplished by vertically integrated firms (producers of goods or services that are inputs to their own further production of goods or services) may be natural monopolies because economies of scale exist in one stage of production, and there are important economies that arise because of vertical integration (Joskow and Schmalensee, 1982:37).

Electricity supply is usually an integrated enterprise. A single firm frequently generates the electricity, transmits it in bulk to distribution centers, and then distributes the electricity to end users. There are, then, at least three stages that one might examine for natural monopoly characteristics; if natural monopoly characteristics exist at any stage, then the three-stage enterprise of electricity supply might be fairly characterized as a natural monopoly. Conversely, the existence of natural monopoly charactistics at one stage in the enterprise would not in itself be theoretical justification for imputing natural monopoly characteristics to other stages or to the enterprise as a whole.

The emergence since World War II of highly integrated transmission and distribution grids and of multiple utility power pools has led to an electricity supply process that is radically different from that which existed during the early 1900s. Before the power distribution and transmission grids had been fully developed, electricity supply was obviously a natural monopoly activity at the distribution and transmission stages. At the production stage, there were technology-determined scale economies (note the efficiency improvement history in Exhibit V.5). However, it is possible that somewhere on the range of scales between the small, generally hydroelectric units of the 1900s and the 800 to 1,200 megawatt coal and nuclear units of the late 1970s, the threshold for power generation scale economies was passed.

There was no resolution during the 1970s or early 1980s as to whether some or all of the stages of electric power supply had lost their natural monopoly characteristics. Certainly, there was no resolution as to how, if at all, to modify electricity price regulations in areas where natural monopoly characteristics no longer existed. However, rising electricity prices did focus debate on the price-setting process. During a time when price deregulation of trucking, airline, and interstate communications enterprises was increasing, previously suspect suggestions that some form of electricity price deregulation might be in order came to be taken seriously (Joskow and Schmalensee, 1982; Bohn et al., 1982; Niskanen and Zych, 1982; Killian and Trout, 1982).

D. FUTURE FINANCIAL CONCERNS

The final section of this chapter explores some financial concerns expected to arise because of future electrification in this country. These concerns are explored against the backdrop of this assessment's six scenarios. The concerns on which this last section focuses are electricity price, electric utility diversification, and electric utility price regulation.

Electricity price. The price of future electrification is discussed here as a function of six variables:

- Plant utilization
- Average fossil fuel heat rates
- Average fossil fuel costs
- Generating plant capital costs
- Suspension or cancellation of power plant construction
- Allowed return on equity investment

The discussion of the price of future electrification concludes with a comment on regulatory treatment of construction work in progress. This consideration primarily affects the timing of electricity price increases.

"Plant utilization," as defined by the earlier discussion, measures the number of kilowatt hours produced annually for each kilowatt of capacity installed at year-end. Increasing plant utilization is consistent with decreasing electricity prices because it reflects efficient use of capital assets (KH:V-3). In the long run, plant utilization can be improved by the growth of electricity demand, by a period of limited capacity expansion, or by a leveling of seasonal and daily peak load patterns. Except in rare cases where technology-related maintenance times are large, short-run plant utilization can be improved only with the advent of increased electricity demand.

Since none of the scenarios postulates rapidly increasing electricity demand during the 1980s, improved plant utilization would be unlikely to constrain electricity prices during this decade. Thereafter, two of the scenarios, the "Nuclear Resurgence" and "Mega-plant" Scenarios, contemplate rather robust (by the standards of the late 1970s) demand growth. Improved plant utilization might, therefore, differentially encourage lower electricity prices in these two scenarios, at least during the 1990s. For the two low total energy demand scenarios, the "Post-industrial" Scenario and the "Economic Malaise" Scenario, a marked difference in electricity demand is postulated. However, the impact on utility plant utilization might not vary much between those scenarios because much of the demand postulated in the "Post-industrial" Scenario is met by non-utility generation. In both these scenarios, utility plant utilization could be expected to improve until the mid-1990s because of limited capacity expansion and, thereafter, to stabilize as demand for electricity (or utility electricity) slackens.

In the long run, implementation of time-of-use rate schedules should also have a beneficial impact on plant utilization; total generation could increase without a commensurate increase in plant capacity because the timing of consumer demand should become more uniform. These schedules should also have positive effects on fuel costs.

Physical deterioration over time of electricity-generating plants also leads to lower plant utilization. This could be a problem in the central scenario, the "Average Future," around the turn of the century because that scenario postulates both modest electricity demand growth and modest growth in real national wealth (a requisite to replacement of deteriorating capital stock). One would expect some instances of plant deterioration in the "Economic Malaise" Scenario during the first decade of the next century because that scenario postulates unpredictable cycles of economic expansion and contraction. Utilities that planned in anticipation of higher electricity demand than is postulated by that scenario for the 2000-2010 period would presumably have unnecessary capacity with which to supply electricity. Utilities that anticipated demand more consistent with that ultimately postulated by the scenario would presumably have spared themselves the burden of unnecessary capacity, but would be forced to rely on older plants for which maintenance time could be important.

If future fossil fuel heat rates at utility-generating plants improve, electricity costs, hence, prices, would experience downward pressure. Better heat rates could be realized through technical improvements in generating equipment. They could also be realized were environmental protection measures made less stringent or less energy intensive. Conversely, more stringent environmental control measures could operate to worsen fossil fuel heat rates.

Technical improvements in conversion efficiencies seem most probable under the circumstances of the "Mega-plant" Scenario and the "Small Coal Plants" Scenario. In the context of the "Mega-plant" Scenario, one might expect the combination of a robust economy and a large high-technology manufacturing base to lead to materials breakthroughs that would, in the long run, allow more efficient fuel conversion. The "Small Coal Plants" Scenario postulates both modest economic growth and a modest high-technology manufacturing sector; however, the emphasis that scenario places on coal-fired generation from, by today's standards, atypically small coal-based facilities suggests that fossil fuel conversion technology might receive particular research attention.

All scenarios except the "Economic Malaise" Scenario postulate environmental regulation of acid rain pollution by 1990. These environmental controls would be expected to have a negative impact on heat rates and, therefore, would exert upward pressure on electricity prices. The "Nuclear Resurgence," "Mega-plant," and "Post-industrial" Scenarios postulate more stringent small particulate pollution control regulation during the 1990s. These regulations might also have negative impacts on fossil fuel heat rates. (Note that even the "Nuclear Resurgence" Scenario postulates almost 50 percent utility reliance on fossil fuels in 2010.) Under the conditions postulated for the "Economic Malaise" Scenario, fossil fuel heat rates would be expected generally to improve, relative to rates in other scenarios, because of somewhat more lenient environmental standards. Of course, plant deterioration also undermines heat rates, so utilities with this problem could suffer worsening heat rates even under the "Economic Malaise" Scenario.

The cost of utility fuel has a substantial impact on electricity prices. High fuel costs were the dominant reason for electricity price escalations during the 1970s. All scenarios postulate more moderate escalation in oil prices during the next 30 years than has been the case in the past 10. Oil prices in 2010 are postulated to be the lowest in the "Small Coal Plants" Scenario and highest in the "Mega-plant" Scenario and the "Post-industrial" Scenario. One would anticipate that per BTU prices of coal will continue to be much lower than per BTU prices for oil and will continue to rise and fall in a pattern approximating the rise and fall in oil prices. As all scenarios contemplate only modest real price increases in oil during the 1980s, one would likewise expect almost flat prices for coal during the decade.

Because of the heavy reliance on nuclear fuel in the "Nuclear Resurgence" Scenario, the long-term impact of fuel price on the cost of generating electricity likely would be lower for that scenario than for any other scenario except the "Post-industrial" Scenario. Note that the "Nuclear Resurgence" Scenario postulates a satisfactory long-term solution to the handling of nuclear wastes; in the absence of such an assumption, nuclear fuel prices might well become significant. The cost of even a satisfactory solution will likely increase nuclear fuel costs somewhat.

The "Post-industrial" Scenario postulates a technological breakthrough in solar photovoltaics during the 1990s. As solar photovoltaics displace other utility-generating capacity, the impact of fuel costs on overall electricity prices would decline. Similarly, the advent of the mega-plants postulated for the "Mega-plant" Scenario would, later in the twenty-first century, make electricity price relatively insensitive to fuel cost. Finally, note that implementation of time-of-use rates also tends to lessen the price impact of fuel costs because these rates will presumably lead to more level demand for utility electricity and thereby make it possible for utilities to reduce the use of high fuel cost peaking facilities.

Increases in capital costs per kilowatt of new generating capacity translates, over time, into increased electricity prices. From previous discussions of power plant capital costs, it appears that increases in environmental control and accident containment equipment have in recent history most contributed to escalations in capital costs.

Projections of real cost escalation for coal and nuclear plants during the 1980s are extremely difficult to reconcile, largely for reasons explained in this chapter's earlier discussion of historic capital costs. However, there seems to be general agreement that coal and nuclear capital costs will escalate over the next 20 years (Appendix B, Item 55; CONAES, 1980:264; EIA, 1981b:304). The previously cited Komanoff study projects approximately 3 percent annual escalation in coal plant capital costs and approximately 4.5 percent escalation in nuclear plant capital costs during the 1980s; Komanoff's hypothesis is that capital costs in the nuclear sector expand as the size of that sector expands (because the importance of accident prevention expands as the size of the sector expands).

Because all scenarios except the "Economic Malaise" Scenario contemplate more stringent acid rain regulations by 1990, capital costs for hydrocarbon plants would be expected to escalate for environmental reasons in each of these scenarios; to the extent that the coal-based plants that come on line around the turn of the century in the "Small Coal Plants" Scenario are of an unconventional design (e.g., fuel cells or fluidized beds), the capital cost influences of pollution control measures could be reduced. For the scenarios ("Nuclear Resurgence," "Mega-plant," and "Post-industrial") that postulate small particulate emission control during the 1990s, coal plant capital cost escalation could be expected to further increase.

If the Komanoff hypothesis is correct and nuclear plant capital costs escalate as the nuclear sector expands, then the "Average Future," "Nuclear Resurgence," and "Mega-plant" Scenarios would call forth escalation in nuclear power plant capital costs. Such a hypothesis would be inconsistent with the marked increase in nuclear power generation postulated for the "Nuclear Resurgence" Scenario; a hypothesis more consistent with the postulated facts in the "Nuclear Resurgence" Scenario is that capital cost escalation in nuclear power plants slows or ceases as plant standardization and other learning curve phenomena become evident.

Cancellation of power plant construction prior to completion reflects a misallocation in society's resources, a misallocation that must be financed by one or more segments of society. Except in cases of extraordinary negligence on the part of utility management, one would expect that the costs of construction abandoned prior to completion would be borne by utility ratepayers (Robinson, 1981). Hence, cancellation of power plant construction prior to completion will generally exert upward pressure on electricity prices. (Prices may be less upwardly influenced than would have been the case were costly and perhaps unneeded capacity completed. However, prices will be higher than they would have been had construction never been undertaken.)

Suspension of construction on power plants pending increases in electricity demand may in some instances result in less waste of capital resources than would outright cancellation. Even this, however, increases overhead and would also be expected to put upward pressure on utility electricity prices.

Particularly as to nuclear plants, construction cancellation and suspension became common phenomena during the late 1970s and early 1980s. As of early 1983, there were approximately 55 nuclear power units actually under construction; these units represent approximately 60 gigawatts of nuclear capacity (Raloff, 1983:12). Of the six scenarios, apparently only the "Nuclear Resurgence" Scenario contemplates enough demand for nuclear electricity during the 1990s to justify completion of all the plants currently scheduled. Within the contexts of the other scenarios, it should be expected that some of the plants currently under construction will be cancelled (particularly in the "Small Coal Plants," "Post-industrial," and "Economic Malaise" Scenarios) and that others will have their completion schedules extended because of construction suspension (particularly in the "Average Future" Scenario) (Congressional Research Service, 1982b).

Note that in the "Mega-plant," "Small Coal Plants," and "Post-industrial" Scenarios it is postulated that currently unconventional sources of electricity supply come to be seen as the technological wave of the future. Under these circumstances, one would expect accelerated decline in the values of more conventional generating facilities. Utility recovery of capital invested in these facilities might also reasonably be accelerated by regulators, another circumstance that would exert an upward influence on electricity prices.

As previously discussed, real rates of return on equity investments on electric utilities suffered during the 1970s but showed signs of marked improvement during the early 1980s. One would expect this improvement be maintained throughout the 1980s because of the already mentioned investor incentives, moderate fuel price increases, decreased utility plans for capital expansion, and the addition to rate bases of large blocks of capitalized AFUDC as plants under construction during the 1970s come on line throughout the 1980s.

Further, the rate of return allowed equity investors in electric utilities, in theory, should be comparable to the rates of return available to investors in other enterprises of similar risk. One would expect, therefore, that rates of return would be maintained or even improved because of general awareness that future electricity demand is uncertain and that the technologies for supplying future electricity demand are uncertain. These uncertainties imply risk, for the assumption of which equity investors should theoretically be allowed higher rates of return. The "Post-industrial" Scenario provides an archetypal example of a future in which an investment in electric utilities should be viewed as a high risk investment; the advent of solar photovoltaic technologies could ultimately obviate the need for conventional electric utilities. Further, in the context of the "Post-industrial" Scenario, the possibility exists for rather rapid loss of regulatory control of electricity prices and, hence, of control of investor rates of return.

Regulatory treatment of CWIP and AFUDC can affect both total future prices for electricity and the timing of changes in electricity prices. As previously discussed, regulatory admission to the utility rate base of the cost (or partial cost) of construction work in progress moves nearer in time the ratepayer obligation to finance the "return" on capital, but it decreases the number of periods over which the costs of construction money compound. Whether the consumer price impact of one method is greater than the other over the long term, say 30 years, depends on such complex variables as the annual rates at which AFUDC is compounded, the rates of return allowed utilities on rate base items, and the real dollar discount rate of the average consumer.

Within the contexts of the six scenarios, one would generally expect that the CWIP/AFUDC controversy would be less important in the future than it has been in the past. Limited plant construction programs or less expensive real plant construction costs subject less capital to delayed return; expansive, long-term capital construction projects are not contemplated for this century in any scenario. The "Mega-plant" Scenario presents a possible exception to this statement, depending on the extent to which costly and long-term solar power satellite construction programs are privately funded (KH:V-4).

There is another matter related to the CWIP/AFUDC problem which may become, however, a point of controversy during the 1980s under each of the scenarios. This is the matter of price increases that would be expected to accompany entry into the rate bases of the 30 to 55 nuclear plants now under construction and actually destined for completion. These will be very expensive plants, costing billions of dollars each, and, because rate base treatment of CWIP is still generally limited or disallowed, their completions will cause the rate bases on which ratepayers pay returns to balloon. At least in the scenarios that postulate less vigorous economic growth, such as the "Average Future" Scenario, the "Small Coal Plants" Scenario, and "Economic Malaise'' Scenario, ratepayer resistance to rapid rate increases might focus debate on schedules that "stretch out" payment of shareholders' earnings. To the extent that plants entering the rate bases provide utilities with uncommitted ("excess") capacity, there may be ratepayer resistance even to their inclusion in rate bases.

From the foregoing discussion one can understand why there is a broad range of opinion about the exact price of a kilowatt hour of electricity at any point in the future. Respondents to the expert opinion survey conducted as part of this assessment (Appendix B, Item 50) estimated prices at the turn of the century that, at the low end, reflect no real escalation in electricity prices and, at the high end, reflect real escalation of more than 3.5 percent annually. Upon reviewing the determinants of electricity price and considering the price affecting inertia of electric power decisions already made, it is difficult to believe real electricity prices could decline for any meaningful periods this century. There exists the very strong possibility that generating plant capital costs during at least the next 20 years will affect electricity prices in the way that (but less dramatically than) fuel costs affected those prices during the 1970s. Because it appears that any reasonable electric power scenario for this country encompasses heavy reliance on coal-fired generation, containment of environmental risks potentially associated with coal combustion could exert strong upward pressure on electricity prices. The prospects for long-term decline in the real price of oil are, of course, uncertain; such an eventuality would have beneficial effects on electricity prices and could, depending on the size and duration of oil price declines, offset the effects of other cost increases.

Electric utility diversification. Under the mandate of such federal legislation as the Public Utility Regulatory Policy Act of 1978, electric utilities in the late 1970s and early 1980s began to provide services that were not conventionally within the realm of accepted or expected utility activities. The following discussion is concerned with this trend to a more broad definition of appropriate "utility" functions; it does not address utility diversification into fields unrelated to energy use.

California's policy on utility diversification into other energy services is probably the country's most developed. The California Energy Commission states that its fundamental premise for broadening the scope of permissible electric utility functions is that "utility customers do not demand energy in kilowatt hours and therms; they demand heat, light, and mechanical power" (California Energy Commission, 1981:107). To the extent then that California utility actions make it easier for utility customers to get the heat (refrigeration), light, and mechanical power they demand, those actions are at least potentially appropriate to the business of an electric utility. Thus, such services as energy auditing, financing of energy conservation investments, and, eventually, the brokering of electric power are specific examples of previously unconventional means by which utilities might acceptably earn revenues in the future.

It is not obvious that utilities will aggressively pursue these diversification avenues in the absence of regulatory mandate (KH:V-5). Offering services such as loans and audits would presumably place utilities in competition with less-regulated firms, such as banks and engineering firms. Further, many potential services would have no constituency within the existing electric utility management (KH:V-6).

The degree to which energy services become elements of future utility business depends largely on society's needs for these services and on the other options available to electric utilities. Modest expansion of utility energy service activities, such as energy auditing and data collection, should be contemplated for each of the six scenarios. Within the contexts of the "Nuclear Resurgence" and the "Mega-plant" Scenarios, however, one would not expect to see aggressive utility promotion of unconventional energy services. In the "Small Coal Plants" Scenario and the "Post-industrial" Scenario, on the other hand, one would expect to see utilities more vigorously undertaking conservation services and services directly related to the sale of electric power, such as the common carriage and, ultimately, brokerage of electric power. The national commitment implied by the economic restructuring hypothesized for the "Post-industrial" Scenario and the important energy role played by electricity in that scenario strongly suggest wide-ranging business opportunities for electric utilities. One would expect very little voluntary diversification within the context of the bleak economic times hypothesized in the "Economic Malaise" Scenario.

Electricity price deregulation. In this chapter's previous discussion of recent responses to changed electric power trends, it was observed that the theoretical justification for electric power price regulation was increasingly subjected to critical examination in the late 1970s and early 1980s.

What are the future prospects for deregulation of electricity price? The most thoughtful analyses of this question recommend that movement to price deregulation, if at all, should be undertaken in a very gradual, experimental fashion, probably beginning at the bulk power sales stage of electricity supply (Joskow and Schmalensee, 1982:255; Bohn et al., 1982:62). It is important to remember that the choice of pricing mechanisms is not between perfect regulation at one end of the spectrum and perfect competition at the other end but, rather, is between various degrees of imperfect regulation and competition. Were price regulation in the future seen to be theoretically unjustified, it still might be more efficient, as a practical matter, than various degrees of realizable competition.

Because of the probability that electricity prices will continue to rise in real terms throughout this century and considering the high level of physical integration and development already attained by electricity supply systems, one should expect continuing debate about the wisdom of and the need for pervasive regulation of electricity price. The extent to which electricity price might ultimately be deregulated varies among the six scenarios. In particular, the prospects for price deregulation seem most constrained within the context of the "Mega-plant" Scenario, if for no other reason than the visibility of scale implicit in systems evolving to reliance on solar power satellites. One would expect more movement to price deregulation in the "Nuclear Resurgence" Scenario, the "Average Future" Scenario, and the "Economic Malaise" Scenario. These three contemplate electricity supply almost entirely from utility sources and from technologies rather similar to those in use today. Price deregulation of bulk power sales seems feasible today, and improving communication technologies also improve the feasibility of more general generation price deregulation.

The "Small Coal Plants" Scenario and the "Post-industrial" Scenario offer the greatest likelihoods of lessened electricity price regulation. In the "Small Coal Plant" Scenario, this likelihood is promoted by the growth of non-utility power generation. Presumably, in the context of this scenario, generation of electricity by cogenerators and small power producers offers economic benefits; otherwise, these groups would not undertake power generation. It seems probable, therefore, that successful utility competition for capital will turn, in part, on utilities' abilities to secure for themselves some of the benefits enjoyed by the non-utility generators. Initially, utilities might be allowed to sell power from small units to themselves at incentive prices, assuming power producers and cogenerators retain this advantage. However, incentive pricing is not likely to be an indefinite phenomenon. If cogeneration and small power production is viable without incentive pricing, then price competition among these producers—some of which would, presumably, be utilities—should be expected.

The potential for price deregulation is also pervasive in the "Post-industrial" Scenario. This scenario postulates technological breakthroughs in power generation that could negate utility scale economies in power generation. One might see not only small entrepreneurial power producers supplying wholesale electricity, but also many end users supplying their own electricity. These could also sell directly to others located nearby. This circumstance and the information society aspect of the "Post-industrial" Scenario suggest that there is a greater possibility here than in other scenarios for the proliferation of supply sources and end use price deregulation.

E. SUMMARY

The precise cost of future electrification is unknowable. It is a function of multiple factors for which future values are themselves unknown.

The 1970s saw trends describing key cost factors reverse previously favorable directions; fossil fuel price was the principal of these but power plant cost, fuel conversion efficiency, and utilization factors were also important. During the 1970s, electrification ceased to be a declining cost activity and became an increasing cost activity. Analyses of cost-influencing factors suggest electrification will remain an increasing cost activity throughout this century at least. Capital costs of future power generation facilities will probably have the most influence on future electrification costs. Environmental and health protection measures could well continue to increase capital costs for conventional facilities. It is expected that truly unconventional facilities would initially be costly and, because of the embedded capacity of conventional facilities, have little opportunity to influence electrification costs. In the absence of technological events that remove from regulators the power to allocate future electrification cost burdens, ratepayers will bear most of these burdens.

Continued utility diversification into more broad energy services is likely in each future considered by this assessment. The degree of diversification is principally a function of demand growth for utility electricity and the need of future society for more broad energy services.

Some degree of electricity price deregulation is more probable in the future than it has been in the past. Rising power supply costs, the current *de facto* national market for electricity, and the potential for cost-competitive smaller scale power generation each increase the likelihood of at least limited price deregulation. The likelihood is greater in some of the futures considered by this assessment than in others. Generally, one would expect price deregulation, if it occurs at all, to begin near the generation stage of supply, probably with bulk power sales. In any event, one would expect very gradual deregulation.

BIBLIOGRAPHY

Baumol, William J. "Productivity Incentive Clauses and Rate Adjustments for Inflation." *Public Utilities Fortnightly* (July 22, 1982), 2-9.

Bohn, Roger E.; Richard D. Tabors; Bennett W. Golub; and Fred C. Schweppe. "Deregulating the Electric Utility Industry." In *Electric Power Regulation, Deregulation and Structural Reform*. Massachusetts Institute of Technology Energy Laboratory Symposium. Cambridge, Mass.: MIT Energy Laboratory, October 1982, Chapter VIII.

Brigham, Eugene F. "Public Utility Finance." In *Financial Handbook*. Edward I. Altman, editor. New York: Wiley, 1981, Chapter 36.

Bryson, John E., and Jon F. Elliot. "California's Best Energy Supply Investment: Interest-Free Loans for Conservation." *Public Utilities Fortnightly* (November 5, 1981), 10-16.

Budwani, Ramesh N. "Power Plant Scheduling, Construction, and Costs: 10-Year Analysis." *Power Engineering* (August 1982), 36-49.

Bureau of Economic Analysis. *Survey of Current Business*. U.S. Department of Commerce, October 1982.

Bureau of the Census. *Historical Statistics of the United States*. U.S. Department of Commerce. Washington, D.C.: U.S. Government Printing Office, 1975.

Bureau of the Census. *Statistical Abstracts of the United States*. U.S. Department of

Commerce. Washington, D.C.: U.S. Government Printing Office, 1979 and 1981.

California Energy Commission. *Energy Tomorrow*. Sacramento, Calif.: Governor's Office, 1981.

Committee on Nuclear and Alternative Energy Systems (CONAES). *Energy in Transition 1985-2000*. San Francisco: W. H. Freeman and Co., 1979.

Congressional Research Service (CRS). *Construction Work in Progress in Electric Base Rate*. For the Subcommittee on Energy and Power of the Committee on Energy and Commerce of the U.S. House of Representatives. Washington, D.C.: U.S. Government Printing Office, 1982a.

Congressional Research Service. *Do We Really Need All Those Electric Plants?* Washington, D.C.: U.S. Congressional Research Service, Library of Congress, 1982b.

Edison Electric Institute. *Statistical Yearbook of the Electric Utility Industry*. New York: Edison Electric Institute, various years.

Energy Information Administration (EIA). *Impacts of Financial Constraints on the Electric Utility Industry*. U.S. Department of Energy, DOE/EIA-0311. Springfield, Va.: National Technical Information System, 1981a.

Energy Information Administration. *1980 Annual Report to Congress*, vol. 3. U.S. Department of Energy. Washington, D.C.: U.S. Government Printing Office, 1981b.

Energy Information Administration. *1981 Annual Report to Congress*, vol. 2. U.S. Department of Energy. Washington, D.C.: U.S. Government Printing Office, 1982.

Joint Economic Committee. *1980 Supplement to Economic Indications*. U.S. Congress. Washington, D.C.: U.S. Government Printing Office, 1980.

Joskow, Paul L., and Richard Schmalensee. *Deregulation of Electric Power: A Framework for Analysis*. Massachusetts Institute of Technology Energy Laboratory for the U.S. Department of Energy. Springfield, Va.: National Technical Information System, 1982.

Killian, Linda R., and Robert R. Trout. "Alternatives for Utility Deregulation." *Public Utilities Fortnightly* (September 16, 1982), 34-39.

Komanoff, Charles. *Power Plant Cost Escalation*. New York: Van Nostrand Reinhold, 1981.

"The Latent Voltage in Utility Stocks." *Fortune* (December 27, 1982), 108-9.

Moody's Investors Service. *Moody's Public Utility Manual*. New York: Moody's Investors Service, Vol. II, 1982.

Navarro, Peter. "Our Stake in the Electric Utility's Dilemma." *Harvard Business Review* (May-June 1982), 87-97.

Niskanen, William A., and Benjamin Zych. Letter to the Editor. *Harvard Business Review* (September-October 1982), 180, 185.

Raloff, Janet. "Nuclear Power at 25." *Science News* (January 1, 1983), 12, 13.

Robinson, Frank S. "Utility Fiascos—Who Should Pay?" *Public Utilities Fortnightly* (December 17, 1981), 17-23.

Thompson, Arthur A. "The Strategic Dilemma of Electric Utilities." *Public Utilities Fortnightly* (March 18, 1982), 19-28.

CHAPTER VI

Health and the Environment

A. OVERVIEW

Energy production and environmental conservation and protection have often been cast in conflicting roles, especially as the federal and state governments have struggled to find policies to ensure adequate energy supplies and to protect the environment and public health. There is little doubt that government decisions and private action involving energy production significantly affect the nation's environment and public health and that efforts to protect the environment and health do have major implications for energy production. A tough challenge for decision makers in the 1980s will be to find a proper mix of energy and environmental policy both for short- and long-term needs.

The 1970s was an era of increased environmental awareness and public demand for protection of health and the natural resources. Major laws aimed at protecting air, water, land, and worker health were enacted by Congress and many state legislatures. Despite these efforts and the resulting protection, public demand for further protection is likely to remain strong in the future (Appendix B, Item 31; Continental Group, 1982; Milbrath, 1981).

These changes came at a time when the stresses on our electrical energy systems were growing. Some of those stresses can be linked to the change in public attitudes and to the government policies for protecting the environment. For example, regulations on the removal of sulfur dioxide from power plant emissions have raised the cost and complicated the process of burning coal. Protection of certain scenic lands and rivers has limited options for siting hydroelectric and other power plants. Although the stresses on production and regulation of electricity in the 1970s were more directly linked to the economy and events such as the oil embargo by the

Middle Eastern countries, many of the tough environmental issues, such as acid rain, allocation of scarce water resources, and protection of workers from exposure to toxic emissions, remain to be coordinated with electrical energy policies in the 1980s.

To complicate many of these decisions, policy makers in the 1980s will also be faced with many technical uncertainties. The effects of a buildup of carbon dioxide from fossil fuel combustion are not now predictable, even the extent of the buildup is uncertain. If the effects were to raise atmosphere temperatures significantly, the results could be devastating. Yet the choice to continue to rely heavily on coal must be set in the face of such uncertainties and with a recognition of the broad implications. This chapter will discuss the implications for land and water resources, occupational health, and the management of residuals.

B. LAND AND WATER RESOURCES

Few systems for electricity production can exist without a heavy reliance on land and water resources. It has often been the availability of one or both of these resources that dictated whether a particular choice of fuel supply or generating facility would be economically feasible in a particular area of the country. The availability of a resource is itself often a complicated question of competition for or government allocation of the resource.

Since the early generation of electricity at Niagara Falls, water has been critical to increased electrification. Whether the electricity is generated directly from the energy of the "head" or height of water or indirectly from steam-powered generators, power plants have historically been sited next to large sources of water. That situation is not likely to change since considerable reliance on traditional nuclear and fossil fuel power plants is expected for the next 30 years. While demand for the water needed for cooling at conventional power plants could be reduced by improved technology or substitute coolants, the extent of the decrease is not likely to be great.

Similarly, demands for land for mining and transporting fuels, for facility siting, and for transmission of electricity are likely to increase. Existing sites and corridors can often be reused for new power plants and added transportation and transmission requirements; however, additional lands will be needed under most of the foreseeable futures.

Western economic theory suggests that scarce resources could be allocated by means of a competitive market system. Thus, if demand for its output is great, a fuel production operation should be able to pay a price sufficiently high to induce other users, such as farms, to sell their water rights. Similarly, acquisition of lands for surface mining or facility siting could often be viewed as simply a matter of a real estate transaction.

Competitive market systems, however, are not always the appropriate means for the allocation of these resources. Goals of society may favor maintaining certain resources for farming at the expense of electricity production. Moreover, individual owners of resources may not make purely economic decisions. On the other hand, public needs may demand that advantages be given for electricity production. For example, electricity producers are often provided with condemnation authority for transmission lines.

In the future, both land and water resources suitable for electricity production are likely to become scarcer. Predictions of national or regional water shortages are common, and the public demand to protect unique public lands remains high. With the competition for the available resources becoming tighter, increased government involvement in the allocation of the resources appears to be likely.

Water. A crisis in water resources has become almost an expectation of the American public. News articles of regional shortages and demands for new water projects have elevated public concern over a water crisis to a level not far behind concern over an energy crisis. In some areas there are likely to be serious water shortages that could affect electricity production (EPA, 1980a:475-87).

The nation has, on an average, 30 inches of rainfall a year. That rainfall, however, is not evenly distributed throughout the United States. Precipitation varies from less than 20 inches per year in most of the West to over 40 inches in most of the East (CEQ, 1981:212). In many of the western states, surface water has to be allocated during periods of drought since sufficient water for all current uses at current prices is not available. Groundwater withdrawals can make up for some of the shortages, although a number of aquifers are already being taxed beyond their capacity to recharge. Again, the problem areas lie generally in the West (CEQ, 1981:215), where population migration is increasing the demand for electricity.

Irrigation is by far the greatest user of water supplies. Exhibit VI.1 shows the consumption rates and the increase in these rates by all sectors. While the steam electric utilities take only a small percentage of the total consumptive use, they show the greatest percentage increase. This increase is likely to continue especially if production of synthetic fuels were to increase significantly (EPA, 1980a:487). Consumption of water, however, is not an accurate measure of the utilities' demand since only a small percentage of the water withdrawn is actually "consumed," primarily through evaporation from cooling towers and from reservoirs. Additional water is withdrawn and used temporarily or shared on a limited basis. Exhibit VI.2 shows the total water demand for conventional power plants. The requirement of a single plant can often be sufficiently large to prevent siting in areas of water shortages.

Hydroelectric power generation creates one of the largest demands for water. There has been a substantial increase in hydroelectric power production, from 100 billion kilowatt hours in 1950 to almost 260 billion kilowatt

Exhibit VI.1
Water Consumption, by Use, 1960–1975

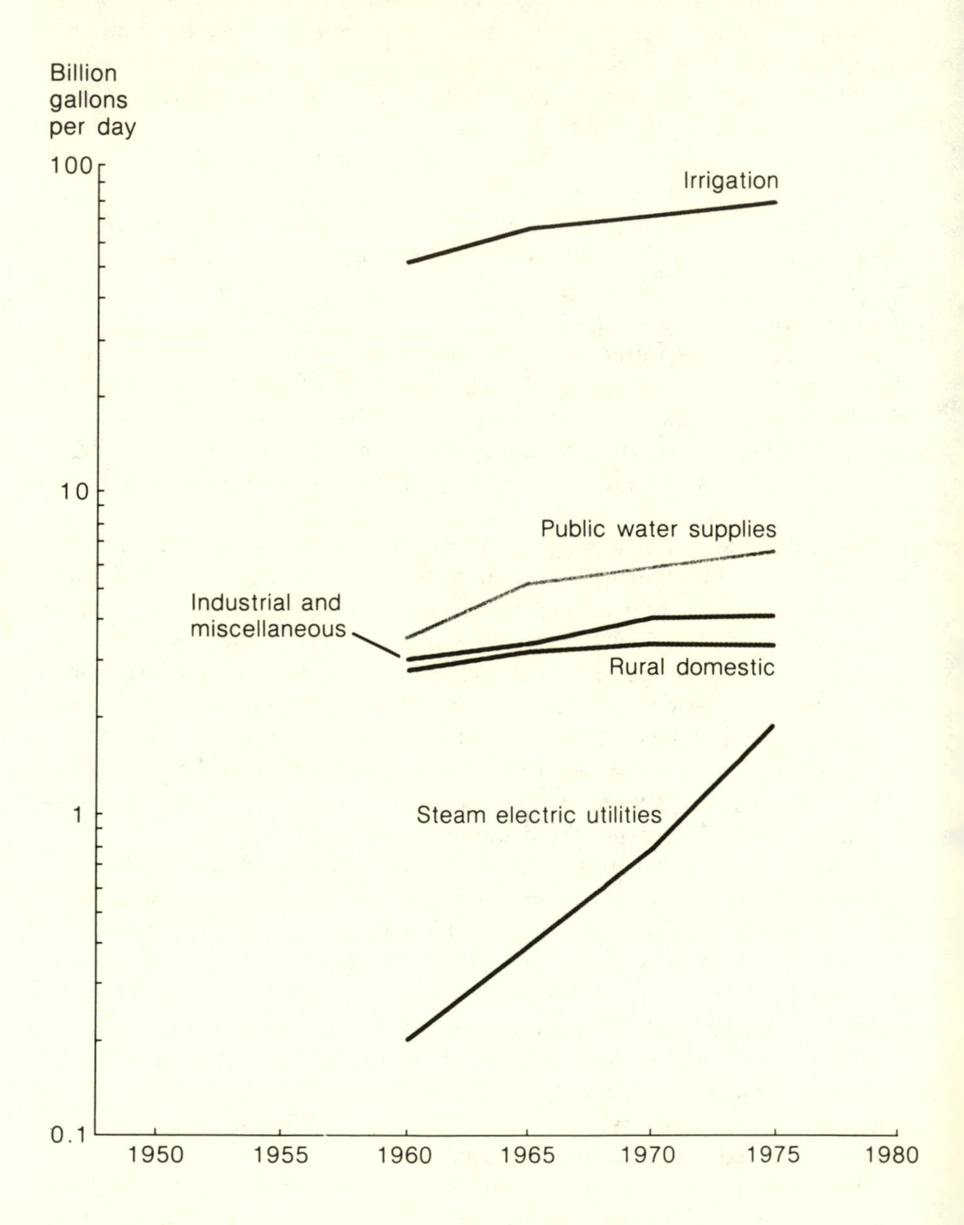

Source: CEQ, 1981:221.

Exhibit VI.2
Water Withdrawal, by Use, 1950–1975

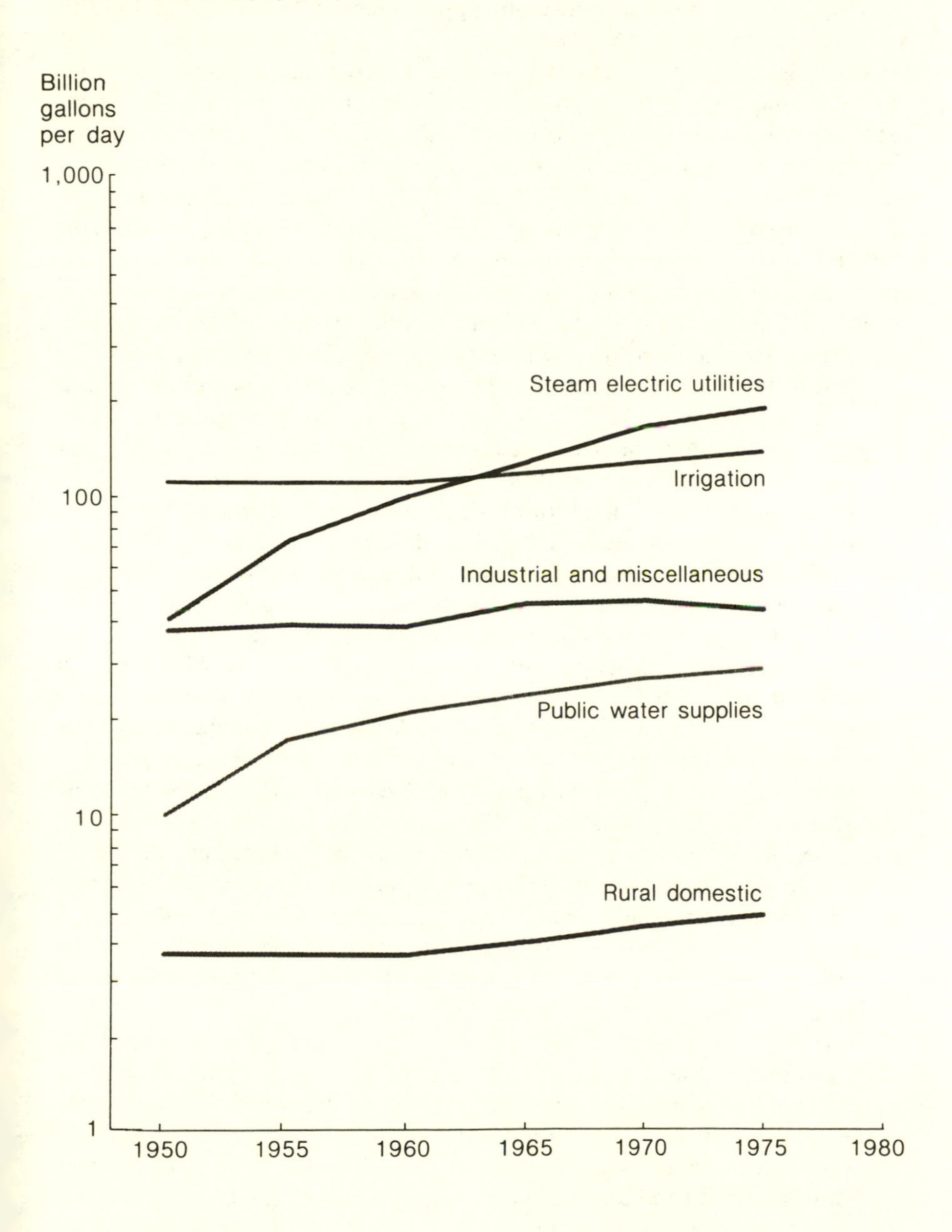

Source: CEQ, 1981:220.

hours in 1981. The percentage of electric power generation contributed by hydro, however, has dropped from approximately 25 percent in 1950 to 11.5 percent in 1981 (EIA, 1982:153).

While hydroelectric power is generally viewed as non-polluting and renewable, the destruction of areas of scenic beauty and of particular recreational opportunities unique to rivers has brought serious opposition to a number of hydro projects. Reservoir projects also have some controversial impacts on land uses. Between 1967 and 1975, the construction of new lakes inundated over 1.5 million acres of prime farmlands (CEQ, 1981:120). Other impacts include reductions in the flow of needed fresh water and nutrients in rivers and into bays and estuaries downstream of reservoirs. Many of these impacts will result from a policy of encouraging many low head hydroelectric dams as well as a policy for several larger projects.

Most of the major sites suitable for large hydroelectric power generation are being used or have been withdrawn from future development by government preservation policies such as the Wild and Scenic Rivers Act (Fowler, 1975:318-19). If significant increases in hydroelectric power generation are to occur, a number of new hydro sites will be needed. Acquiring these sites will not be easy. Increased efficiency in turbines, low head generation, and added pumped storage will also need to be developed.

Water used for cooling in coal, oil, natural gas, and nuclear power plants is the other major demand for water in electricity generation. In a future with low growth in electricity generation, such as "Economic Malaise" and "Post-industrial" Scenarios, existing reservoirs together with new technologies for reduced water consumption could almost eliminate the need for substantial new water supplies for generation. In a future of mega-plants or power parks, however, water availability determinations limit the siting of the facilities. Finding even the new sites postulated in the "Mega-plant" Scenario will not be easy. In a future with many small coal plants, the overall size of the water demand will be less important than the number of suitable sites for reservoirs. Again, reuse of existing sites will help, but a large number of new reservoirs will be needed. Greater reliance on salt water for cooling should be expected, since such a shift would leave fresh water available for uses that require fresh water.

Whether the utilities and fuel producers can get the water they need has normally been a function of the availability of the water, the degree of competition for the water, and state water laws. In areas of the West where water is in short supply, most states have a regulatory system for allocation of water during periods of drought (Hutchins, 1971:1-19). Under these systems, municipal and domestic uses usually have the top priority and will be the last to be cut off. Thus, in states where water shortages are expected, the states generally have established mechanisms to allocate the available water.

The federal government also has a role in the allocation of water resources. While not yet directly setting national priorities for use, the federal government has made water available for certain uses through construction and subsidy of major water projects. In the future, if water crises have a greater national impact, the federal government's role will likely shift away from that of a supplier to that of an allocator of water in and among states. For example, if coal slurry pipelines are installed, the interstate transport of the slurry water will require federal policy decisions.

In any future that requires an increase in water supplies for electricity production, some utilities and fuel suppliers will be faced with water shortages. The solution to these shortages may depend on the ability of utilities to develop new water conservation technologies or to convince policy makers of the need for special treatment. Utilities also may be faced with mandatory conservation requirements and may have to develop ways to ensure that other users such as farmers have the incentives to conserve in order to make additional supplies available.

Land. As with water resources, there appear to be, on a national average, sufficient lands for greatly increased electrification. Again, however, the competing interests for use of the land and the requirements for the suitability of the sites have created and will continue to create some shortages in the land resources needed for electricity generation.

Corridors for transmission lines are examples of the allocation problem that is usually solved with special treatment of utilities by state governments. Eminent domain powers are given to utilities as a public policy to avoid forcing the utilities either to pay higher prices for rights of way or to route transmission lines inefficiently if landowners refuse to sell easements at market value.

Use of coal for electricity generation has a more significant land resource implication because of the effects of mining. The availability of lands for mining of coal will not be a limitation on any of the scenarios described in Chapter II; however, the costs of reclaiming the mined lands are likely to put some limits on coal production.

Reliance on a solar power system, such as a thermal station, could also require a considerable commitment of land (Fowler, 1975:334-38). Thousands of acres may be required at a single site, and, if desert lands were used, the impacts on those fragile ecological systems could be significant. Similarly, biomass production could require a dedication of a significant amount of land and water for cultivation, competing directly with the cultivation of food crops. With wind generation, the problems do not involve amounts of land but rather the acquisition of suitable locations.

While the federal government has an important role in the allocation and protection of privately owned lands that are needed for electricity production, it has an additional role where it is the landowner (GE:VI-1). Much of the western coal resources and a large percentage of the oil and gas resources lie under federally owned lands and under the outer continental shelf. Other demands on federal lands, however, have limited and will continue to limit energy development. Approximately 25 percent of all federal lands have been removed totally or partially from use for mineral production and set aside as national parks, wildlife refuges, and wilderness areas (CF, 1982:296-97; see also Appendix B, Item 33). In contrast, a July 1982 decision by the Department of the Interior has opened up essentially all federal outer continental shelf lands to oil and gas leasing (BNA, 1981:371). Thus, for U.S. public lands, the federal government can determine not only the manner in which mining or oil production can be carried out, but also whether there will be any such production.

In meeting their needs for land and water resources, fuel suppliers and electricity generators are not competing in a purely free market. Governmental actions, such as development and allocation of water resources, authorization of use of eminent domain, and leasing of public lands, have conferred competitive advantages and disadvantages to the industries. The trend appears to be an increasing need for government policies, especially at the federal level (GE:VI-2).

C. OCCUPATIONAL HEALTH

Human resources needed for mining, transportation, operation of power plants, installation of rooftop solar units, and the many other jobs associated with the generation of electricity are as critical to power production as the land and water resources. While mechanization has reduced and is likely to continue to reduce the labor needed to produce a kilowatt hour of electricity, an increase in electrification will require substantial numbers of workers in all aspects of the fuel cycles.

As fertility rates decline and as the average age of the population increases, there is concern in many industries over the availability of an adequate labor pool (GE:VI-3). As a general rule, electricity production is not labor intensive. There could be short-term problems if a sudden increase in power plant construction caused a shortage in engineers or other workers with specialized skills. In addition, labor shortages could affect the electric power industry by affecting production of necessary components, end use products, or by raising costs. None of the types of futures described in Chapter II, however, creates a serious concern over shortages of most types of workers for electricity generation.

The major implications for workers from increased electrification are more likely to involve health effects. Within the existing fuel cycles, job-related injuries and disease have always been a problem. In all exploration, drilling and production of oil and gas in 1978, approximately 75,700 workers were injured, and 22 of the workers were killed (API, 1979; Brown, 1979:25-37). Millions of dollars are paid annually to compensate victims of black lung from coal mining (Ramsay, 1979:108-11).

If a future with increased electrification allows greater mobility and an increased opportunity for personal choices of jobs and life-styles, utilities or other industries with positions that pose serious health risks may have difficulty filling the jobs. There is also the possibility of disruption in one aspect of electricity production as a result of a major walkout over health risks. While several such scenarios could be proposed, one possibility involves a gradual buildup of concern over an uncertain risk, worker frustration over unexplained illnesses, and a sudden release of "proof" of the connection between the illness and the job. The probability of such an occurrence is low, but the impacts could be high.

Many decisions concerning the acceptability of a risk of occupational injury and the methods for reducing the risks of occupational injury will be complex, especially those dealing with chronic health risks with long latency periods. While policy decisions concerning the use of land and water resources will involve issues of the availability of the resources or distributing the resources in some equitable fashion, decisions concerning many health problems continue to be complicated by the scientific uncertainties.

Some of the occupational health risks associated with electricity generation are common to most fuel cycles. An increase in the number of transformers will likely result in an increased exposure of workers to the aromatic hydrocarbons and other chemicals used as or in the insulating liquids (U.S. Department of HEW, 1978b). New transformer fluids should, however, reduce the risks of exposure to the toxic chemical polychlorinated biphenyls (U.S. Department of HEW, 1978b). Hearing loss from noise pollution is a significant problem in many industries, and the problem is likely to demand increased attention in construction of new power plant facilities and in retrofitting old ones (EPA, 1980a:696). Furthermore, construction of any generating facility is likely to result in common problems of exposures to organic solvents and other toxic chemicals (U.S. Department of HEW, 1979).

In assessing the health risks associated with increased electrification, it is tempting to compare the risks associated with the various types of fuel cycles. Such comparisons, however, may not be very appropriate or helpful. First, the mix of fuel cycles is likely to continue, and emphasis on reducing the risks associated with each cycle will be necessary. In addition, the types of risks associated with the different fuel sources are not easily compared. For example, risks of immediate bodily injury from drilling for oil cannot readily be compared to delayed effects of exposure to a toxic chemical used in solar cell manufacturing or to fears that a uranium mine worker might have about adverse health effects from exposure to radiation. Further, a quantitative or even qualitative assessment of the risks associated with each type of fuel source is extremely difficult in light of the uncertainties surrounding the causes of some diseases (GE:VI-4).

Protection of worker health in the nuclear fuel cycle illustrates some of the difficulties in assessing the risks. It appears that most of the associated occupational diseases occur as pulmonary malignancies among mine and mill workers (Ramsay, 1979:109). Exposure of workers to radiation at nuclear power plants also is expected to result in some injuries (EPA, 19080a:651-57). An actual quantification of the risks or even an accurate description of the nature of the risks associated with exposure to radon gas, its radioactive "daughters," and other sources of low levels of radiation, however, is the subject of considerable debate (NAS, 1972).

In addition to the limitations on the evaluation of occupational health risks associated with fuel cycles, other difficulties exist when attempting to compare the future health impacts of alternative electricity supplies. A comparison of ranges for predictable injuries of past coal and nuclear operations is likely to show that the coal fuel cycle is a greater source of injuries than the nuclear cycle (Ramsay, 1979:108-18). Historic comparison, however, cannot safely be used to predict accurately the risks associated with the technologies that are likely to be used in the future. Increased surface mining and improved boiler design should significantly reduce risks associated with coal. *In situ* uranium mining has the potential to reduce radiation exposures for workers. Thus, decisions concerning future choices of fuels cannot be based solely on a comparison of historic risks.

Finally, a comparison of the risks to workers involved in the different fuel cycles may not give much guidance for future choices of power sources unless there is an equal degree of knowledge of risks. While coal has been used as a fuel for a much longer time than uranium, health effects associated with the exposure to coal combustion products have not been as thoroughly studied as the health effects of the nuclear fuel cycle. This is partly a result of traditional public acceptance of coal. The situation again suggests that historic comparisons may have limited value (KH:VI-1).

While uncertainties limit the value of comparing worker health risks when evaluating the implication of alternative futures, the assessment of the types of risks associated with each fuel cycle does suggest some important areas for policy decisions. Most fuel cycles create health risks that are difficult and expensive to reduce and for which major uncertainties exist. Creating mechanisms for addressing these problems may be critical to the continued public acceptance of a particular fuel cycle.

As discussed above, worker exposure to radon gas and other radioactive materials associated with the nuclear and, to some extent, the coal fuel cycles will demand additional monitoring and research. Combustion of most hydrocarbon fuels, including oil, gas, coal, and biomass, can result in a range of risks from boiler explosions or fire (Bliss et al., 1976:228-36) to toxic air emissions exposure. Many of the chronic health implications of exposure to these chemicals are unknown or uncertain (MITRE 1980:1-7).

Burning synthetic fuels, including gasified and liquified coal and oil shale, pose most of the same problems associated with combustion of other fossil fuels. In addition, workers will be exposed to risks at the mining and processing facility. While there has not yet been adequate research on the subject, it appears that workers at coal liquification or gasification facilities will run the risk of exposure to carcinogens, such as aromatic amines and polycyclic hydrocarbons, and acutely toxic materials, including hydrogen sulfide and carbonyl sulfide (Brown and Whitter, 1979; Brown, 1981). There are, however, indications that new high temperature and pressure gasification technologies could significantly reduce these exposures. Finally, oil shale operations can result in additional risks of exposure to numerous potential toxicants in the fugitive dust caused by mining (Brown, 1979).

Exposure to toxic chemicals is not eliminated by reliance on renewable sources of electricity. Geothermal energy production can result in exposure to hydrogen sulfide, mercury, and sulfuric acid (U.S. Department of HEW, 1978a; DOE, 1982a). Manufacturing of solar voltaic cells carries with it use and exposure to chemicals such as silicon and arsenic (DOE, 1982b).

For any of these fuel cycles, monitoring of worker health and research on the mechanism and effects of chemical exposures are needed to reduce the uncertainties and determine the risks. Undoubtedly, new chemicals with uncertain health effects will enter the fuel cycles and create new risks. With the higher levels of public awareness of health risks that began in the 1970s, workers and the public are likely to demand that more emphasis be placed on reducing the chronic health risks associated with exposure to chemicals.

Additional research is often the logical first step in dealing with the uncertainties in health risks. Money spent by utilities for monitoring and research, however, is not likely to be viewed favorably by the public or regulators who are concerned with increasing productivity or reducing costs of electricity.

The federal government might, therefore, be the only significant source for major research or research funds in this area. The potential health effects of occupational exposure to chemicals will not differ greatly from region to region or from state to state. Regulatory exposure limits can, therefore, be determined on a national basis.

New research in health risks may not provide the answers or do so in a timely fashion. For that reason, money should not be spent entirely on research on health effects. Research and development for minimizing human exposures or on maximizing the substitution of machines for workers in positions where the risks are high or unknown also should be funded.

In any case, attention to the occupational health aspects of increased electrification will be a key factor in setting electric power policy in the future.

D. RESIDUALS MANAGEMENT

Before the 1973 oil embargo and resulting "energy crisis," the United States had begun to react to a perception of a long-range environmental crisis. Doomsday forecasts were interspersed with the calmer, yet no less emphatic, statements that the United States was slowly destroying its air, water, and land and was reducing the life expectancy of its citizens. Reaction by the federal and state governments was often swift, even with a threatening energy crisis. Some major improvements resulted.

The policies of the early 1970s to control residuals or wastes were at times made without full recognition of their implications for other policies, including electric power policies. Efforts were being made to coordinate energy and environmental policies, but in some cases the new environmental laws were enacted primarily to get the needed programs started to protect public health and the environment. Further, the laws and policies often addressed the more obvious environmental problems and, as a result, many decisions of the early 1970s left the tougher issues unresolved. It will not be surprising if some of the decisions that were made to get waste management programs started come to be viewed as misdirected with the hindsight of the 1980s. In addition, some of the decisions of the 1970s are likely to limit or to complicate efforts to establish more appropriate policies in the 1980s. Regardless of the "correctness" of the 1970s decisions, they form the foundation for environmental policies.

Many difficult environmental policy decisions remain for the 1980s. Some decisions, such as those dealing with acid rain and nuclear waste, raise questions of equitable distribution of risks over both large geographic areas and over present and future generations. Other decisions, such as those involved with carbon dioxide buildup, may have to be made before significant uncertainties are resolved.

These policy decisions may become even more difficult as they are integrated with electric power policy. For example, acid rain problems might be better resolved by forcing utilities that intend to rely on coal for electricity generation to use coal gasification technologies rather than using traditional coal-fired plants (KH:VI-2). Use of the traditional coal technologies with the related pollution control equipment can result in long-term commitments to a technology that does not solve many of the pollution problems and that significantly reduces the efficiency of the power generation. For example, the scrubbers that are used to reduce sulfur dioxide emission in coal-fired plants can add considerable expense to the cost of a power plant and can greatly reduce the efficiency of operations (Brown, Ouellette, and Cheremisinoff, 1983). New technologies, such as gasification, may not only eliminate the problems of reduced efficiency, but may also provide better pollution controls.

Likewise, resolution of electric power policy issues and environmental policy issues needs to be better coordinated with other public policy issues.

Increased electricity rates that result from increased waste management regulation can create serious problems for the poor and those on fixed incomes (Hinkle, 1978). On the other hand, failure to reduce pollutants, such as sulfur oxides, can have economic ramifications for agriculture and food production (Ramsay, 1979:22, 137). Although the importance of coordinating different public policies is well recognized, the actual coordination is difficult. Efforts to interrelate the different policies will have to be dealt with in the 1980s.

Air pollutants. Since the middle of the century, major air pollution incidents such as the ones in London, England, and in Donora, Pennsylvania, which resulted in a number of illnesses and deaths have demonstrated the serious need to control emissions of waste in the air. In the 1970s, a number of federal and state programs were established or revised to reduce pollution. The clearest proof of the benefits of these efforts is in the air. Using an index tied to the five major air pollutants, the number of unhealthy days in 23 metropolitan areas appears to have dropped 39 percent from 1974 to 1980 (CF, 1982: 47-49). This trend is expected to continue (Appendix B, Item 34).

In spite of these efforts, however, serious health and environmental impacts remain. In a recent study of the Ohio River Basin, a comparison of the mortality rates that can be linked to air pollution in the basin with those rates on a national average suggested that, in 1975, a number of deaths in that region could be tied to coal combustion emission (EPA, 1981b:22).

A major contributor to air pollutants has been the burning of fossil fuels for heat and power (NAS, 1981). The combustion can result in the emission of hydrocarbons, nitrogen oxides, sulfur compounds, particulates, and a number of toxic materials. Many of those emissions are from the tailpipes of automobiles; many others are from the stacks at power plants and other industries. With the multiple sources, multiple pollutants, multiple reactions of the chemicals in the atmosphere, multiple transport mechanisms, and difficulties in monitoring the pollutants, the exact source of an air pollution problem or the proper solution for cleanup are often not easily identifiable.

Fossil fuels are not the only source of air contamination. Emissions of radioactive materials are associated with the nuclear fuel cycle, although combustion of many fossil fuels can also result in emission of radioactive materials.

With the passage of the Clean Air Act in 1970 (USCA, 1973, vol. 42:7401), regulation of several major pollutants increased on a national level. Monitoring of ambient levels of these "criteria pollutants" shows an average decrease in all but nitrogen oxides over the last decade, although in certain areas some of the levels for other criteria pollutants are on the rise (CF, 1982:48-49). It may be that this general success will continue, but the reduction of these pollutants will not guarantee a reduction in the risks from air pollutants. Many of the more complex and less understood air pollution problems remain.

For example, while sulfur dioxide and nitrogen oxides have been regulated as criteria pollutants with resulting reduction or leveling off of emissions in the immediate area of the sources, the increased use of coal and other fossil fuels is apparently causing long-range and remote impacts. Emissions of sulfur dioxide and nitrogen oxides have been blamed as the primary cause of acid rain (NAS, 1981). The increase in acidity of many lakes and resulting loss of the aquatic life in New York, Canada, and Sweden exemplify the seriousness of this problem. Questions involving the transport of the acid compounds over great distances leave the cause-and-effect relationship not fully resolved. Nevertheless, public demand for solutions has forced the question into the political arena. Rather than waiting to resolve all the uncertainties, requirements for major reductions in emissions of the pollutants soon may be required by Congress.

Less certain is the likelihood of new programs for reduction of the fine or inhalable particles that result from combustion of fuels. Particulates are criteria pollutants under the Clean Air Act, and major progress has been made at reducing the overall quantity of particulate emissions to the benefit of visibility and public health. This reduction has often been simply a result of processes that filter out the larger particles that make up the bulk of the particulate emissions.

Since the larger particles have been the easiest to remove, they may have been the logical class of particulates to control first. As a result, however, the more serious threat to public health may have been allowed to slip through uncontrolled. It is the fine or inhalable-size particles which travel deepest into the lungs and which appear to be linked to many serious respiratory difficulties and diseases (NAS, 1980:8).

Dealing with the fine particles will require the resolution of a number of uncertainties. For example, the type and extent of risk to human, animal, or plant life from many of the trace chemicals are not well understood. With the expected high costs of developing and installing the additional monitoring and control equipment, the inability to quantify the risks reduces the incentive for control. Moreover, the fact that there is an existing program to regulate particulates may make it more difficult to implement a program for fine particles.

While the problems of control of particulates will be difficult to solve, they can be dealt with on a national or state level. The potential greenhouse effect of increased carbon dioxide buildup, however, creates a controversial problem of global scale (EPA, 1980a:173-95; NAS, 1981). It has been suggested that unless increased carbon dioxide emissions from combustion of any hydrocarbon—coal, oil, gas, or biomass—are absorbed by the oceans and vegetation, the increased atmospheric concentration of carbon dioxide could raise the temperature of the atmosphere. The results could range from small climatic changes to the melting of the polar ice caps and dramatic rises in sea levels.

The lack of knowledge concerning this problem is ominous; for, without convincing proof, the likelihood of action to reduce carbon dioxide emissions by a reduction in the burning of fuels in the United States or elsewhere around the world is doubtful. The possible buildup of carbon dioxide could present the dilemma that, by the time the problem is verified, it may be too late to rectify. More optimistically, if major global effects such as changes in weather patterns were to begin, they would occur slowly and provide time for a change in energy production patterns. Shifting from fossil fuels in the United States would still be difficult, especially if other nations were not also willing to make the needed changes.

Finally, a new type of air pollution problem is demanding attention in the 1980s. As energy conservation is increased by sealing and insulating homes and offices, trace elements can build up in the air (EPRI, 1981:164; CF, 1982:77-78). If public exposure to these indoor pollutants is determined to be serious, energy conservation programs could be limited. Some research efforts have begun, but resolving these pollution problems may be limited by available construction materials and consumer preferences for certain types of household and office products.

Water pollutants. Waste discharges into water courses have, like air emissions, long been a common waste disposal practice. Some discharges, like those associated with wastewater from power plant operations, are intentional discharges; others, such as mine drainage, often are unintentional and uncontrolled. In instances such as discharge of cooling tower blow down, the pollutant will end up in surface waters; in other cases, as with oil field brine disposal, the risk can be to groundwater. In many fuel cycles, waste heat is discharged to water courses.

While less progress is generally recognized in cleaning up water pollution problems than those with air emissions (CF, 1982:96-98), striking examples of successes are well known. The occurrences of fish returning to bodies of water that previously could not support such life are always noted events.

As with air pollution, industry and government have tended to pursue the easier and more obvious water pollution problems first. As a result, the problems of heavy metals contamination in power plant effluents often remain. Recent examples of major fish kills in cooling lakes for coal-fired plants have resulted from discharges that have concentrated trace levels of selenium from the coal (Cumbie, 1978; Sorenson et al., 1981). The buildup of trace contaminants, such as heavy metals, over a period of operation can create significant risks to the environment. Solving many of these problems has just begun.

While the cleanup of surface waters received major attention with the passage of the Federal Water Pollution Control Act in 1972 (USCA, 1973, vol. 33:1251), groundwater contamination has only recently been a focus of federal action. With the 1981 promulgation of regulations for injection wells for the disposal of chemical wastes and oil field brines (CFR, 40:146) and with the 1982 publication of rules for hazardous waste disposal (CFR, 1982:264), more protection of groundwater should result. Recently, public awareness of the problem of groundwater protection appears to be increasing and the likely result is a demand for stricter protection (EPA, 1980b).

The potential sources of groundwater contamination associated with electricity production include brine disposal from oil and gas production, both in surface pits and through injection wells; leaching of wastes at coal and uranium mines; and the seepage of contaminants through liners at solid and liquid waste storage and disposal facilities. Groundwater contamination is often not obvious. Slow migration rates often mask the problems. Yet once contaminated, an aquifer or source of groundwater is extremely difficult to restore.

As with other pollution problems, the extent of the risks or the perceived degree of the risks will determine when and how much attention is given to the problems. In the case of groundwater contamination, the development of regional water shortages will likely add to the demand for water protection. The traditional coal and uranium cycles will be hardest hit by new regulation for groundwater protection, although concern for leaching of pesticides and fertilizers with biomass farming may also require special consideration.

Land disposal of wastes. As the Clean Air and Clean Water Acts have required controls on emissions and effluents, the resulting wastes have often become a land disposal problem, especially for coal-fired plants. The nuclear fuel cycle, of course, has its own unique waste problems.

The buildup of radioactive wastes has highlighted the difficult issue of burdening future generations with the cost and risk from today's electric power generation. High-level radioactive wastes are extremely hazardous and, if not properly disposed of, could have devastating effects. Even if it were agreed that the wastes could be disposed of for thousands of years with no risk, the political problems are still controversial since few communities will readily accept the disposal in their "backyard." The federal Nuclear Wastes Policy Act of 1982 has not totally eliminated the political problems. Until both the technical and political questions are fully resolved, spent fuel will continue to be stored in pools and the possibility of a resurgence in nuclear power generation is reduced.

Associated with the nuclear waste issue are questions of decommission and disaster cleanup. Decommissioning of nuclear power plants creates another future waste problem, and in most instances no mechanisms for financing the decommissioning have been provided. The Three Mile Island incident has raised the issue of who will pay for costs for spill cleanup and decommissioning. In the cases of waste disposal, decommissioning, and disaster cleanup, the trend may be a reliance upon federal funds (AK:VI-1).

Although such reliance might give the industry an economic advantage, it might also stimulate political opposition to increased use of nuclear energy.

Hazardous chemical waste has the potential for becoming the type of problem for the coal cycle that nuclear waste has been for the nuclear cycle. With the Love Canal incident, the public became acutely aware of the problems of hazardous waste management, and a complex federal program for hazardous waste regulation has been established.

Ashes and sludges from coal burning are a significant solid waste problem for utilities because of the large quantities of the wastes (EPA, 1980a:158). In some cases those wastes are hazardous, although currently these utility wastes are exempted from the federal hazardous waste program. The federal Environmental Protection Agency is studying this inconsistency, and utilities could be faced with significant expenses from new land disposal regulation for their hazardous wastes.

Whether the utility wastes are reclassified, they still will be buried in the ground. Even under strict hazardous waste regulation, however, the risk of escape of hazardous chemicals and the resulting threat of groundwater contamination will be present. In the same fashion that the elimination of all air pollutants or associated health risks would be economically infeasible, requirements for zero risk for waste disposal have not been proposed for land disposal (EPA, 1981a). Thus, the utility wastes could create burdens for future generations. The extent of the burden left for the future will be a decision that must be faced in the 1980s as part of electric power policies.

E. SUMMARY

In the short run, growth in electrification is likely to result in significant impacts on health and the environment. Increased demand for water and lands for electricity production will cause some significant controversies both at the federal and state policy-making level. Occupational health risks will attract increased attention. Greater reliance on coal will assure greater problems of residuals management unless new technologies are developed. The continued public demand for protection of health and the environment will require that these controversies, risks, and problems be addressed.

Turning attention to these problems often means facing significant uncertainties, and policies will have to be set in light of these uncertainties. In addition, determining which areas of the country, which segment of society, or which generation will bear the risks and burdens will not be easy. Finally, integrating environmental policies into electric power policy analyses will create new challenges.

A wide range of decisions must be made in the 1980s in order to allow increased electrification while protecting the environment and public health. Those decisions that are characterized by larger uncertainties and tough equity questions will raise the greatest debates. The natural tendency of government may be to delay such decisions for additional research or simply to avoid a decision that would burden one area or one segment of society. Delaying decisions will at times also keep options open by not foreclosing a change in direction. Nevertheless, failure to make tough decisions can exacerbate the problems. An integrated approach to solving environmental and energy problems must be found so that policies will be set that allow both increased electrification and improved environmental protection.

BIBLIOGRAPHY

American Petroleum Institute (API). *Summary of Occupational Injuries and Illnesses in the Petroleum Industry*. Washington, D.C.: American Petroleum Institute, 1979.

Bliss, C. *Accident and Unscheduled Events Associated with the Non-nuclear Energy Resources and Technology*. McLean, Va.: The MITRE Corporation, M76-68, December 1976.

Brown, Richard D. *Health and Environmental Effects of Synthetic Fuel Technologies: Research Priorities*. McLean, Va.: The MITRE Corporation, MTR-80W348, April 1981.

Brown, Richard D. *Health and Environmental Effects on Oil and Gas Technologies: Research Needs*. McLean, Va.: The MITRE Corporation, MTR-81W77, 1979.

Brown, Richard D.; R. P. Ouellette; and A. N. Cheremisinoff. *Pollution Control at Electric Power Stations, Comparisons for U.S. and Europe*. Ann Arbor, Mich.: Ann Arbor Science, January 1983.

Brown, Richard D., and Alice Whitter. *Health and Environmental Effects of Coal Gasification and Liquefication Technologies*. McLean, Va.: The MITRE Corporation, M78-58, May 1979.

Bureau of National Affairs (BNA). *Environmental Reporter*, 12 (July 24, 1981), 371.

Code of Federal Regulation (CFR), 40. Washington, D.C.: U.S. Government Printing Office, 1982.

Conservation Foundation (CF). *State of the Environment*. Washington, D.C.: U.S. Government Printing Office, 1982.

The Continental Group, Inc. *The Continental Group Report: Toward Responsible Growth*. Stamford, Conn.: The Continental Group, Inc., 1982.

Council on Environmental Quality (CEQ). *Environmental Trends*. Washington, D.C.: U.S. Government Printing Office, 1981.

Cumbie, Peter M. *Belews Lake Environmental Study Report: Selenium and Arsenic Accumulation*. Charlotte, N.C.: Duke University, 1978.

Department of Energy (DOE). *Health and Environmental Effects Document on Geothermal Energy*. Washington, D.C.: U.S. Government Printing Office, 1982a.

Department of Energy. *Health and Environmental Effects Document for Photovoltaic Energy Systems*. Washington, D.C.: U.S. Government Printing Office, 1982b.

Electric Power Research Institute (EPRI). *Overview and Strategy 1982-1986*. Palo Alto, Calif.: Electric Power Research Institute, 1981.

Energy Information Administration (EIA). *1981 Annual Report to Congress*, vol. 2. U.S. Department of Energy. Washington, D.C.: U.S. Government Printing Office, 1982.

Environmental Protection Agency (EPA). *Environmental Outlook 1980*. EPA 600-18-80-003. Washington, D.C.: Environmental Protection Agency, August 1980a.

Environmental Protection Agency. "EPA Supplemental Notice of Reproposed Rules on Permits for Hazardous Waste Disposal Facilities." *Federal Register*, 46. Washington, D.C.: U.S. Government Printing Office, (January 1981a).

Environmental Protection Agency. *Ohio River Basin Energy Study*. EPA 6007-81-008. Washington, D.C.: Environmental Protection Agency, January 1981b.

Environmental Protection Agency. *Proposed Groundwater Protection Strategy*. Washington, D.C.: Environmental Protection Agency, November 1980b.

Fowler, John M. *Energy and the Environment*. New York: McGraw-Hill, 1975.

Hinkle, L. E., Jr. "Health Benefits and Health Costs of Controlling Sulfur Oxides in Air." *Bulletin of the New York Academy of Medicine* , 52 (December 1978), 1257-67.

Hutchins, W. A. *Water Rights Laws in the Nineteen Western States*, vol. 1. Washington, D.C.: U.S. Department of Agriculture, 1971.

Milbrath, L. W. "Environmental Values and Beliefs of the General Public and Leaders in the United States, England, and Germany." In *Environmental Policy Formation: The Impact of Values, Ideology, and Standards*. Dean Mann, editor. Lexington, Mass.: Lexington Books, 1981.

The MITRE Corporation. *Health Effects of Coal Technologies: Research Needs*. MTR-79W15 902. McLean, Va.: The MITRE Corporation, September 1980.

National Academy of Sciences (NAS). *Atmosphere-Biosphere Interactions: Toward a Better Understanding of Ecological Consequences of Fossil Fuel Combustion*. Washington, D.C.: National Academy of Sciences, 1981.

National Academy of Sciences. *Controlling Airborne Particulates*. Washington, D.C.: National Academy of Sciences, 1980.

National Academy of Sciences. *The Effects on Population of Exposure to Low Levels of Ionizing Radiation*. Washington, D.C.: National Academy of Sciences, 1972.

Ramsay, William. *Unpaid Costs of Electrical Energy*. Published for Resources for the Future. Baltimore: Johns Hopkins University Press, 1979.

Sorenson, E.M.B., C. W. Harlan and J. S. Bell. *Renal Changes in Selenium-Exposed Fish*. Memphis: Department of Biology, Memphis State University, 1981.

U.S. Code Annotated (USCA). Volumes 33 and 42. St. Paul, Minn.: West Publishers Inc., 1973.

U.S. Department of Health, Education, and Welfare (HEW). *Health Hazard Evaluation Determination Report* 77-121-490. Cincinnati: National Institute for Occupational Safety and Health, 1978a.

U.S. Department of Health, Education, and Welfare. *Health Hazard Evaluation Determination Report* 79-130-645. Cincinnati: National Institute for Occupational Safety and Health, 1978b.

U.S. Department of Health, Education, and Welfare. *Health Hazard Evaluation Determination Report* 79-130-645. Cincinnati: National Institute for Occupational Safety and Health, 1979.

CHAPTER VII

Power Systems Reliability

A. OVERVIEW

The functioning of America's homes, offices, and industries is dependent upon a reliable and constant flow of electricity. Any interruption to that supply is not just an inconvenience—it can potentially affect the well-being and security of the nation. Like any complex system, however, the electric network, from fuel source to wall socket, is open to disruption.

The system's level of "security," as opposed to "adequacy" (the term used to define sufficiency of installed capacity), is determined by the degree to which it is vulnerable to operating disruptions and the extent to which it can survive such disturbances without cascading loss of facilities leading to load loss (NERS, 1981a:25). Historically, the electric utility industry in the United States has maintained a high degree of service reliability. However, despite the overall robust nature of power systems in the United States, certain conditions will occasionally occur which result in customer interruptions over widespread areas.* Potential failures can be foreign or domestic,

*According to the National Electric Reliability Study (1981a:9) there was an average of 45 bulk power system interruptions per year for the 1970-1979 period. (A bulk power failure is defined as a service interruption affecting either half system load or at least 100 megawatts during a fifteen-minute period or more.) These disruptions involved an average of 250 megawatts of load per failure. The study found that the interrupted loads were increasing slightly (14.5 percent) per year and also increasing in frequency at a rate of 44 percent per year.

The generic causes of outages based on the ERA data base (Economic Regulatory Administration 1970-1979) were broken down as electric system components—48 percent; electric system operation—11 percent; weather—27 percent; and miscellaneous (including fires, accidents, and sabotage)—14 percent. Most of the bulk outages are coupled to the transmission system, with the distribution sector responsible for over 80 percent of all interruptions to end users.

civil or military, deliberate or accidental in origin. For the sake of convenient discussion, these causes will be grouped into four categories:

1. Natural disasters, such as storms, lightning, fires, and earthquakes
2. Aggressive physical acts, such as terrorism, sabotage, or war
3. Technical failures of mutually dependent systems—and/or their controls—due to intrinsic factors or human error
4. "Political" disruptions, such as strikes, embargos, moratoriums, and changes in law

Any of these disruptive events can affect one or several of the essential components of the electric system. Disturbances can take place in the fuel supply, in the generating facilities, and in the transmission and distribution networks. (Individual end use failures will not be discussed since these are local and temporary in nature.)

This chapter analyzes a range of potential disruptions, their causes, unexpected effects, and interactions with each other—not only as existing vulnerabilities, but as factors operative to varying degrees in future projections. An attempt will be made to identify those features in electric systems that promote vulnerability and those that provide greater recovery capability in the face of potential disruption. Starting with examples of significant power interruptions and their causes over the last decade, the chapter proceeds with an analysis of present and future aspects of power system vulnerability.

B. A SAMPLING OF DISRUPTIONS

Over the last decade or so, just about all the possible combinations of disturbance to our electric supply outlined above have occurred. A partial listing follows.

Disruptive natural events. Under this heading fall not only rare, cataclysmic natural disasters, such as earthquakes, tidal waves, and volcanic eruptions, but the more common manifestations of severe weather—a condition that occurs with some frequency in a country as vast as the United States. Each region has its own characteristic features, such as hurricanes in Florida, droughts in Texas, tornadoes in Oklahoma, and blizzards in Wisconsin.

Since most fuel is transported outdoors over long distances, all forms of fuel shipment are subject to inclement weather and natural disaster (CRS, 1977, vol. 3:189). In 1976 through 1977, for example, Midwestern utilities experienced a severe coal shortage because of an especially severe winter. Barge traffic carrying coal was blocked due to the Ohio River's freezing over from bank to bank. Other coal shipments by rail were frozen solidly in box cars (the coal had been wetted down at the mine to suppress dust) and had to be blasted out for removal.

Another weather system that can affect a power system is rainfall. From 1975 to 1977, California experienced 60 percent less rainfall than the 1931-1977 average. As a consequence, the region's hydroelectric output was cut almost in half. Hydro-dependent Pacific Gas and Electric Co., one of the state's largest utilities, used an additional 50 million barrels of oil, raising its operating expenses by a third (Quirk and Moriarty, 1980:90-92).

Generating plants are generally not affected by natural disturbances, but if a stray cat can be classified as such, the blackout in Managua, Nicaragua, on August 28, 1981, can be cited as an example. In this case, a cat wandered into a central power station and accidentally caused a short circuit (*New York Times*, August 29, 1981). And as long as organisms are being listed, the young of the introduced Asiatic clam *Corbicula fluminea* have been known to clog up the fresh water-cooled steam condensers in power stations. Too small to be stopped by screens, these bivalves proliferate in the warm, artificially created environment, and some stations have to shut down twice daily to shovel out the tiny clams (*Wall Street Journal*, August 12, 1980).

Transmission line failures due to meteorological causes are much more common, not a surprising fact in view of the over 365,000 circuit miles of overhead transmission lines in the United States (NERS, 1981b:2-9). Of the twelve worst interruptions in U.S. bulk power supply during 1974-1979 attributable to failures in transmission and distribution, seven were initiated by bad weather (NERS, 1981b:44,45).

One of the worst and most dramatic recent blackouts, the 1977 New York City power outage, was triggered by lightning striking an imperfectly grounded transmission line tower in Westchester County, New York, short-circuiting two high-voltage lines (Boffey, 1978). (The resulting cascading power failure was a story of cumulative technical and human error.)

The April 5, 1979, Miami blackout (which also affected Fort Lauderdale and West Palm Beach) was caused by a buildup of dust and salt spray on insulators, short-circuiting the output of three generating stations (*International Herald Tribune*, April 6, 1979:5).

Fires have also taken their toll on transmission and distribution facilities. On January 8, 1981, an accidental trash fire at the Utah State Prison produced enough heat to cause arcing in an adjoining switchyard. The resulting cascading transmission failure blacked out all of Utah and parts of Idaho and Wyoming. (*Electric Light and Power*, 1981). In Southern California, brush fires often cause extensive damage to wooden-poled lines, sometimes ionizing the air sufficiently to short-circuit high-voltage conductors (Chenoweth et al., 1963:34).

In addition to so-called "natural" disasters affecting transmission lines, accidents involving helicopters and even flying kites have caused shorting out of overhead lines (Clapp, 1978:41).

Aggressive physical acts. Unlike the previous category, power failures under this heading are deliberate in nature. Destructive acts of war, terrorism, and sabotage purposefully seek out points of high vulnerability to produce maximum damage.

During World War II, for example, highly centralized German power stations (80 percent of the electrical output came from only a few hundred plants) formed an attractive target. However, Allied strategists assumed—erroneously—that grid interconnections were sufficiently flexible to make bombing ineffectual. As a consequence, the stations were not bombed until the massive raids of 1944. After the war, this was found to have been a costly mistake. German officials told U.S. authorities that the war could have been shortened by at least two years if the power stations had been eliminated earlier as German war production was almost entirely dependent on electricity (Clark and Page, 1981:3, 52-54; Clark and McCosker, 1980:19-29).

In contrast, the Japanese power grid during World War II was dominated by numerous dispersed small hydroelectric plants. Strategic bombing of power facilities was only possible for the relatively few central steam plants near urban areas, leaving the grid almost totally operative.

During the Korean conflict, the huge hydroelectric dams on the Yalu River became a primary target, whereas during the Vietnam War, North Vietnamese energy plants were too small and dispersed to make attacking them worthwhile (Clark and Page, 1981:52-54; Barnet, 1981:67).

Incidents of sabotage or terrorism affecting energy facilities ranging from fuel supply systems to transmission lines and generating stations have been increasing in all countries over the last decade. Clark and Page (1981:57) have compiled a table (see Table VII.1) based on Department of Defense data listing 174 domestic and 191 foreign incidents. Topping the list are attacks on power lines and electric central stations and substations. And, by the count shown in Table VII.1, attacks are occurring at the rate of one every three weeks in the United States and one every ten days throughout the world.

During 1972-1980, the FBI reported a total of more than 1,500 actual or attempted bombings in the United States (FBI, 1973-1981). Electric utilities represented about 1 percent of the total—or some 20 bombings per year.

Table VII.2 is a partial listing of some typical examples drawn from data compiled by the Energy and Defense Project and listed in a report for the Federal Emergency Management Agency (Clark and McCosker, 1980:16). As can be seen, most of these attacks are relatively minor in their effects, but several were of significant consequence, such as the blackout of the eastern half of Puerto Rico and the outage to Sausalito, California. Also, although no deliberate attacks have equalled the accidental cascading transmission failures that blacked out whole states and regions of the United States over the past several years, planned sabotage could readily duplicate and even increase the extent of such disasters. After an investigation of the

Table VII.1
Attacks Against Energy Facilities, 1970–1980

Target	Domestic	Foreign
Power Line	55	48
Power Station or Substation	43	21
Pipeline	27	54
Oil or Gas Storage Facility	15	15
Nuclear Energy Support Facility	15	32
Oil Refinery	6	12
Oil Well	5	1
Hydroelectric Facility	4	2
Mine (Coal or Uranium)	2	2
Coal Train	1	—
"Nuclear Weapon Association"	1	—
Oil Tanker	—	3
Nuclear Waste Freighter	—	1
Total	174	191

Source: Clark and Page, 1981:57.

1977 New York City blackout, a U.S. Department of the Interior expert stated that "a relatively small group of dedicated, knowledgeable individuals . . . could bring down [i.e., by disrupting the power grid] almost any section of the country" (Joint Committee on Defense Production, 1977, vol. 1:87).

Such planned power outages affecting vital areas of a country have become common incidents overseas. Angola, Argentina, Afghanistan, Bolivia, Brazil, Chile, Cyprus, Ecuador, El Salvador, Egypt, the Federal Republic of Germany, France, India, Iraq, Iran, Ireland, Japan, Kuwait, Lebanon, Mozambique, The Netherlands, Nicaragua, Portugal, Qatar, Saudi Arabia, Singapore, Spain, South Africa, Sweden, Taiwan, Turkey, Uganda, and the United Kingdom have all had their share of guerrilla- or terrorist-provoked power failures.

Here are some examples: Red Brigade terrorists caused $600,000 worth of damage and blacked out parts of Rome for several hours on June 14, 1978, with a series of bombs in a central power station. In 1980, the Soviet-controlled airport at Jalalabad, Afghanistan, was blacked out by guerrilla sabotage. During 1981, important sections of El Salvador (including the capital) were repeatedly blacked out by sabotage to power facilities. In 1969, Israeli commandos cut power lines from the Aswan High Dam to Cairo, severely crippling that city (Lovins and Lovins, 1982:70-72).

Nuclear power facilities, because of their potential for "melt-down" and radioactive fallout, are of particular concern when it comes to war, sabo-

Table VII.2
Incidents of Energy-Related Terrorism, 1968–1978

Facility	Reference/Date	Who	Reason	Damage	Interruptions
Transmission Towers Owned by Public Service Electric and Gas in Cedar Grove, N.J.	*New York Times*, 11/6/68, 40:1	Unknown	Unknown	Tower footing damaged	None
Shell Oil Co. Gasoline Pipeline in Oakland, Calif.	*New York Times*, 3/19/69, 28:4	Unknown	Unknown	Fuel carried by creek to nearby community, three injuries	None
Four Transmission Lines in Colorado	*New York Times*, 4/16/69, 54:1	"Campus Revolutionary"	Electricity ran to local defense plants	Not available	Not available
Refinery Owned by Humble Oil in Linden, N.J.	*New York Times*, 1/27/70, 1:5 and 56:4	United Socialist Revolutionary Front	Get political prisoners freed	"Millions of dollars" to four units	Production halted but no interruption
Transformer in Puerto Rico	*New York Times*, 1/1/75, 36:1	Assumed Puerto Rican Nationalists	Assumed in protest of visit by Henry Kissinger and Nelson Rockefeller	App. $100,000	East part of island without power
Pipeline in Puerto Rico	Same as above	Same as above	Not available	Not available	

Table VII.2 *(continued)*

Facility	Reference/Date	Who	Reason	Damage	Interruptions
Six Transmission Towers Near Oakland, Calif. Owned by Pacific Gas and Electric	*Los Angeles Times*, 3/22/75, I, 25:4	New World Liberation Front (NWLF)	Rate protest	Slight	None
Substation Owned by Pacific Gas and Electric Near San Jose, Calif.	*Los Angeles Times*, 4/19/75	NWLF	Rates	App. $15,000	12,000 homes without power
Substation Owned by Seattle Light in Seattle, Wash.	*New York Times*, 1/2/76, 45:2	George Jackson Brigade	Unknown	Not available	2,000 without power
Pacific Gas and Electric Substation Near San Jose, Calif.	*New York Times*, 1/2/77, 17:1	NWLF	Rates to low-income consumers	Not available	Not available
Pacific Gas and Electric Substation Near Cupertino, Calif.	*Los Angeles Times*, 1/28/77, I, 3:6	NWLF	Same as above	"Substantial"	Power out to 21,000 for 30 minutes
Pacific Gas and Electric Substation Near Oakland, Calif.	*Los Angeles Times*, 4/16/77, 28:3	NWLF	Same as above	Not available	Power out to 5,000

Pacific Gas and Electric Transformers in Sonoma, Calif.	*Los Angeles Times*, 4/19/77, I, 2:5	NWLF	Same as above	Not available	Power out to 8,000
Alaskan Oil Pipeline	*New York Times*, 7/29/77	Local miner	"Disgruntled by line's construction"	Insulation damaged	Flow halted for repairs
Pacific Gas and Electric Substation in Sausalito, Calif.	*New York Times*, 8/30/77, 10:6	Assumed NWLF	Rates	Not available	Blackout in Sausalito
Alaskan Oil Pipeline	*New York Times*, 2/18/78, 18:6	Unknown	Unknown	One-inch hole in line; 8,000 barrels lost; unassessed environmental damage	None
Pacific Gas and Electric Substation in Concord, Calif.	*Los Angeles Times*, 3/16/78, I, 23:3	NWLF	Rates	Not available	Power out to 50,000

Source: Clark and McCosker, 1980.

tage, or terrorism. Security precautions for nuclear plants are among the most stringent in the power industry; however, most of these measures are specifically designed against inside sabotage and not against military action or damage by foreign agents or subversive organizations (Turner et al., 1970; Wagner, 1977).

The first precedent for an open act of war against a nuclear facility was the Iranian bombing of Iraq's Tuwaitha nuclear research center outside Baghdad on September 30, 1980. The second incident was even more dramatic. On June 7, 1981, an Israeli air raid using one-ton bombs deliberately destroyed the Osirak "research" reactor at the same site—fortunately, just before it was loaded with fuel. The Israelis considered the reactor to be a cover for the manufacture of plutonium for nuclear weapons use (*New York Times*, June 8, 1981).

Although not common, terrorist attacks on nuclear facilities have occurred all over the world. So far, none of these attempts has resulted in any major disaster—the overall intent of such attacks being political, aimed mainly at stopping nuclear construction. The following are examples of nuclear sabotage.

When the Argentinian Atuchi-1 reactor was built in 1973, it was taken over by guerrillas for publicity (Burnham, 1975:32). A May 1975 bombing damaged the Fassenheim reactors in France. Two months later, the gas-cooled reactor at Monts D'Aree, Brittany, had to be shut down due to Breton separatists causing damage to a cooling water inlet and air vent (Burnham, 1975:29). In January 1982, five Soviet-made rockets were fired at the construction site of the French Super-Phoenix fast-breeder reactor, causing only minor damage due to bad aim (*Los Angeles Times*, January 19, 1982). Basque separatists have repeatedly bombed the Lemoniz reactor under construction near Bilbao in northern Spain, killing workers and causing heavy damage (*Nucleonics Week*, February 18, 1982). Armed terrorists broke into a nuclear facility in Italy in 1982 (Bass et al., 1980:74). In the U.S., the visitor center of the Trojan reactor in Oregon was bombed (Kupperman and Trent, 1980:36), and reactor guards at several U.S. installations have been fired at (Subcommittee on Energy and the Environment, 1977:247). Unexploded bombs have been found at several sites, such as the Point Beach reactor in Wisconsin in 1970 (Comptroller General of the U.S., 1977:2). In 1976, the Nuclear Regulatory Commission issued a security alert to all U.S. power plants on the basis of "inconclusive information" the nature of which was never disclosed (*Los Angeles Times*, June 1, 1976).

In addition to acts attributable to outsiders, inside sabotage at nuclear plants has also occurred. The 1971 fire at the Indian Point Two reactor in New York, which caused millions of dollars worth of damage, was set by a maintenance man at the plant (Bass et al., 1980:40). Worker sabotage has also damaged seven other U.S. plants—the two most recent incidents being at Brown's Ferry in Alabama in 1980 and Beaver Valley in Pennsylvania in 1981 (*Nucleonics Week*, February 11, 1982:3-4). The Surry reactor in Virginia was extensively damaged by two control room workers in a protest over what they called lax security and unsafe working conditions at the plant (*New York Times*, October 18, 1979:A18).

Technical failures of electrical system components, their controls, or operation. System component failure and operator error account for 60 percent of all bulk outages reported for 1967-1979. When the ERA data base for the above statistics is broken down by generic subsystems (i.e., generation, transmission, and distribution), failures in transmission account for the vast majority of these outages in every year for which figures have been gathered.

While severe weather conditions were the initiating cause of all the worst outages based on duration (see "Disruptive Natural Events"), equipment problems rank as the highest cause of the worst blackouts based on load lost or number of customers. Table VII.3 lists five of these recent major outages. In addition, San Diego, Butte, Montana, and Boston have all experienced massive outages over the last five years due to system and operator errors (NERS, 1981a:10-14, 294).

Table VII.3
Worst Bulk Power Interruptions Due to Technical Failure, 1974–1979 (in order of lost capacity)

Utility or System	Date	Mega-watts Lost	Customers	Duration	Cause
Los Angeles Department of Water and Power and Other Western Interconnected Systems	9/10/76	3,632	1,769,500	1 minute to 1 hour, 28 minutes	Equipment trip
Florida Power and Light	5/16/76	3,227	1,300,000	4 hours, 36 minutes	Transmission line trip
Northwest Power Pool Area (Idaho, Montana, Washington, Utah)	11/27/79	2,400	350,000	Momentary to 2 hours	Computer misoperation
Western Systems Coordinating Council Area	3/21/75	1,665	1,300,000	4 to 28 minutes	Equipment failure
Florida Power and Light	4/25/74	1,415	1,000,000	15 minutes	Testing error/ human

Source: NERS, 1981b.

More recent examples include the following power interruptions: On September 24, 1980, most of Montana was blacked out for about an hour due to equipment failure (*Great Falls Tribune*, September 26, 1980:6A). On July 12, 1981, a Con Ed substation fire initiated by transformer failure caused an interruption of the electric supply via thirteen underground distribution cables to most of lower Manhattan, including the financial district, and disrupted the world's greatest concentration of financial computers for several hours (*New York Times*, September 11, 1981:1).

Overseas, cascading transmission failures due to equipment problems blacked out most of France on December 19, 1978; most of Israel on February 5, 1979; and over half of Great Britain on August 5, 1981 (Lovins and Lovins, 1982:129).

Wide-spread blackouts caused by generator failure are uncommon. For the 1967-1975 period, the Department of Energy/ERA data base shows that only 10 percent of bulk outages were attributable to generation component failure. Exceptions are a few municipal utility systems with old and badly maintained central stations and with no backup from neighboring facilities in case of a generating unit failure (NERS, 1981b:268). However, generating units are frequently "tripped" (due to loss of synchronism among the generators) during cascading transmission problems and grid instabilities adding to the problem (e.g., the 844 megawatt, Con Ed Ravenswood no. 3 generator during the 1977 New York City failure) (Clapp, 1978:17).

Technical failures at nuclear plants, although extremely rare and to date of no major direct consequence, hold such a vast threat for immediate and long-lasting disaster that the impact is far greater than any actual or potential disruption of electric supply (see Table VII.4).

Three Mile Island (March 28, 1979) is, of course, a well-known story of technical and human error. And, in 1975, at the TVA's Brown's Ferry reactor in Alabama, a technician testing for air leaks with a candle caused a fire that burned 1,600 electric cables in the control room. Other incidents involve the forced shutting down of reactors because of external grid instabilities (reactors are dependent on off-site and backup on-site sources of electricity for cooling and control) as well as common-mode failures involving the electric supply, leaking floats, water damage, wiring errors, contaminated oil, incorrect installation of equipment, and frozen pipes due to faulty thermostats (Pollard, 1979:11-75).

Political disruptions. Included in this category are embargos, strikes, moratoriums, and regulatory changes. The most potent example to date is the 1973-1974 Arab oil embargo during which the United States—and the rest of the world—became painfully aware of its dependence on a source of fuel shipped halfway around the world. Five years later, in 1979, the Iranian revolution with its consequent 1 percent reduction in world oil availability triggered gasoline shortages and more than doubled oil prices (Ebinger,

Table VII.4
Reactor Accident Consequences

Plant/Location	Peak Early Deaths	Peak Early Injuries	Peak Fatal Radius in Miles	Peak Injury Radius in Miles	Scaled Financial Consequences in Billions
Beaver Valley 1(O), Shippingport, Pa.	24,400	271,000	20	55	122.0
Braidwood 1(C), Joliet, Ill.	6,750	63,300	15	60	127.0
Callaway 1(C), Callaway, Mo.	11,300	31,200	17.5	35	110.0
Calvert Cliffs 1(O), Lusby, Md.	7,090	25,900	20	55	87.4
Catawba 1(C), Rock Hill, S.C.	41,900	87,400	20	30	101.0
Commanche Peak 1(C), Glen Rose, Tex.	1,210	13,800	25	35	117.0
Cook 1(O), Bridgman, Mich.	2,060	92,700	15	70	91.9
Fermi 2(C), Laguna Beach, Mich.	8,150	349,000	15	70	136.0
Fitzpatrick (O), Scriba, N.Y.	1,520	28,800	12.5	40	103.0
Haddam Neck (O), Haddam Neck, Conn.	60,600	144,000	17.5	50	74.1
Indian Point 2(O), Buchanan, N.Y.	56,600	227,000	17.5	50	274.0
Lasalle 1(O), Ottawa, Ill.	14,900	13,900	15	60	118.0
Limerick 1(C), Montgomery, Pa.	77,700	710,000	20	55	213.0
Marble Hill 1(C), Jefferson, Ind.	12,300	158,000	15	60	87.2
McGuire 1(O), Cornelius, N.C.	11,600	19,400	15	35	106.0
Peach Bottom 2(O), Peach Bottom, Pa.	74,500	50,900	20	55	119.0
Salem 2(O), Salem, N.J.	102,000	75,700	20	55	135.0
Summer (O), Fairfield, S.C.	6,280	110,000	20	30	68.5

(O) indicates the plant is complete.
(C) indicates the plant is under construction.

Table VII.4 (*continued*)

Plant/Location	Peak Early Deaths	Peak Early Injuries	Peak Fatal Radius in Miles	Peak Injury Radius in Miles	Scaled Financial Consequences in Billions
Susquehanna 1(O), Berwick, Pa.	70,600	53,900	20	55	143.0
Waterford 3(C), St. Charles, La.	90,400	253,000	20	40	131.0
Zimmer (C), Moscow, Ohio	12,300	196,000	20	30	84.5
Zion 1(O), Zion, Ill.	12,300	181,000	15	70	146.0

Source: *Washington Post*, November 1, 1982.

1982:78). The oil crisis over the last seven years has been at the root of the energy conservation movement and the search for alternative sources of energy that have so deeply affected the electrical industry.

The coal strike of 1977-1978 caused severe shortages to many coal-fired generating plants, threatening region-wide brown- and blackouts. Attempts to ship coal from facilities with large reserves were nullified because there was a shortage of equipment for *reloading* coal for shipment elsewhere; all the available equipment was for *unloading* (Subcommittee on Energy and Power, 1978:9).

The 1974 coal miners' strike, producing power shortages all over Great Britain, helped to bring down the Heath government, and British power workers have repeatedly threatened strikes as a means of extracting pay increases (*Daily Mail*, April 10, 1979:2). In Israel, power plant employees blacked out large areas of the country during their 1981 strike (*Los Angeles Times*, May 27, 1981:1, 2), and, in Australia, electric workers frequently threaten blackouts (*Straits Times*, March 28, 1980:5).

The political repercussions of nuclear accidents, coupled with environmental concerns over radioactive wastes, have been far reaching and threaten to throttle the nuclear power industry. Unless the nation can be reassured on both counts—namely, that reactors are accident proof and that their wastes can be disposed of safely—it is unlikely that future electric demand will be filled by building more nuclear plants. The political climate at the moment is, in effect, almost equivalent to a nuclear moratorium.

Increasingly stringent environmental regulations may also make it cumulatively more difficult and financially unrewarding to strip-mine coal or to burn that fossil fuel in the quantities necessary to meet electric demand, creating a possible shortfall in generating capacity with resulting cutbacks in the use of electricity.

C. ANALYSIS OF VULNERABILITY OF POWER SYSTEMS

The five major components that keep a power system operational are (1) the fuel supply, (2) power stations, (3) transmission lines with their associated switch gear and transformers, (4) distribution systems with their transformers and switches, and (5) control and communication systems that coordinate the above components. The following discussion reviews the "vulnerability profile" of each component.

Fuel supply vulnerabilities. Any electrical systems predominantly dependent on a single fuel source are open to disruption when that fuel becomes unavailable. A system with a varied generating mix or one whose plants can use substitute fuels is more flexible in the face of a particular fuel's being in short supply. The impact of a given fuel's unavailability will also vary with the extent of the affected system's load diversity, dependence on adjacent power grids, levels of stockpiles, and applicable regulatory policies.

The nation as a whole appears to have a reasonably varied generation mix—50 percent coal; 14 percent oil; 13 percent gas; 12 percent nuclear; 10.5 percent hydroelectric; 0.5 percent geothermal, refuse, wood, and other; and 1 percent imports. On a region-by-region, state-by-state basis, however, this view changes in the direction of much less diversity (NERS, 1981b:209).

Of the nine National Electricity Reliability Council (NERC) regions, the Western Systems Coordinating Council (WSCC) is one of the more isolated divisions. It is dependent on the following mix: 36 percent hydro, 10 percent gas, 20 percent oil, 25 percent coal, and 6 percent nuclear. The region is therefore vulnerable to extended drought conditions affecting hydropower generation. Within the region, the Pacific Northwest with its 80 percent dependency on hydropower is especially vulnerable, as was the case in 1974 and 1977. California, on the other hand, is not as open to drought disruptions; 62 percent of that state's generation is oil fired, making it vulnerable to oil cutbacks.

On the other side of the continent, the U.S. portion of the Northeast Power Coordinating Council (NPCC) depends heavily on oil (52 percent). The rest of the mix consists of 13 percent hydro, 11 percent coal, and 24 percent nuclear. The Northeast Power Pool (NEPOOL) is not only especially open to oil interruption but also to a nuclear moratorium—New England uses 60 percent oil-fired and thirty-seven percent nuclear plants.

The Mid-Atlantic Area Council region (MAAC), although leaning to coal, has a relatively balanced mix (23 percent oil, 51 percent coal, 14 percent nuclear) and on a state-by-state level exhibits even greater diversity.

The Southeastern Electric Reliability Council region (SERC) depends heavily on coal (55 percent) with nuclear and oil generation accounting for 21 percent and 16 percent, respectively. The balance is provided by hydro (6.5 percent) and gas (2.5 percent). However, there are wide state-by-state differences. Florida is 50 percent oil dependent, Alabama has 60 percent coal generation and almost 40 percent nuclear, and, in North Carolina, 50 percent of the electric supply comes from nuclear reactors and 40 percent from coal-fired plants.

The East Central Area Reliability Coordinating Agreement region (ECAR) is overwhelmingly coal dependent with almost 90 percent of its capacity in coal, oil and nuclear splitting the balance.

The Mid-American Interpool Network (MAIN) is another region heavily reliant on coal (70 percent), but its nuclear capacity is greater (24 percent). Most of its component states follow the regional pattern; Michigan, however, has proportionally more nuclear and oil plants.

The Mid-Continent Area Reliability Coordinating Agreement region (MARCA) is also predominantly coal dependent (60 percent) with the balance primarily in hydro (14 percent) and nuclear (25 percent).

The Southwest Power Pool (SWPP) has nearly 55 percent of its capacity in gas-fired (domestic natural gas) plants; the remainder is evenly divided between coal and oil. Within the region, Louisiana and Oklahoma are the heaviest users of gas (68 percent and 87 percent, respectively), whereas Kansas shows a 40 percent dependency on coal.

The most isolated region in terms of connection to other regions is the Electric Reliability Council of Texas (ERCOT). Here, generation is mainly (75 percent) gas fired from domestic supplies with the balance provided by coal.

As a generalization, the following conclusions can be made as to present regional fuel vulnerabilities:

- Disruptions in coal supply will have the worst effects in Midwestern areas, with the Southeast also badly affected.
- The West—and especially the Northwest—is particularly vulnerable to drought and dam damage.
- Any interruption in oil will cripple much of Northeastern power generation.
- Natural gas disruptions will have the most adverse impact on Texas and the Southwest in general.
- Disruptions in the nuclear fuel supply will be felt mostly along the East Coast and portions of the Midwest.

Coal. As we have seen, coal supplies are mostly open to disruption from bad weather affecting transportation and from strikes (in both the mining and transportation sectors) but could also become scarce due to regulatory restrictions regarding mining practices.

Coal becomes increasingly important for national power generation in all the projected scenarios, but especially in the case where coal is the overwhelming dominant fuel in use by numerous 200-400 megawatt plants ("Small Coal Plants" Scenario). Electric consumption is almost doubled here, and, with coal as the main fuel, this could mean mining about 2 billion tons per year (Lovins and Lovins, 1982:171). Currently, coal for domestic use is transported mainly by railroads (50 percent), water carriers (25 percent), and trucks (10 percent) (since all of these modes are dependent on diesel fuel, interruptions in oil supply automatically threaten coal supply lines), with coal slurry pipelines being considered as an important alternative (one is already in operation). The rest of the coal (15 percent) is consumed by plants at the mine mouth (CRS, 1977:35-36).

Increasing demand for coal will be filled mainly by surface-mined western coal (especially from Montana and Wyoming), transported by mile-long special freight trains and coal slurry pipelines. In one projection (Lovins and Lovins, 1982:171), a handful of single rail corridors from the Powder River Basin to the Midwest and Southwest would carry more coal than is now mined in all the rest of the country. The vulnerability of such concentrated corridors of transportation to natural disasters, sabotage, or acts of war is apparent. Also, under the increased coal demand scenarios, the impact of coal worker strikes, especially if they are coupled with sabotage of mines or slurry lines, becomes correspondingly significant.

Oil and natural gas. Although oil and natural gas consumption is down in all scenarios, a disruption of supply would nevertheless have a major regional impact on electric system integrity. Oil and gas combustion turbines, which provide the majority of system peaking capacity, may be inoperable. In scenarios with considerable cogeneration—such as in the "Small Coal Plants" and "Post-industrial" Scenarios—coal- and shale-derived liquid fuels would be diverted to the transportation and home heating sector, shutting down cogenerating small liquid fuel-fired power plants and fuel cells.*

Water for hydropower. This form of power production is not especially significant on a national level for any of the scenarios, even if there is little doubt that some areas of the country will continue to be heavily hydro dependent. Much as at present, these areas will be affected by droughts or decreased water availability due to other demands. In the case of the larger dam systems, deliberate damage becomes a factor.

Nuclear fuels. The light water reactors in use at present require enriched uranium hexafluoride (UF_6) pellets for their functioning. The initial source for this fuel is uranium ore. Six basic steps are involved in the nuclear fuel process.

1. Uranium ore is shipped from the mine to a milling facility where it is refined to uranium oxide (or yellowcake). New Mexico, Wyoming, and Utah are the major domestic sources of uranium ore. Canada, Australia, South Africa, Zaire, and

*There is vast literature on the subject of national vulnerability to future oil (and gas) shortages caused by new embargos, Middle Eastern crises, and so on. The Lovinses (1982) also devote considerable space to the openness of domestic oil pipeline and natural gas pipeline systems to sabotage.

Gabon are major foreign sources. Yellowcake is produced by some 30 separate facilities in the United States.

2. Two conversions plants in the United States (at Metropolis, Illinois, and Sallisaw, Oklahoma) convert yellowcake to uranium hexafluoride.
3. Three "gaseous diffusion plants" produce "enriched" UF_6 along with depleted uranium tailings.* The enrichment facilities, all federally owned, are located at Portsmouth, Ohio, Oak Ridge, Tennessee, and Paducah, Kentucky. The latter plant performs only the initial step in the enrichment process and ships its material to the other two plants for final enrichment.
4. The enriched UF_6 requires fabrication into pellets before it can be inserted into the fuel rods of light water reactors (LWRs). Seven companies in the United States fabricate these pellets by converting the UF_6 into powdered uranium dioxide (UO_2), which is then pressed into pellets. Some of these companies fabricate whole fuel assemblies which are then shipped to reactor sites.
5. The highly radioactive spent fuel assemblies are either stored at the reactor site or shipped to a fuel repository. Two of the latter are at sites intended for fuel-reprocessing facilities, although no such reprocessing is at present taking place. The three commercial reprocessing plants that have been built but which are inactive are in West Valley, New York, Morris, Illinois, and Barnwell, South Carolina. The last site is now being considered for federal operation (*Wall Street Journal*, August 6, 1982:14).

   After use in the reactor for about eighteen months, the spent fuel is initially stored at the reactor site to allow for intense, short-lived radioactive decay. After that, most wastes—since repository facilities are inadequate—are stored in temporary quarters at the reactor site. Even these temporary facilities are presently filled to overflowing, and although additional on-site storage pools can always be constructed, a long-term solution to the spent fuel storage/disposal issue is essential.
6. Other wastes containing relatively low levels of radioactivity, such as contaminated radiation suits, gloves, and tools, are buried at one of six commercial sites.

Transport of nuclear materials is predominantly by commercial trucking on highways. (Enriched UF_6 is shipped directly by the federally owned enrichment plants to the pellet and fuel assembly fabricators.) The general pattern of nuclear fuel transportation is mainly from west to east. Yellowcake (U_3O_8) shipments move from Colorado, Wyoming, and New Mexico to the two conversion plants in eastern Oklahoma and Illinois. From there, gaseous UF_6 moves farther east to the three enrichment facilities in Kentucky, Ohio, and Tennessee. The majority of the enriched material now

*LWRs need a fuel that is about 3 percent fissionable U-235. This requires enrichment of the original yellowcake, leaving tailings still containing about 1.5 percent of the original U-235. The tailings are stored at the enrichment site as a potential fuel for the advanced breeder reactors still in development.

travels farther eastward to pellet and fuel assembly fabricators in North and South Carolina, the nuclear portion returning west for fabrication in Oklahoma and Washington. Thus, the concentrated and crucial intermediate steps in the nuclear fuel cycle center in the lower mid-Atlantic and Appalachian regions. Fuel consumption, in turn, is concentrated in relatively few states: Illinois, New York, Connecticut, Pennsylvania, South Carolina, Virginia, and Florida. The resultant wastes are presently stored in pools at the reactor sites.

Truck transport of nuclear fuels is light compared to other commodities. According to one study the heaviest recorded flow between two points of the nuclear cycle averages out to one truck per day. Interstate Highway 40, between Nashville and Knoxville, is the heaviest nuclear materials roadway (Clark and McCosker, 1980:82-85).

The vulnerabilities of such a fuel cycle, dependent as it is on just a few crucial links, are apparent. Any malfunctions or deliberate sabotage of one of the two conversion plants would seriously disrupt the initial phase of the cycle. Failure of one or more of the three enrichment facilities would be even more critical to the flow of nuclear fuel. And, ultimately, fuel assembly can be seriously interrupted* by accidental or deliberate problems at some or all of the seven fabrication plants.†

The "Nuclear Resurgence" Scenario postulates the greatest amount (about 40 percent) of nuclear generation. Whether the nuclear fuel cycle here would remain more or less the same, other than being increased tenfold in the quantities of fuel involved, depends on the technological nature of the reactors. If LWRs remain the primary generating units (and this seems to be the most likely case), the same processing sequences, with their inherent vulnerability to disruption, would be taking place.‡

Of potential importance to any nuclear resurgence is the question of uranium resources. Current domestic reserves are estimated at about 700,000 tons with another 1 million tons in the "undiscovered but probable" category and some 2 million tons being considered as "speculative" by the United States Geological Survey (USGS) (Clark and McCosker, 1980). Under the higher nuclear projection, the nuclear power industry could be consuming as much as 2 million tons of uranium ore over the next 30 years (extrapolation from USGS data). Even with improved "once-through"

*Most nuclear plants stockpile one year's worth of fuel. Thus, interruptions in the fuel cycle would have to be repeated or long lasting to have the most impact.

†Only the effects of fuel interruptions on power generation are discussed here. The possibilities of highly damaging explosions and fallout from accidents or sabotage to some of these facilities are not under consideration and neither are the chances for diversion of enriched fuels for nuclear weapons production.

‡If breeder reactors have become at all significant, then fuel reprocessing for plutonium extraction becomes essential and active reprocessing facilities will be very much part of the cycle. Such reprocessing sites will constitute an additional crucial link that will need to be carefully secured from disruption.

reactor technology and new mining techniques, a shortfall in domestic resources seems possible. This could mean that the United States will be dependent on uranium imports in a manner analogous to the country's current reliance on Middle Eastern oil (DD:VII-1).*

Generating vulnerabilities. Perhaps the most important consideration here is plant size and density of generating distribution throughout the system. Natural events, accidents, and component and operator failures, as well as sabotage and strategic bombing can disrupt the generating sources of an electrical system. The impact of a power station's failing is proportional to its overall generating role within the grid. Other variables entering into this somewhat simplistic formulation are the degree of interconnection to other systems, the precise function of the transmission lines directly affected, and so forth. Nonetheless, the bigger the power output interrupted by accident or malicious intent, the more significant the outage.

Fine, modular, redundant structure is one of the main criteria in use by analysts when they consider resiliency to disruption in complex systems. The historical trend toward greater and greater centralization of power systems has resulted in larger and larger plants that are proportionately less dispersed and more critically interconnected to each other and to end users. The result is anything but the kind of network that can easily withstand unexpected failure of just a few or even a single major generating component (KH:VII-1).

When a big centralized power system "cascades to failure" (because of generator outage or any other reason), it pulls down all its interconnected customers unless there is a way for local areas or establishments to stand alone and disconnect from the system. Residents of Coronado, California, for instance, did not even know that the San Diego grid had crashed due to operator error creating a chain reaction of generating cutouts on March 8, 1978. This community was dependent for its power on an independent co-generation plant (*San Diego Union*, December 26, 1978). Similarly, Holyoke, Massachusetts, was able to escape the 1965 Northeast blackout by isolating itself from the regional grid and operating on its own gas turbine installed for just such an emergency (*Electrical Construction and Maintenance*, 1965).

As mentioned earlier, strategic bombing and sabotage are most effective when highly centralized power stations are the target. Defense preparedness studies (e.g., Clark and McCosker, 1980) emphasize this present national vulnerability† and advocate a more dispersed system or at least one with

*The uranium resource shortfall might push reactor development strongly in the direction of breeder units. However, financial and security consideration regarding the high costs of reprocessing and the increased risk of nuclear proliferation could be strong enough deterrents to inhibit breeder development (*Wall Street Journal*, August 6, 1982:14).

†Fewer than 300 generating stations are presently supplying about half of the nation's electricity. Most of these plants are located in or near major urban areas (Joint Committee on Defense Production, 1977:I:1).

more independent and optionally interconnected backup generation in case of emergency.

In future projections, "centralization," meaning large units, often geographically clustered and sparsely linked at a few nodes over long distances either to each other or the end users has not lessened except for the "Small Coal" and "Post-industrial" Scenarios. The size of plants is greater in the "Average Future" Scenario; even larger in the "Nuclear Resurgence" projection and at least equally as large in the "Mega-plant" Scenario (where, in addition, there are several 5,000 megawatt giant generating facilities). In the "Small Coal Plants" era, a degree of decentralization has taken place—despite the emphasis on one main fuel—since the coal plants are relatively small (200-400 megawatts), dispersed geographically, and connected over shorter distances and less critical for end use application. There is also a significant amount of industrial and commercial cogeneration. The "Post-industrial" projection, however, seems to incorporate to the greatest extent most of the features commonly associated with decentralized, dispersed, and varied, as well as optional electric-generating sources. This is a society with a low energy consuming but highly "electrified" profile. End use efficiency, auto- and cogeneration have been emphasized to the utmost. Dependence is increasingly on solar technologies both for residential as well as more encompassing municipal and local regional use. Accidental and deliberate disruptions of the electric system would have the least end use impact under these conditions. The system under the "Economic Malaise" Scenario is about as centralized as at present, with sabotage vulnerability perhaps increased due to social discontent fermented by general economic hardship.

Transmission and distribution vulnerabilities. The electrical transmission system comprises over 365,000 circuit miles of overhead cables, including lines of voltages ranging up to 765 kilovolts* (DOE/Energy Data Reports, 1978). In addition to transmission within the United States, high voltage interties with Canada exist for the importation (or exchange) of electricity. (The 1965 Northeast blackout was attributable to a Canadian electric component failure precipitating a cascading series of overloads and outages across the border by way of just such an intertie.)

The distribution system comes into effect after the electricity has been transmitted to the local service area. Substations transform transmission to lower voltages for distribution over subtransmission lines and over 4 million miles of retail (industrial, commercial, and residential) distribution lines. As previously mentioned, transmission failures are responsible for the majority

*High voltage lines carry a tremendous amount of energy. Lines rated at 500 kilovolts, a common size today, handle about 1,000 to 2,000 megawatts, the output of one to two large power stations; and a 765 kilovolt line handles about 3,000 megawatts (Joint Committee on Defense Production, 1977, vol. 2:36). In the future, thousands of miles of ultra high voltage (UHV) lines—1,000 kilovolts and above—may also be in operation.

of bulk outages, while the distribution network accounts for 80 percent of more localized kinds of blackouts.

Overhead transmission is the most economical and efficient method of moving electricity. Also, other things being equal, the per unit cost of transmitting large amounts of electric energy over significant distances is greatly reduced by using the highest voltage lines available. Because power lines are not easy to site (15 to 20 acres of land are required per mile of overhead transmission line), the higher the voltage the less the land use and consequent costs. As the power grid expands, especially in builtup areas, newer and higher-capacity lines tend to be built alongside already existing ones. New York City and Southern Florida are good examples for supposedly independent transmission lines being squeezed into single narrow corridors (Joint Committee on Defense Production, 1977, vol. 1: 7-8).

Within the various NERC regions, the constituent power pools do not have unlimited (or 100 percent redundant) capacity to interchange power with each other. That ability depends mainly on a few vital extra-high voltage transmission segments such as the Wisconsin-Missouri-Illinois intertie, a single 700 megawatt corridor. Bulk power transmission is also dependent on specially manufactured switch gear and transformers at both ends, equipment that, when damaged, takes at least a year to replace (Kupperman and Trent, 1980:71-72, 106).

The high degree of existing interconnection and the creation of power pools is intrinsic to the historical growth of centralized power systems. Such systems have proliferated because of the economies of scale involved and because interconnection allows a given amount of generating capacity to meet a larger amount of scattered demand—since not all demand occurs simultaneously. But, although the ability to interchange power can make for greater reliability in case of failure of one component, the same capacity can have the reverse effect: "crashing" a whole grid (or power pool) when a critical linkage or node malfunctions and "fail-safe" provisions are intrinsically or mistakenly inadequate.

Substations with their associated distribution networks, although not usually as dramatic and far reaching in their failures as the outages caused by transmission problems, have historically been even more vulnerable to malfunction (NERS, 1981a:15-17). Also, as recounted earlier, substations and associated transformers have been a high priority for sabotage. Also, since the typical distribution network is radial (and often underground), in contrast to the more "meshed" network of a transmission system, restoration of power to the affected end users generally takes a longer time because of the complex series of switching operations involved.

As far as future scenarios are concerned, similar comments about respective vulnerabilities can be made here as in the previous section, "vulnerabilities of generating facilities." Highly centralized facilities cannot exist without long-distance transmission and distribution linkages. Therefore, the most vulnerable and highly impacted scenarios are the same as before, with the "Post-industrial" Scenario emerging as a front runner in any stakes for flexibility in the face of disruption.

Control and communication vulnerabilities. The stability of today's complex power grids depends on control centers that communicate with each other and with field equipment (switches, relays, generators, etc.) to balance loads and route power as the occasion demands. When component failure occurs, it is the control center that re-stabilizes the grid by shutting down circuits and shifting loads.

Communication takes place by telex, telephone, signals over transmission lines, radio, and private microwave circuits. These methods, even though they are backed by standby power supplies, are open to disruption (DEPA, 1962:31). Planned sabotage of utility communication systems would be more effectively disrupting to most utility operations than any interruption of the generation and transmission components.

The centralized control stations are a vital link in the whole operation. Their computer facilities are usually redundant and, in at least one center, triply redundant, and their power supplies are generally doubly backed by standby generators (NERS, 1981a:2-8). Yet, there are devices that are similar in effect on a local scale to the electromagnetic pulse (EMP) produced by high-altitude nuclear explosions (Dircks, 1981). Activated by a saboteur outside a grid control center, such an effect could conceivably render most of its computers and other equipment permanently inoperable (Lovins and Lovins, 1982).

Aside from cutting or otherwise physically disrupting communications, a saboteur might be able to take over their control. "Phone phreaks" are a common phenomenon today, tapping into public and private telephone lines and manipulating supposedly secure data and programs in computers all over the world. Some of the more sophisticated "phreaks" have devised methods for fooling systems into giving away services and products (such as free phone calls, telex, water, electricity, and gas) by tapping into accounting computers (*Los Angeles Times*, December 16, 1981, II:1; *New York Times*, August 3, 1981). If the utilities' accounting computers are open to decoding, what about the control computers? It might be possible for advanced "phreaks" (or saboteurs) using a portable dish to take over a whole utility grid, cutting out power stations, rerouting connections, changing voltages, or doing whatever they please (Lovins and Lovins, 1982).

Complex control and communication systems are part and parcel of centralized power networks. Again, as in the previous two sections, the "vulnerability index" here is highest for those scenarios with the greatest amount of large unit generation and regional transmission interdependence; it is lowest for the more dispersed and diversified systems such as exist partially under the "Small Coal Plants" Scenario and more fully under the "Post-industrial" projection.

Dynamic vulnerabilities. Having looked at potentially weak points in the separate components that make up an electric system, it is time to regard the whole system and consider its internal dynamics. In the case of complex systems, any valid assessment of vulnerability cannot rest on a mechanical stringing together of assessments of the vulnerability of separate parts (Dresch and Ellis, 1966:3).

Contemporary electric grids rely on great delicacy of balance and timing for stable functioning. Transient surges of power (from any number of internal and external causes) can break down insulation in a cable or transformer, creating a secondary fault that can propagate new transients throughout the network. Normally, automatic circuit breakers open and reclose within a fraction of a second. However, if the fault has not cleared or if the breaker malfunctions, the circuit stays broken. Generally, an alternate transmission path is available and the electric flow redistributes itself within a few cycles. This redistribution may overload other lines that, although they are built to withstand considerable overloads, must be disconnected —by shedding loads or re-routing power—in time to return them to safe limits. Operator error and communication failures can, however, prevent proper load adjustments. The whole grid will then "crash" as interconnected generators "pull out" of synchronism and system frequency drops below the 60 cycles per second required for stable operation. Once the grid has crashed, re-starting procedures involve so many complex operations in order to re-establish generator synchronism and "reactive balancing" of loads and lines that they generally require twelve or more hours. That amount of time without electricity in an "electrified" society can be very disturbing—economically as well as psychologically.

In 1976, the assistant director for systems management and structuring in the U.S. Energy Research and Development Administration stated that "Our interconnected electric energy systems seem to be evolving into a new condition wherein 'more' is turning out to be 'different.' As [these systems] become more tightly interconnected over larger regions, system problems are emerging which are neither presaged, predicted, or addressed by classical electrical engineering and which are no longer amenable to *ad hoc* solutions." He goes on to say that "The larger more tightly interconnected system is behaving in a fashion *qualitatively* different from that of earlier smaller systems." Examples mentioned are new "subsynchronous resonance" effects along with difficulties in determining control strategies for information transfer and decision making in the case of multiple independent but interconnected control centers. His final point is that "The conceptual tools and underlying theory required for the effective solution [of these rising problems] have not yet been developed" (Fink, 1976:20-21).

The understanding of some of these difficulties has advanced since 1976, but as grids become increasingly complex and stringently intertied under the more "centralized" scenarios, new problems concerning dynamic interactions within the systems are likely to arise. These will, of course, require vigorously undertaken solutions if the nation wishes to have a stable, predictable supply of energy.

D. SUMMARY

U.S. utility performance has been excellent, if measured by an index of reliability based on total number of customer hours per year compared with the total number of customer hours interrupted per year (over 99.9 percent). Nevertheless, major bulk power system failures do occur. Over the last twelve and one-half years, there have been 136 such interruptions, each of which has affected about 10,000 customers, and 14 interruptions that have affected about 100,000 customers. Also between 1974 and 1979, a few interruptions affecting up to several million people have been recorded (see Table VII.5). Initiating causes for these bulk power interruptions include inclement weather as well as component and operating failures. However, of the average 100 minutes of service loss experienced per U.S. customer every year, only 10 minutes are attributable to bulk power failure.

Although no major catastrophic outages can be attributed to sabotage or terrorism, such incidents directed at power systems components have been increasing all over the world. Fuel supply disruptions, too, are potential sources of electric supply shortfall. Depending on the generating mix of particular regions and states, different vulnerabilities exist.

The weaknesses of the present system are those inherent in any large, complex, highly interconnected network. Transmission problems head the list for bulk outages, with distribution malfunctions accounting for the majority of smaller disturbances. Generator outages are relatively minor factors in initiating interruptions but add their share of problems once a grid begins to "crash."

For any complex technological system to survive unexpected stress, a number of elements need to be present. Among these are the following (RB:VII-1):

- Fine-grained modular structure
- Early fault detection
- Redundancy and substitutability
- Optional interconnection
- Diversity
- Dispersed components
- Hierarchical embedding
- "Forgivingness" while failing
- Limited societal demands and social acceptability

Table VII.5
Worst Power System Interruptions, 1974–1979

Date	Utility or System	Lost Megawatts	Customers	Duration	Index (Megawatts-Customer-Hour, in Billions)	Initiating Event
7-13-77	Consolidated Edison Company	5,750	2,700,000	25 hours	388	Thunderstorms
5-16-77	Florida Power and Light Company	3,227	1,300,000	4.5 hours	18.9	Transmission line trip
1-13-78	Long Island Lighting Company	700	340,000	72 hours	17.1	Ice storm
3-1-76	Wisconsin Power and Light and Municipal Cooperatives	790	270,000	1 to 9 days	15.2	Winter storm
3-21-75	Western Systems Coordinating Council	1,665	1,300,000	4.5 hours	9.74	Equipment failure
4-8-79	Detroit Edison Company	500	230,000	50 hours	5.75	Ice storm
9-10-76	Los Angeles Department of Water and Power	3,632	1,769,505	0.75 hours	4.82	Equipment trip
6-27-78	Potomac Electric Power Company	1,000	100,000	35 hours	3.50	Thunderstorm
3-24-79	Illinois Power Company, Central Illinois Power Company, etc.	323	200,000	24 hours	1.55	Ice storm
12-31-78	Dallas Power and Light Co.	237	86,000	48 hours	0.98	Ice Storm
3-8-78	San Diego Gas and Electric	856	318,000	3.5 hours	0.95	Operator error
7-3-78	Montana Power Company	600	207,000	7.5 hours	0.93	System Failure

Looking through this list, it seems that electric system reliability would be least served by centralized systems consisting of widely separated clusters of gigantic single-fuel plants interconnected over long distances along critical, interruptible nodes. Conversely, a diversified mix of evenly dispersed, relatively small generating units, richly (but optionally) interconnected appears to fill the requirements best for overall strength and elasticity in the face of accidental and deliberate disruption (KH:VII-2).

BIBLIOGRAPHY

Barnet, R. *Real Security: Restoring American Power in a Dangerous Decade.* New York: Simon and Schuster, 1981.

Bass, G., B. Jenkins, K. Kellen, J. Krofcheck, G. Petty, R. Reinstedt, and D. Ronfeldt. *Motivation and Possible Actions of Potential Criminal Adversaries of U.S. Nuclear Programs.* Santa Monica, Calif.: Rand Corporation, February 1980.

Boffey, P. M. "Investigators Agree NY Blackout of 1977 Could Have Been Avoided." *Science*, 201 (September 15, 1978), 994-98.

Burnham, S., ed. *The Threat to Licensed Nuclear Facilities.* MTF-7022. MITRE Corporation Report to NRC. Project #2770. Department W-50. McLean, Va.: MITRE Corp., September 1975.

"The Case for Emergency Power." In *Electrical Construction and Maintenance.* New York: McGraw-Hill, December 1965.

Casper, B. M., and P. D. Wellstone. *Powerline: The First Battle of America's Energy War.* Amherst: University of Massachusetts Press, 1981.

Chenoweth, J. M., L. A. Hoh, R. C. Hurt, and L. B. McCammom. *A Method for Predicting Electric Power Availability Following a Nuclear Attack*, vol. I. April Report #5N108-1 to Office of Civil Defense. Pasadena, Calif.: National Engineering Science Company, 1963.

Clapp, N. M. *State of NY Investigation of NYC Blackout.* July 13, 1977. Report to Governor Hugh Carey. Albany, N.Y.: Governor's Office, January 1978.

Clark, W., and J. McCosker. *Dispersed, Decentralized and Renewable Energy Sources: Alternatives to National Vulnerability and War.* Energy and Defense Project. Report to Federal Emergency Management Agency (FEMA). Washington, D.C.: U.S. Government Printing Office, December 1980.

Clark, W. and J. Page. *Energy, Vulnerability and War: Alternatives for America.* New York: W. W. Norton, 1981.

Comptroller General of the United States. *Security at Nuclear Power Plants—At Best, Inadequate.* EMD-77-32. Washington, D.C.: U.S. General Accounting Office, April 7, 1977.

Congressional Research Service (CRS). *National Energy Transportation.* 3 volumes. Science and Transportation. Washington, D.C.: U.S. Government Printing Office, May 1977.

Defense Electric Power Administration (DEPA). *Protection of Electric Power Systems.* Research Project 4405. Washington, D.C.: Department of the Interior, June 1962.

Department of Energy. *Energy Data Reports.* Washington, D.C.: U.S. Government Printing Office, 1978.

Dircks, W. J. *Electromagnetic Pulse (EMP)—Effects on Nuclear Plants.* Memo to Commissioners, Secretary-81-641. Washington, D.C.: Nuclear Regulatory Commission, November 1981.

Dresch, F. W., and Ellis, H., *Methodology for Assessing Total Vulnerability.* Menlo Park, Calif.: Stanford Research Institute, 1966.

Ebinger, Charles K. *The Critical Link: Energy and National Security in the 1980s.* Center for Strategic and International Studies. Ballinger, 1982.

Federal Bureau of Investigation (FBI). *Bomb Summary 1972 through 1980.* Uniform Crime Reports. Washington, D.C.: FBI, 1973-1981.

Fink, L. H. "Systems Engineering Challenges Emerge as Electric Energy Network Increases in Complexity." *Professional Engineer*, 47, no. 12 (December 1976), 20-21.

Joint Committee on Defense Production. *Civil Preparedness Review.* 2 volumes. 95th U.S. Congress. First Sess. Washington, D.C.: U.S. Government Printing Office, August 1977.

Kupperman, R., and T. Trent. *Terrorism: Threat, Reality, Response.* Stanford, Calif.: Hoover Institution Press, 1980.

Lovins, A. B., and L. H. Lovins. *Brittle Power.* Andover, Mass.: Brick House Publishing Company, 1982.

National Electric Reliability Study (NERS): *Final Report.* DOE/EP005. Washington, D.C.: U.S. Department of Energy, April 1981a.

National Electric Reliability Study: Technical Study Reports. DOE/EPP05. Washington, D.C.: U.S. Department of Energy, April 1981b.

Pollard, R. D., ed. *The Nugget File.* Cambridge, Mass.: Union of Concerned Scientists, 1979.

Quirk, W. J., and J. E. Moriarty. "Prospects for Using Improved Climate Information to Better Manage Energy Systems." In *Interactions of Energy and Climate.* Bach, W. et al., editors. Reidel, Dordrecht, Netherlands, 1980, pp. 88-99.

Subcommittee on Energy and the Environment. *Allegations Concerning Lax Security in the Domestic Nuclear Industry.* July 29. 95th U.S. Congress. First Sess. Serial #95-23. Washington, D.C.: U.S. Government Printing Office, 1977.

Subcommittee on Energy and Power. *Energy Impacts of the Coal Strike.* February 16. 95th U.S. Congress. Second Sess. Serial #95-132. Washington, D.C.: U.S. Government Printing Office, 1978.

Turner, S. E., C. R. McCullough, and R. L. Lyerly. "Industrial Sabotage in Nuclear Power Plants." *Nuclear Safety*, 11, no. 2 (1970), 107-14.

"Utah Governor Blacks Out Data on State-Wide Energy Outage." *Electric Light and Power*, 59, no. 3 (March 1981), 1-7.

Wagner, N. R. *A Survey of Threat Studies Related to the Nuclear Power Industry.* SAND 77-8254. Sandia Laboratories, August 1977.

CHAPTER VIII

Demography

A. OVERVIEW

This chapter is concerned with the demographic implications of increased electrification. Specific demographic areas are set forth and potential impacts on them of increasing levels of electrification are discussed. In considering each of the demographic factors, at least two important underlying assumptions must be kept in mind. First, we assume that electrification will increase over time. Indeed in many instances, over half of the energy input is projected to be for electricity generation. Second, we assume that, in the future, coal will play a much larger role as a fuel source for electricity generation in absolute and relative terms than has heretofore been the case.

The topics to be covered in this chapter on demographic implications of increased electrification are the distribution of population and industry, sectoral transformation of the labor force, and the potential for divisiveness among regions in the United States. Since numerous possible paths exist for achieving increased electrification based on differential levels of growth and demand for power as well as variation in fuel mix, cost of fuels, and overall economic growth, a good deal of attention is given to specific demographic implications that emerge due to movement along alternative paths. However, the underlying context is that indicated by the two assumptions mentioned above. The implications for population distribution of increased electrification are discussed first.

B. POPULATION DISTRIBUTION

Background. Three main trends have historically characterized the distribution and redistribution of population in the United States: (1) rural to urban redistribution, (2) migration of the population from the South to the North, and (3) the outward expansion of cities.

Although recent striking deviations from these trends have occurred, it will be useful initially to consider these long-standing patterns in greater detail. Through recognition of the causes of these historical trends and of the reasons for recent departures from them, it may be possible to gain leverage in attempts to anticipate future variations more adequately.

First, consider the redistribution of population from rural origins to urban destinations. Between 1910 and 1970, population increases in metropolitan areas exceeded those in non-metropolitan areas in both absolute and relative terms in every decade. This is true irrespective of whether one relies on official definitions of metropolitan areas or on constant geographical areas. Indeed, if official definitions are the criterion, non-metropolitan areas actually recorded negative growth in the 1960-1970 decade (Hawley, 1971:154). However, the population size of rural, non-farm areas leveled off during the 1950-1970 interval, thereby partially setting the stage for the post-1970 reverse of this historic trend.

With respect to the redistribution of the population from the South to the North, the South experienced net migration losses to all other regions throughout much of the history of this country. In fact, this was the experience of the South as far back as the decades immediately following the Civil War. With the approach of the 1950s, migration from the South to the North (and West) took on special significance because it was primarily a migration of blacks (Lee, 1964:124). A foreshadowing of future events, however, occurred between 1949 and 1950 when the South received a very slight net in-migration (Poston, et al., 1981:11). On balance, the South was a donor of population to the rest of the country up until 1960. Except for the depression of the 1930s, the net out-migration of whites was in the vicinity of a half million persons for each decade ending in 1930 and 1950. During the 1950s, however, there was a modest net in-migration of whites. It is probably the case that the net out-migration of blacks was spurred by occupational opportunities in the North during both of the world wars with some diminution of the trend in the depression years. The absolute volume, however, has been declining since 1940.

Turning to the third aspect of population distribution, the outward expansion of cities, it is clear that since 1920 the growth rate of central cities has fallen below that of metropolitan areas as a whole; and after 1930, the growth of central cities fell below that of the country as a whole. This deconcentration seems to have been primarily the result of the diminution of the relationship between central location and accessibility. That is, in the twentieth century, improvements in both inter- and intraurban transportation and in manufacturing technology, plus the extension of utilities and other services to the periphery of cities, created a situation in which the benefits of central location physical propinquity began to fade. No longer was it necessary for urban populations and functions to cluster near the city's core. This along with the "push" created by deterioration, crime, and congestion near the city's center and later the advent of large shopping malls in suburban rings gave great impetus to the suburbanization of population and industry, which peaked in the 1950s but which continues to the present. Especially hard hit by population loss have been the oldest (and largest) central cities in the Northeast.

Recent trends. Recently, however, important reversals have occurred in two of the above trends. There has been a historically unprecedented net migration turnaround and a complete reversal of regional trends such that in the 1970s the South has experienced net migration gains from all other regions. At the same time, there has continued to be a massive suburbanward movement. Let us examine each of these recent developments in more detail.

The historic trend in movement toward metropolitan areas was reversed in the 1970s. "Between 1970 and 1978, more than 2.7 million more people moved out of metropolitan areas than moved into them. Whereas one-sixth of all metropolitan areas lost population, three-fourths of all non-metropolitan counties gained population" (Kasarda, 1980:380). Perhaps even more surprising, the greatest rate of net migration gain was in the most rural counties, that is, those not adjacent to metropolitan areas and those with no incorporated urban place in them. Virtually all socio-demographic groups seem to have been part of this turnaround in migration, with the exception of blacks (cf. Beale, 1978; Zuiches and Brown, 1978; Tucker, 1976; Wardwell, 1977).

Reversing the trend of net loss to all other regions, the South gained population from the Northeast and North Central regions in the 1960s. In the 1970s decade, the South gained population from all other regions, including a gain of about 200,000 persons from the West between 1970 and 1978. From 1965 to 1978, the South netted over 1.8 million persons in exchange with the Northeast and close to 1.5 million from the North Central region (Kasarda, 1980:374-75). The West continued to receive migrants from the Northeast and North Central regions though the absolute gains were somewhat smaller than those recorded by the South. Specifically, the net gain for the West was about .75 million and 1.2 million from the Northeast and North Central regions, respectively, in the 1965-1978 interval.

The trend in movement of population and industry away from the core of the city to suburban areas has continued. Although the single heaviest decade of suburbanization was the 1950s, the increase in suburban population in the 1960-1970 decade was close to 24 million, while central city populations grew by a little less than 7 million (unadjusted for annexations: Berry and Kasarda, 1977:Table 8.3). Moreover, during the 1960s the percentage of the labor force living in suburbs increased by about 40 percent, and the growth of employment in suburban areas increased by 48 percent (Logan, 1976:335).

Impacts of increased electrification. What are the implications of increased electrification for population distribution? It may be the case that the primary mechanism underlying the relationship to population distribution of increased electrification is that of locational advantage. "Population settlement has always been sensitive to the cost and availability of energy" (Wardwell and Gilchrist, 1980:567). That is, energy needs and costs have remained of crucial importance, but changes in these factors have "reduced and, in some cases, reversed the previous locational advantages" (Kasarda, 1980:378).

A situation in which there is an increase in the fraction of energy consumed for production of electricity based on heavy reliance on coal might well shift at least part of the locational advantage back to the Northeast and North Central regions and help maintain the currently existing advantage of the West, all at the expense of the South (SM:VIII-1). Kasarda (1980:378) points out that during the 1960s there was a substantial conversion of industrial plants and some utilities in the North from coal to oil or natural gas. Then with the rapid rise in the price of petroleum products in the 1970s, severe problems were created for these northern areas. However, counterpart facilities in the South were helped because of more ready access to oil and natural gas in those regions.

With a shift toward greater nationwide reliance on coal, the Northeast, North Central, and West regions might be in a stronger competitive position since the industrial structure of the northern states has been oriented to the use of coal, and the West has large deposits of low sulfur coal. In addition, coastal utilities now dependent on delivery of oil directly from tankers have little storage capacity for coal, so that conversion to coal-fired generation of electricity would be extremely difficult if not impossible (EPRI, 1981:2). Thus, heavy use of coal, as depicted under the "Average Future" and "Mega-plant" Scenarios, would be apt to have a positive impact on the economy and population growth of the West and perhaps diminish to some extent the redistribution of population out of the Northeast and North Central states.

On the other hand, it would be naive to expect that increases in the generation of electricity, even based largely on coal as a fuel source, would, in and of itself, lead to huge dislocations of population and industry. It may be that the costs of transporting coal by rail will prohibit wide-spread development of generation facilities at locations far removed from the source of the raw material (DD:VIII-1). But this situation can probably be offset since it is likely that transmission of electricity for distances of up to 1,000 miles can effectively substitute for the transport of coal. Further, coal-oil and coal-water slurries may be a solution for utilities currently without adequate storage capacity for coal itself. In any event, the point is not that major reliance on coal will under any and all circumstances either provide an overriding stimulus for the North or lead to a complete reversal of Sunbelt growth. The point is simply that a coal-oriented future might mitigate or dampen the recent trends described above. For example, the economies of coal- and lignite-producing regions ought to experience some positive "multiplier" effects in terms of the growth of ancillary or support industries that serve the extracting industry.

Under conditions of high levels of energy consumption and moderate fuel and electricity costs ("Small Coal Plant" Scenario), there would seem to be little to reverse the recent trends in population redistribution. This would be particularly true where power stations themselves are fairly widely distributed.

To the extent that "soft energy" alternatives (solar, wind, hydro) achieve prominence, the impetus to wider dispersion of population and industry into non-metropolitan areas might be expected to continue. Wardwell and Gilchrist (1980:575) note that "soft energy alternatives may be neutral to scale; hard energy alternatives are not. . . . The latter demand production centralization and long-distance transmission. The former permit decentralization of production and less reliance upon transmission." It also appears that the importance of locational advantages may be heightened by moves toward generation via soft energy sources since such sources "are more dependent upon the characteristics of the physical environment at the production site, and lend themselves less readily to long-distance transmission of energy" (Wardell and Gilchrist, 1980:575).

Of course, as noted previously, once generated, transmission of electricity over substantial distances is entirely possible if issues of cost and interface with the soft energy system can be resolved. In sum, migration of population to the South and West might well continue, or even accelerate, due to greater reliance on solar or water power since these areas have the greatest potential in this regard. Again referring to Wardell and Gilchrist (1980:575), it is clear that the current trend in migration "favors locations rich in natural amenities such as mild and sunny climates, access to seashores or fresh-water lakes and rivers, and scenic vistas of mountains and deserts. Many of these locations may lend themselves to the solar, wind and water generation of energy that can provide the residents with the material amenities and conveniences to which they have been accustomed."

Nevertheless, the cost of producing electricity must be constantly borne in mind. To illustrate, conditions of low total energy consumption, a high electricity fraction, and extensive auto-generation (e.g., "Post-industrial" Scenario) would favor increasing dispersion. That is, to the extent that power generation is decentralized, locational advantages of concentration (i.e., economies of scale) diminish. Thus, one might predict more movement to suburban settings, and beyond to non-metropolitan areas, especially in the West and South. But if fuel costs are high ("Post-industrial'' Scenario), this might severely retard expansion in these areas. This could occur not only due to the high price of electricity, per se, but also because of the costs involved in expanding and maintaining the distributive industries in general and transportation facilities in particular that would be necessary to sustain a widely dispersed population. High energy costs, such as those envisioned under the ''Post-industrial'' Scenario might either make such a decentralized society impossible to sustain or result in a highly segmented, uncoordinated society held together by common needs (such as national defense) but moving toward local self-sufficiency. This, in turn, might increase the importance of regional metropolises as trade and administrative centers for a loosely coordinated society—though large national metropolises would certainly retain scale advantages.

A future with low total energy consumption and a relatively small fraction consumed for electricity implies economic stagnation, which historically has inhibited migration (e.g., ''Economic Malaise'' Scenario). A plausible result would be little in the way of locational shifts once the economic malaise is felt. Some reconcentration of population is apt to occur (in central cities, or perhaps more likely, in near-to-city suburbs) as economies of scale become more important as a means of cost reduction.

C. INDUSTRIAL DISTRIBUTION

Background. Basically, the patterns of industrial redistribution are the same as those examined for the population, namely, movement of industry to the Northeast and North Central regions, growth of metropolitan areas, and large scale suburbanization. Early in the history of the United States, the Northeast and the North Central regions became the principal loci of industrial production. Major cities grew up and prospered in these regions at deepwater ports and railheads. Metropolitan areas in general grew rapidly as the rise of commercial agriculture with its increasing mechanization and high volume production meant less manpower was needed on farms. The South, as the leading agricultural region, sent many thousands of migrants, both black and white, North in search of jobs. Industry was fairly well concentrated in central cities in the nineteenth century. But as new technology in manufacturing led to assembly line techniques and thus large, single-story space requirements; as improvements in short-range transportation allowed easy accessibility to all parts of the city; as trucks (instead of railroads) assumed a larger share of long distance transport thereby making the periphery an easier place to receive materials; and as services spread to the periphery, the old constraints were removed and suburban movement of industry began to occur rapidly. Later, large shopping malls began to lead, rather than follow, out movement of population.

Recent trends. In the 1970s, paralleling the movement of population there has been a redistribution of industry resulting in major growth in nonmetropolitan areas, a shift in the direction of the movement of industry to the South, and the continued outward expansion of industry to suburban and exurban rings.

As an illustration of these trends, consider the following. Between 1970 and 1980, "Manufacturing jobs in the South increased by 926,000, whereas the Northeast lost 585,400 manufacturing jobs and the North Central region lost 116,300 jobs" (Kasarda, 1980:378). (In addition to cheaper power, of course, the generally more favorable political, land cost, and tax climate of the South gave impetus to this redistribution to non-metropolitan areas.) The "225 largest Standard Metropolitan Statistical Areas (SMSAs) suffered a net loss of 513,000 jobs in manufacturing between 1970 and 1978, [while] nonmetropolitan counties experienced an increase of 619,000 manufacturing jobs." Further, "An additional 3,452,000 service-sector jobs emerged in the nonmetropolitan counties between 1970 and 1978. By 1980, nearly two-thirds of all non-metropolitan workers were employed in the expanding service-performing sector" (Kasarda, 1980:381, citing Haren and Hollins, 1979).

Figures on the movement of industry from central cities to suburbs are presented for the largest central cities in a later section. However, to indicate the direction of effect and the relatively long-standing nature of the suburbanization trend, note that "the 49 central cities that had attained 50,000 or more population before 1900 showed losses of 1,370,000 in manufacturing, 750,000 in retail and 805,000 in wholesale trade" (Kasarda, 1980:384). Not only have the South and West gained industry in general, but most of the gains have come in high growth service sectors. It is also the case that more desirable industries, such as electronics, have begun more and more to concentrate in the Sunbelt.

Impacts of increased electrification. What are the implications for industrial redistribution of increased electrification? As was argued in the preceding section, the primary mechanisms through which many impacts might be felt would seem to depend on locational advantage. However, there are differences in the forces that can be expected to influence industrial redistribution as compared to population. High taxes in the central city may push out both population and industry, but shifts to non-metropolitan areas and to the South and West as related to taxes would seem more pertinent to industry than to population. The point is twofold. Cost-imposing factors are more important to industrial relocation than are life-style differentials, while the latter may be decisive for population shifts as long as decent employment opportunities exist at potential destinations. Second, electricity consumption and overall energy consumption are only two among a large number of factors that have an impact on the redistribution of industry and population. These other determinants must be factored in at least by the way of caveats regarding the magnitude of impact of those energy-related variables that are our focus.

Any scenario that does not incorporate disincentives, including those related to electricity and energy consumption, would suggest a continuation of industrial redistribution out of the North and large metropolitan areas. As pointed out previously, the South and West already have numerous advantages in pulling industry to these regions: lower taxes, right-to-work laws, aggressive "boosterism," and so forth (Kasarda, 1980). Thus, a scenario in which power generation is decentralized, as might be realized with many small, dispersed coal plants, coupled with moderate reliance on oil and gas, hydro, and nuclear power (as suggested in "Small Coal Plant" Scenario) would appear particularly conducive to the continuation of recent trends in the migration of industry—the more so if fuel cost and the price of electricity are relatively cheap.

Future electricity generation scenarios based on the assumption of a strong and/or growing use of coal as a primary fuel source might shift some of the locational advantage back to the Northeast and North Central states. In this regard, "Average Future" and "Mega-plant" Scenarios are similar, with the major exception that the "Mega-plant" future assumes lower power costs and a robust economy which might make it economically feasible to ignore partially the economies of locational advantage and scale. In that event, little stifling of the trends favoring non-metropolitan areas and the South and West would seem likely.

A resurgence of nuclear power could produce an "eastward tilt" in industrial movement. The vast majority of all nuclear plants currently being planned or in operation are east of the Mississippi. If this were to continue, certain states along the eastern seaboard and in the Southeast could become the "power centers" of the nation. Again, though, it is certainly not necessary for industry to be located in close proximity to power generation facilities, if the costs of long-range transmission are as reasonable as the Electric Power Research Institute (EPRI) suggests. It is possible that a nuclear resurgence might give particular advantages to non-metropolitan areas if, as seems reasonable, many nuclear plants would be sited in less populated areas.

Economic stagnation (i.e., "Economic Malaise" Scenario) would inhibit recent trends in industrial redistribution for reasons similar to those discussed in the population distribution section. Finally, it is conceivable that certain combinations of economic and technical factors might in the future lead to emigration of substantial numbers of manufacturing plants to other countries. It is possible, for example, that low levels of electricity generation coupled with relatively high costs of fuel, could result in the movement of energy-intensive industries to Mexico in numbers large enough to be economically disruptive of the economies of regions where such industry has long been a mainstay.

D. CENTRAL CITIES

Background. The migration trends discussed in the preceding sections, particularly the non-metropolitan turnaround and the movement to the

Sunbelt, are, in some respects, a continuation of urban deconcentration initiated around the turn of the century (Berry and Kasarda, 1977; Hawley and Mazie, 1981). Particularly after World War II, the United States experienced major shifts from the central cities to the suburbs, and these moves, along with others related to the overall deconcentration patterns, have been selective by income level. Indeed, Kasarda has observed that "the automobile and commuter train—along with rising real incomes, state and federally funded radial and suburban highway systems and federally insured home mortgages—encouraged and facilitated the mass suburbanization of the middle and upper class in metropolitan America" (1980:383).

Recent trends. Regarding job shifts in the metropolitan area, an especially applicable illustration should suffice: "The central cities of the 33 SMSAs with over one million population in 1970 lost more than 880,000 manufacturing jobs between 1947 and 1972, while manufacturing jobs in their suburbs grew by over 2,522,000. These same 33 central cities lost more than 565,000 retail jobs and 302,000 jobs between 1947 and 1972 while retail and wholesale jobs in the suburban rings grew by 2,360,000 and 798,000, respectively" (Kasarda, 1980:384). But these trends do not apply to SMSAs with populations less than 250,000 in 1970. The central cities and the suburbs in these smaller SMSAs showed gains in these employment categories.

Another factor strongly impacting the relative growth and economic position of central cities compared to their suburbs is the emergence of peripheral market and production centers. Kasarda observes that "to escape the costs and headaches of a central city location, yet maintain important agglomeration and scale economies, many manufacturers and wholesalers have banded together in modern industrial parks along major suburban and exurban expressways" (1980:385). Over 2,000 of these types of locations have been formed since 1960. Also, we have witnessed in the past quarter decade the proliferations of suburban shopping centers and regional malls. "Between 1954 and 1978, more than 15,000 shopping centers were constructed" (Kasarda, 1980:386). By 1975, these kinds of shopping configurations accounted for over half of the total annual retail sales in the United States (Muller, 1976).

Impacts of increased electrification. What are the specific implications for central cities of the various scenarios dealing with increased electrification? In the first place, it may be argued that current redistribution trends in which both population and industry have displayed a marked preference for small communities (especially in the South and West) would be enhanced by increased use of soft energy alternatives as fuel sources for, among other things, electricity generation. Unfortunately, it is not known "what optimal settlement densities are, how they are affected by the scale economies of soft energy production and the extent to which they correspond to the density gradients of growing rural and nonmetropolitan communities" (Wardwell and Gilchrist, 1980:576). Wardwell and Gilchrist (1980:577) point out that soft energy alternatives (solar or hydropower for example) can be pursued on a smaller scale than hard energy alternatives. This, along with the potential for substituting labor for energy, may contribute even further to the demographic and industrial growth of small towns and cities and hasten central city decline. "Small Coal Plants" and "Post-industrial" Scenarios, with a fairly large percentage of autogeneration and, compared to other scenarios considered here, more reliance on soft energy sources imply a situation more conducive to further deconcentration of population and industry.

In contrast, scenarios such as "Average Future" or "Mega-plant," which emphasize hard energy sources (particularly coal), might once again place a premium on higher density settlement, since hard energy sources are more tied to economies of scale. High costs of power and/or economic stagnation suggest advantages of concentration in cities, although with a depressed economy little real "re-vitalization" would be expected. On the other hand, there is no assurance that the population will respond very quickly, if at all, to pressures to concentrate as long as it is affordable to live in suburban, exurban, or non-metropolitan areas. Thus, scenarios such as "Nuclear Resurgence," "Mega-plant," and "Small Coal Plants," which incorporate high levels of consumption and moderated costs, may not lead to any significant departures from current trends in urban deconcentration.

E. THE SERVICE SECTOR

Background and trends. This section focuses on the implications of electrification for the continued growth of the service economy. With respect to past events, two major trends characterize the sectoral transformation of the U.S. labor force in the past century: the movement out of agriculture and the growth of services. In the interim, there was movement into the transformative sector and then a stabilization in that sector. To illustrate, "In 1870, 52.3 percent of the U.S. labor force was in the extractive sector. Fifty years later in 1920, this sector had declined to 28.9 percent and by 1970 it had further declined to a mere 4.5 percent. The transformative (secondary) sector for the same periods rose from nearly a quarter (23.5 percent) of the labor force to nearly a third (32.9 percent), but then changed relatively little over the next fifty years, ending at 33.1 percent, in 1970" (Browning and Singlemann, 1975:xv).

The service sector of the labor force can be conceptualized in terms of four components: *distributive services* (transportation, communication, wholesale and retail trade, except eating and drinking establishments); *producer services* (financial, insurance, engineering, law, and business services); *social services* (health, education, welfare, and government); *personal services* (domestic, lodging, repair, and entertainment). Services as a whole have increased every decade since 1870, but the gains in the various components have been uneven. Actually, in one important area, personal services, there has been a negative change. Domestic services comprised 7.4 percent of the total labor force in 1870, but declined to about 2 percent by 1970. In the other three types of services, there have been increases. Distributive services experienced significant gains between 1870 and 1920, with transportation contributing heavily to the increase. With the growing efficiency of transportation since 1930, however, proportionally fewer workers have been needed in this enterprise. Producer services have grown from about 3 percent in 1910 to more than 8 percent in 1970 (prior to 1910, producer services were such a minor part of the total that they were included in the trade category). Even though there has been a decline in domestic services over time, the larger category of personal services has remained just about constant over the past 100 years (9.3 percent in 1870 and 10.0 percent in 1970). The labor force engaged in the provision of social services has increased dramatically. In 1870, social service workers made up 3.4 percent of the labor force. By 1970, this component had grown to 21.9 percent of the labor force. Within the sector, between 1950 and 1970, education increased from 3.7 percent to 4.6 percent (Browning and Singlemann, 1975:xv).

Factors mainly responsible for these sectoral changes in the labor force, and the resultant buildup of the service sector, include productivity gains (movement from lower to higher productivity levels) and increases in per capita income leading to increased demand for a variety of services. The sectoral transformations have clearly not led to decreases in productivity in the extractive and transformative sectors; the number of persons employed in a sector does not necessarily reflect the magnitude of its output. Indeed, agriculture and manufacturing have experienced astounding productivity gains in the past 60 years (Browning and Singlemann, 1975:xv-xvi).

Impacts of increased electrification. In light of the above trends, what are the implications of increased electrification for the continued growth of the service sector? It is certainly the case that the sectoral transformation of the labor force has been under way since the turn of the century. But the kinds of structural changes discussed above pertaining to population and industrial redistribution have given impetus to the transformations already in existence, particularly with respect to the growth of services. For with movement from the central city to the suburbs, from the Northeast and North Central regions of the nation to the South and the West, from the metropolitan areas to the non-metropolitan areas, there have been corresponding increases in the demand for services—particularly producer, distributive, and social services. Growth in personal services has probably not been as great, but there have been increases in this sector also. For instance, looking at the growth of the southern region between 1970 and 1980, the U.S. Bureau of Labor Statistics (1970; 1980) has reported that during this decade more than 7.5 million non-agricultural, non-manufacturing (i.e., service) jobs were added in the South, more than any other region in the country. Availability of services encourages new industry and the movement of old industry as we have seen in the South and to a lesser extent in the West. Also, with the increased growth of manufacturing industry and population in new areas, services accompany and follow these changes.

With regard to the specific electrification scenarios, it would appear that expansion of distributive services (particularly transportation and communications) has been a key element in the redistribution of population and industry. Only as the friction of movement over time and space has been widely reduced have population and productive functions been able to disperse widely. As alluded to above, it was in this manner that the advantages of concentration and centrality were diminished. Thus, any set of power generation relationships that act to increase the friction of movement via high costs of fueling transportation by spreading power generation facilities at intervals too wide to be efficiently used would place considerable strain on the distributive sector. Whether this would mean devoting more economic and human resources to distributive services or would lead to reconcentration of population or would result in technological innovation to meet the increased needs is a matter of speculation.

The "Average Future" Scenario, with its lower level of growth, and the "Economic Malaise" Scenario, which represents economic stagnation, would probably inhibit population and industrial redistribution. One would anticipate a much smaller movement of the labor force into services with "Economic Malaise" Scenario than would be the case with others.

Alternatively, "Nuclear Resurgence," "Mega-plant," "Small Coal Plants," and "Post-industrial" Scenarios suggest the increased growth of services. The "Post-industrial" Scenario is characterized by extensive autogeneration, which favors increased dispersion (see the above discussion under population) and would imply increased services. "Mega-plant" and "Small Coal Plants" Scenarios, with high levels of consumption and moderate costs, would also encourage services expansion since these two scenarios would likely involve the continued redistribution of population and industry. The "Nuclear Resurgence" Scenario does not contain any disincentive related to energy consumption either, so one would expect continued redistribution trends of population and industry under this scenario and thus continued growth of the service sector.

F. REGIONAL DIVISIVENESS

Background. This final section examines the topic of regional divisiveness. To what extent will divisiveness between and among the regions of the United States occur under increased electrification, and how might this divisiveness vary according to the various electrification scenarios?

Generally, the focus here is on the census-defined regions, that is, the Northeast, the North Central, the South, and West. For certain kinds of comparisons, finer distinctions could, and perhaps should, be made. These could be census-defined divisons (which are components of the four larger regions) and have the advantage of being units for which large amounts of relevant data are readily available.

What then are the underlying bases of regional divisiveness? Primarily, the basis lies in a broad range of differences by region in growth, resources, and needs. Specifically, there are at least six main bases of potential divisiveness (some of which are described in EPRI, 1981). They are the following:

1. Low growth versus rapid growth within the regions (directly speaking, the differentials in population and economic growth mentioned above).
2. Differences by region in natural resources, including differences in geographic, geologic, and topographic factors.
3. Differences (1) and (2) lead to regional differences in characteristics that are directly related to electrical generation/consumption and energy use, including variation in transmission requirement, fuel source type and availability, and nature of environmental issues.
4. Demographic composition (e.g., by age or race) also varies by region and may play a part in divisiveness possibilities.
5. Governmental/legal differences (tax rates, right-to-work laws, etc.) also exist among the regions and potentially could lead to divisiveness.
6. There are also obviously partisan political differences, but we view these as the results of, and manifestations of, the variation in the factors listed previously.

Impacts of increased electrification. Given these differences, what are the implications of increased electrification under the various scenarios? Note first that regional divisiveness is at least partially a function of population and economic growth. As EPRI (1981) notes, regions experiencing high growth will have a larger range of power generation options, while low-growth, capital constrained areas will be more interested in extending the life of existing facilities and probably more eager for an expanded use of coal, if projections regarding the cost of coal as the major fuel source for electrical generation are accurate.

Thus, under the "Average Future" Scenario (moderate energy consumption and prices with well over half the generation from coal), low-growth regions such as the Northeast and parts of the Midwest would no doubt lobby for programs designed to make use of coal more profitably and for environmental policies designed to control acid rain. These kinds of activities would not necessarily be in the best interests of the South (and Southwest), where coal is not readily available and where gas and oil might well be preferable as a fuel source. Nor would these policies necessarily coincide with the interests of the West, where coal is of the low sulfur variety and where environmental issues of visibility and water availability and quality are of greater concern. It should be remembered that all scenarios discussed here predict a heavier reliance on coal than is presently the case. Thus, one would expect some degree of divisiveness in each of the scenarios, based on environmental concerns and variation in location of large coal deposits. In 1980, many regions of the country—and, incidentally, these are the same regions with the lowest cost of electricity generation—relied not on coal but on either natural gas (Electric Reliability Council of Texas; Southwest Power Pool) or hydro (Western Systems Coordinating Council—see EPRI, 1981:38). There are two possibilities here. Very heavy use of coal may result in major shifts or dislocations in the economies of the South and Southwest (at least those subregions now relying on natural gas). On the other hand, those regions with special natural resource advantages (gas in the Southwest; hydro in the far West) may continue to take advantage of them, while other regions turn more and more to coal. In either case, differentials in resource base and environmental impacts will probably act to reinforce regional disagreements.

"Nuclear Resurgence" Scenario may have the greatest potential for both intra- and interregional conflict. The issue is so emotionally charged that it is at least conceivable that the populations of some regions with few nuclear power facilities and that are not even being asked to help store or dispose of nuclear waste might still offer resistance in regard to areas that do have or want nuclear generation facilities. This may not seem likely, but then many did not think that the current level of opposition was likely either.

High-priced power may also have divisive effects. In the high-priced, inflationary 1970s, the major shifts of population, industry, and capital took place in earnest from the Northeast and North Central regions to the South and West. As a result, divisions created by conflicting vested interests have deepened. For instance, it appears that many recent votes in Congress have begun to be more along regional than party lines. This might be exacerbated in the "Post-industrial" Scenario by the spatial and organizational segmentation implied in that particular scenario.

If the economy is strong, with plenty of energy available at reasonable prices, there would be no apparent economic, energy-related basis for increased divisiveness. Conversely, the effect of severe nationwide recession is hard to predict with respect to regional divisiveness. In earlier periods of depression on a national scale, the population of the United States has tended to pull together and to subordinate regional and local interests to the federal government, which is presumably better able to deal with truly nationwide problems. It is perhaps the situation in which some regions experience an economic downturn while others remain relatively prosperous that holds the greatest potential for divisiveness.

G. SUMMARY

This chapter describes how the future of the generation availability and demand for electric power affects five national demographic characteristics. These five demographic areas are as follows: (1) population distribution (e.g., a heavy reliance on coal as an energy source might shift population movement away from the South and West and back to the Northeast and Central regions); (2) industrial distribution (e.g., if power generation is decentralized, industry would continue to move away from Northeastern cities); (3) central cities (e.g., greater reliance on soft energy sources that are not tied to large generation plants would hasten central city decline); (4) service sector (e.g., the service sector will follow population and industry movement toward energy supplies); and (5) regional divisiveness (e.g., differences by region in natural resources, including geologic, geographic, and topographic factors, lead to regional preferences for certain national energy policy decisions). Thus, energy policy makers can benefit by calculating the demographic implications of their decisions, along with other socio-economic factors, when deciding on policy alternatives.

BIBLIOGRAPHY

Beale, Calvin L. "People on the Land." In *Rural U.S.A.: Persistence and Change.* Ames: Iowa State University Press, 1978, pp. 37-54.

Berry, Brian J. L., and John D. Kasarda. *Contemporary Urban Ecology.* New York: Macmillan, 1977.

Browning, Harley L., and Joachim Singlemann. *The Emergence of a Service Society: Demographic and Sociological Aspects of the Sectoral Transformation of the Labor Force in the U.S.A.* Springfield, Va.: National Technical Information Service, 1975.

Electric Power Research Institute (EPRI). *1982-1986 Overview and Strategy.* Palo Alto, Calif.: Electric Power Research Institute, 1981.

Haren, Claude C., and Ronald W. Hollins. "Industrial Development in Nonmetropolitan America: A Locational Perspective." In *Nonmetropolitan Industrialization.* New York: Wiley, 1979.

Hawley, Amos. *Urban Society.* New York: Ronald Press, 1971.

Hawley, Amos, and Sara Mills Mazie. "An Overview." In *Nonmetropolitan America in Transition.* Chapel Hill: University of North Carolina Press, 1981, pp. 3-23.

Kasarda, John D. "The Implications of Contemporary Distribution Trends for National Urban Policy." *Social Science Quarterly,* 61 (December 1980), 373-400.

Lee, Everett S. "Internal Migration and Population Redistribution in the United States." In *Population: The Vital Revolution.* Garden City, N.Y.: Doubleday Anchor, 1964, pp. 123-36.

Logan, John R. "Industrialization and the Stratification of Cities in Suburban Regions." *The American Journal of Sociology,* 82 (September 1976), 333-48.

Muller, Peter O. *The Outer City*. Resource Paper no. 75-2. Washington, D.C.: Association of American Geographers, 1976.

Poston, Dudley L., Jr.; William J. Serow; and Robert H. Weller. "Demographic Change in the South." In *The Population of the South: Structure and Change in Social Demographic Context*. Austin: University of Texas Press, 1981, pp. 3-22.

Tucker, C. J. "Changing Patterns of Migration between Metropolitan and Non-metropolitan Areas in the United States." *Demography*, 13 (November 1976), 435-43.

U.S. Bureau of Labor Statistics. *Employment and Earnings, Monthly Reports*. Washington, D.C.: U.S. Government Printing Office, 1970, 1980.

Wardwell, John M. "Equilibrium and Change in Nonmetropolitan Growth." *Rural Sociology*, 42 (Summer 1977), 156-79.

Wardwell, John M., and C. Jack Gilchrist. "The Distribution of Population and Energy in Non-metropolitan Areas: Confluence and Divergence." *Social Science Quarterly*, 61 (December 1980), 567-80.

Zuiches, James J., and David L. Brown. "The Changing Character of the Non-metropolitan Population, 1950-1975." In *Rural U.S.A.: Persistence and Change*. Ames: Iowa State University Press, 1978, 55-72.

CHAPTER IX

Research, Development, and Demonstration Policy

A. OVERVIEW

Of all the means available for providing American society with the capability to meet anticipated demands for electric power, none is more potent or practical than carefully conceived and effectively executed programs of technical research, development, and demonstration (RD&D). If properly managed, such programs will provide society in general, and the electric utilities in particular, with a spectrum of feasible, economic, and safe means for generating, transmitting, distributing, and utilizing electric power. Given this spectrum, policy makers can select the combination of technologies that can best accommodate society's needs and desires. However, this flexibility of choice, like all commodities, has a price, and it will be the test of policy makers, both inside and outside of the government system, to determine the proper level of RD&D support.

Designing a national electric power RD&D policy involves a number of subordinate policy questions. What will be the size and scope of the national RD&D effort? How will the effort be apportioned between technical alternatives? Who shall perform each of the required tasks? How will costs be assessed and recovered? How will electric power efforts be fashioned to serve other national goals? Many of these questions are discussed in other chapters of this report as well as in Appendix G of the NSF report. This chapter's purpose is to present background information from which to discuss national electric RD&D policy in terms of the six alternate scenarios. To accomplish this purpose, an overview of the history and present direction of RD&D in the electric power industry is presented. Next, a typology of RD&D programs is developed. Finally, types of RD&D programs appropriate for each scenario are outlined along with a discussion of environments conducive to particular types of RD&D.

B. RD&D HISTORY AND CURRENT STATUS

Before discussing appropriate RD&D strategies for each scenario, it is helpful to review the history and current state of RD&D. The evolution of RD&D in electricity can be divided into three periods of distinct character. The first period, from the 1790s to the late 1880s, was when experimental scientists developed a simple understanding of the properties of electricity. The second period, from the late 1880s to the early 1960s, was when electric power systems were commercially developed and implemented throughout the United States. And finally, the third period, from the early 1960s to the present, is when RD&D has become increasingly directed toward overall energy and socioeconomic effects of the electric power system.

1790s-1880s. The first period of RD&D, lasting almost to the turn of the nineteenth century, was dominated by scientists pursuing basic electrical knowledge with primitive tools. Their questions concerned: What were the properties of electricity? How could it be produced? Could it be transmitted? There was relatively little interest in application as a basic understanding of the phenomenon of electricity was still to be achieved. Yet, these scientists developed electrical knowledge to a point where there was no question about its usefulness to technology.

1880s-1960s. Around the late 1880s, the beginning of the second period, entrepreneurs with an understanding of the electrical principles began to see commercial potential for electric power. Wide-spread implementation of electric power systems required more commercially directed RD&D than was previously performed. Individuals from many fields made significant contributions to this process, but the two major catalysts were Edison and Westinghouse. The companies they founded in the early 1890s are still the largest generation and transmission device manufacturers today. Once the commercial feasibility of large electric power systems was firmly established, RD&D became more and more centralized in the laboratories of what were to become General Electric and Westinghouse.

During the Great Depression, equipment manufacturers divested themselves of the distribution companies (utilities), which were granted natural monopoly status. At the same time, individual utility systems were entering a rapid expansion phase. Systems operations management and maintenance became significant problems as the individual utility grids grew in size and complexity. The manufacturers supplied the equipment, but it was the utilities' responsibility to keep the system running efficiently and profitably. Performing RD&D for these tasks was the individual utilities' first entry into the research field and remained their primary area of concentration until recently.

Similarly, as the electric power system became an indispensable part of the economy and as the technologies employed became capable of national impact, it became appropriate for the federal government to have a role in industry management and direction of RD&D. Federal involvement was first evidenced during the 1940s when major industries were mobilized to support the war effort. Later, the federal presence became permanent with the Atomic Energy Commission's effort to develop practical methods for using nuclear power to generate electricity (EPRI, 1979:69).

1960s to the present. The direction of RD&D prior to 1965 was not so much rationally planned as financially determined. With the available resources (coal, water, oil, and gas) and technologies, electric power was produced progressively more efficiently through economies of scale. Large generating devices with higher speeds and temperatures had better electricity conversion efficiencies. The higher the transmission voltage, the less the transmission line energy loss. The higher the degree of interconnection among individual utility grids, the less reserve capacity each utility was required to have (see Exhibit IX.1).

The evolution of increasingly larger and more complex utility systems progressed steadily until the 1960s when non-technical issues came to the forefront. The 1965 massive power failure, blacking out the northeastern United States, brought to national attention questions about the reliability of the electric power system. The 1960s also saw the raising of the nation's ecological consciousness and environmental concerns. The 1973 oil embargo encouraged RD&D efforts to develop renewable or non-fossil fuels. Finally, higher fuel costs promoted RD&D toward improving technological efficiency of generation and use of electric power.

Because of such events and public concern, RD&D policies shifted into a reactive rather than its previous expansive mode. RD&D policies responded to the issues of the day, making the policy decision process more complex. Believing federal intervention forthcoming, the National Electric Reliability Council (NERC), a voluntary industry group, was organized to address the reliability problem. NERC was composed of members from nine regional councils. Although not a RD&D organization by name, their mandate was to augment the reliability and adequacy of bulk power supply in the electric utility systems of North America and it led them to make specific recommendations concerning the RD&D needs of the industry. Not surprisingly, systems monitoring and controls were emphasized (NERS, 1981, pp. 17-18).

The Electric Research Council (ERC), another voluntary organization founded in the fall of 1965, published after extended analysis its "Electric Utility Industry Research and Development Goals through the Year 2000" (ERC, 1971). It was a major step for RD&D planning for the industry. Needs in generation, transmission, environment, utilization, and systems management were identified and required expenditures were projected through the year 2000 (see Exhibit IX.2). The Electric Research Council concluded that:

> Historically, our industry has relied heavily on manufacturers to do R&D and pass along the costs in the price of the ultimate product. While this has led to many important advances, it is not an altogether satisfactory arrangement. For example, it

Exhibit IX.1
Trends in Technology and Cost of Electric Energy, 1880–1980

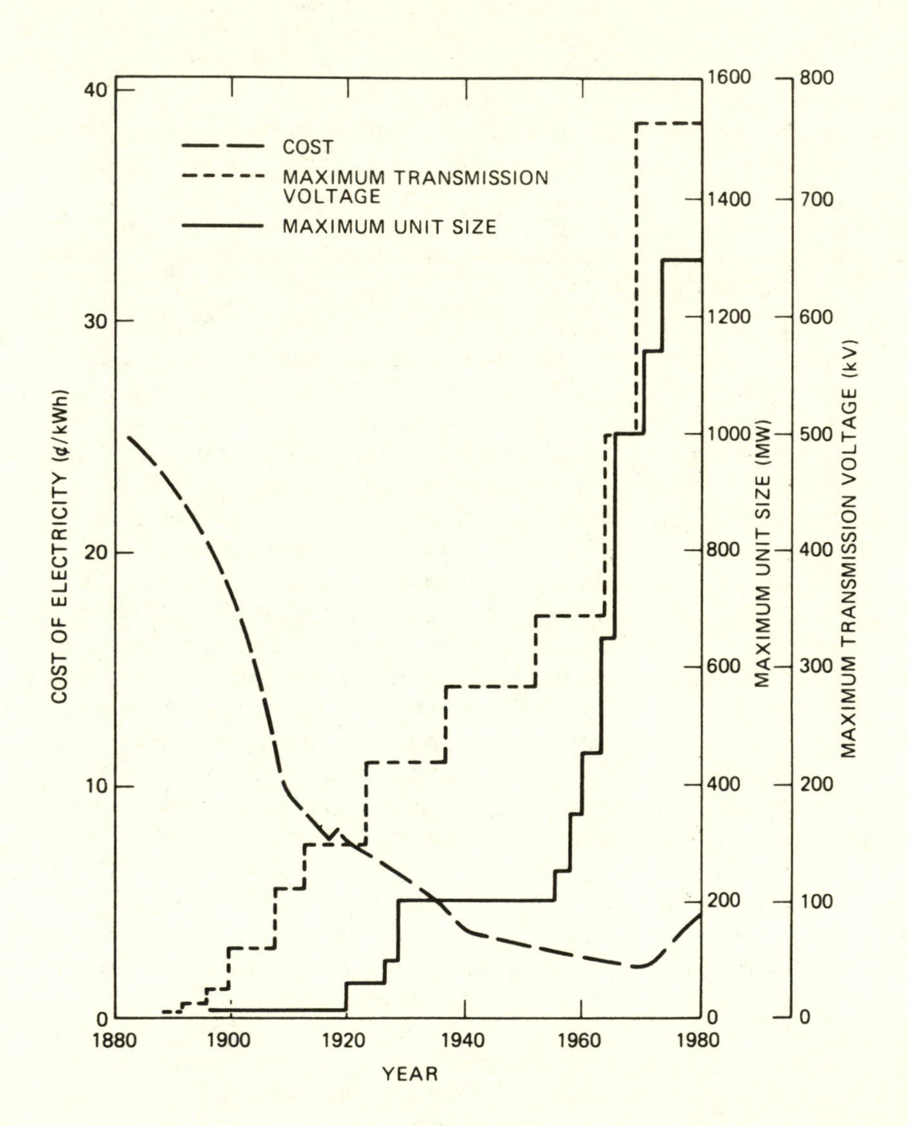

Source: NERC, 1981.

Exhibit IX.2
R&D: Distribution of Total Estimated Annual Cost to Utilities, Manufacturers, and Government

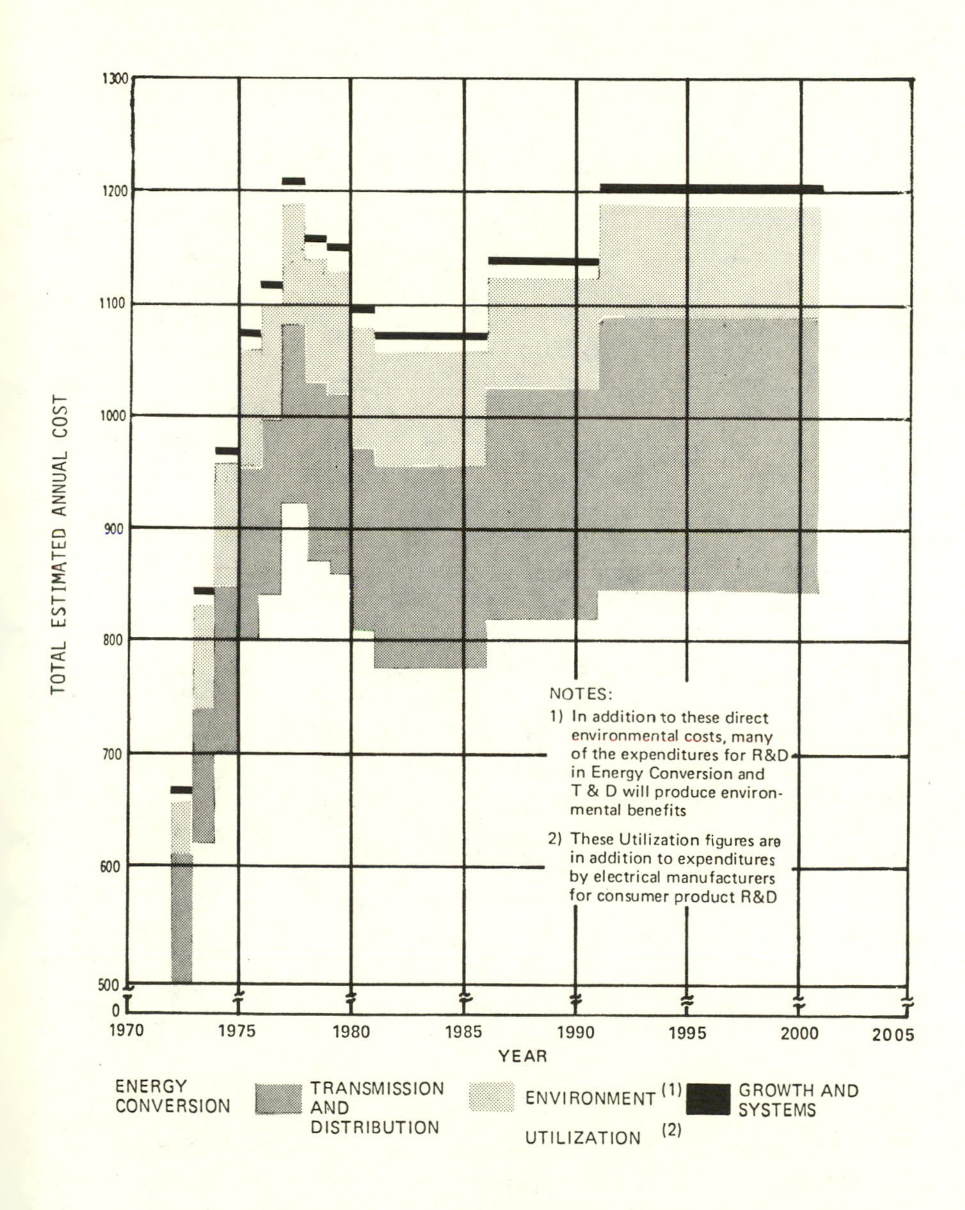

Source: ERC, 1971.

does not provide for comprehensive systems oriented R&D. . . . While we will continue to expect much from manufacturers, we cannot merely take what is given us to work with. We have an obligation to take the future more into our own hands. [ERC, 1971:5]

At approximately the same time, the Federal Power Commission (FPC) published a similar report (1973) with strikingly different conclusions. The ERC targeted only one-seventh of RD&D expenditures to address environmental concerns. While the FPC did not specify funding levels with their recommendations, they clearly emphasized environmental concerns, derived from a different set of needs criteria (see Table IX.1). Conspicuously absent from the FPC list is anything relating to systems operation (a benefit to the utilities in efficiency) or end use (a benefit to the utilities from increased demand). The manufacturers' role in the RD&D area, especially planning, was growing progressively smaller, as concerns other than generation efficiency affected the industry. Probably the largest component of manufacturer RD&D expenditures during the late 1960s to early 1970s was improving the efficiency of reactor vessel design. Being responsible to a smaller subset of society, manufacturers were not required to take as broad a view as the utilities that served a broader range of customers or the federal government whose constituency was broader still and could concentrate on the most profitable areas of RD&D.

Obviously, the differences among manufacturers', utilities', and the government's RD&D reflect the different needs of their constituencies and different capabilities. This is the informal beginning of a tacit allocation of RD&D responsibilities of the industry. Later this tacit allocation would be formally defined by contract among the RD&D participants.

In the late sixties and early seventies environmental RD&D topics related to electricity received more attention from the public, government, and industry. Government-funded research in the early sixties led to regulation and enforcement that required all aspects of the electric power system to take into account its environmental effects. Manufacturing companies began to supply generators that caused less pollution and devised retrofit equipment to reduce emissions from existing plants. Local environmental issues (i.e., plant sitings or effects on immediate biosphere) drew individual utilities further into the RD&D area.

Emphasis on environment control substantially increased the cost of generation facilities construction. For example, equipment to control sulphur dioxide emissions escalated the cost of building a coal plant in 1978 by 26 percent over the cost of building the same plant under standards which existed in 1971. Furthermore, the cost of adding equipment to reduce nitrogen oxide emissions is projected to raise the cost for the same type of plant by another 25 percent (Komanoff, 1981).

In 1973, the focus of RD&D took on a new character. The oil embargo caused the perception of an acute petroleum shortage. At the time, about 75 percent of the nation's total energy was derived from petroleum and natural gas and was used for about one-third of all of electricity production (DOE, 1980). This elevated the energy and electric supply issue to a national priority. The immediate need was for a reduction of the economy's dependence on oil.

In response to the oil crisis the National Power Survey of the Federal Power Commission (NPS-FPC), together with A. D. Little, updated in April 1971 their 1970 version of RD&D needs for the industry. It set the tone for many similarly directed reports. One major change in the policy perspective of the NPS-FPC study was that electric supply came to be viewed as part of the overall energy system. Also, RD&D tasks were subdivided by type and the appropriate funding source was identified. Finally, short-term, mid-term, and long-term goals for the electric RD&D were formulated, while considering the energy system and the national economy. Among other observations, the study noted that, in the short term, energy systems research will play the major policy role because such systems have the longest potential for conservation with the least economic loss. Mid- to long-term recommendations focused on the development of alternative technologies and the creation of an agency within government with sufficient authority to ensure these areas were funded. The government body assigned such a responsibility was the Energy Research and Development Administration (ERDA) from 1974 to 1977 and the Department of Energy (DOE) from 1977 to the present. The proposal for a coordination body in government echoed a sentiment quoted earlier from 1971 ERC report (i.e., that the RD&D task facing the industry was too large to be carried out by any of the present research efforts and that an industry group would be better able to coordinate this function than government). On this recommendation, in 1972 the Electric Power Research Institute was formed (EPRI, 1979:90).

By the mid-1970s, with the push for alternative or renewable sources of energy and electricity, a broad range of potential alternative generation technologies emerged. In 1975, ERDA, in its first analysis of national energy RD&D needs, published a report titled, *Creating Energy Futures.* The study identified five national policy goals relating to energy. These five goals addressed national security, economic health, life style diversity, world stability, and environmental protection. The study concluded that the objectives could best be attained by having a wide choice of energy and electricity supply options to draw upon. To create these options, a broad technical development program was implemented. The plan specified that "the federal government has a responsibility to undertake RD&D in those cases where there is such urgency, risks or magnitude to the effort that private industry cannot reasonably be expected to carry the entire burden." Because of the long lead times required for major new technologies to be

Table IX.1
Major Research and Development Needs in the Field of Electric Power

		Major Areas of Benefit					
Category	**Item**	*Reliability and Safety*	*Cost Reduction*	*Conservation of Fuel Resources*	*Air Quality*	*Water Quality*	*Aesthetic Values*
A. Generation Technology	1. Reduction of sulfur emission from fossil fuel-fired plants through the development of equipment for the removal of sulfur dioxide from combustion gases and of techniques for the removal of sulfur from coal before it is burned.				X		
	2. Reduction of nitrogen oxides emissions from fossil fuel-fired plants through closer control of combustion conditions.				X		
	3. Reduction of particulate emission from fossil fuel-fired plants through the continued refinement of electrostatic precipitators and other stack gas-scrubbing equipment and techniques.				X		
	4. Reduction of radioactivity releases from nuclear plants through continued refinement of plant and equipment design and operating techniques.				X	X	
	5. Reduction of warm-water discharge from thermal power plants in general through advances in the design and use of cooling ponds and towers and also through the development of new generating technologies (see below).					X	

6. Minimization of the effect of warm-water discharge through the development of improved cooling water intake and discharge structures and operating techniques.					X
7. Continued scale up of the size of generating units through advances in equipment and plant design.		X			
8. Improvement in plant dependability through the development and use of more rigorous quality control techniques and procedures, the gathering and analysis of equipment reliability statistics, more sophisticated maintenance programming, and the like.	X				
9. Development of new techniques, such as magnetohydrodynamics (MHD), which will make it possible to achieve higher thermal conversion efficiencies and reduce waste heat rejection.			X	X	X
10. Continued nuclear power safety research including study of accident mechanisms and testing of engineered safeguards.	X				
11. Expeditious development of breeder reactors for use in nuclear power plants.		X	X		X
12. Continued research on fuel cells.			X	X	X
13. Research on controlled fusion.		X	X		
14. Investigation of the possibilities of new approaches to tapping solar energy.			X	X	X

Table IX.1 *(continued)*

Category	Item	Major Areas of Benefit					
		Reliability and Safety	*Cost Reduction*	*Conservation of Fuel Resources*	*Air Quality*	*Water Quality*	*Aesthetic Values*
B. Transmission Technology	1. Improvements in equipment reliability through advances in materials technology and engineering design, and the establishment of needed testing facilities.	X					
	2. Continued improvement in the aesthetics of transmission tower design.						X
	3. Continued advances in extra-high-voltage (EHV) transmission (from 500 kilowatts to 765 kilowatts and above).		X				
	4. Continued development of new transmission techniques, such as the use of superconducting cryogenic cable and direct current transmission methods.		X				
	5. Development of reliable economic methods for undergrounding high-voltage transmission lines.	X					X
C. Distribution Technology	1. Improvement in undergrounding of distribution lines.						X

	2. Improvement in the automation of distribution facilities.		X			
D. Environmental Sciences	1. Continued study and monitoring of individual ecosystems—for example, a particular lake or estuary.					X
	2. Development of improved techniques and devices for tracing the path of effluents from power operations, including constituents of stack gases from fossil fuel-fired plants and radioactivity releases from nuclear plants.				X	
	3. Continued study of the short- and long-term biological effects of subtle changes in environmental conditions resulting from power operations.	X			X	X
	4. Establishment of data gathering and analysis centers and technical information dissemination services in specialized areas of the environmental effects field.				X	X
	5. Continued research in the environmental ''ologies'' (seismology, meteorology, geology, hydrology, etc.).	X			X	X
	6. Development, through the application of the knowledge gained from the foregoing, of improved air and water quality standards and improved environmental protection criteria in general.				X	X

Source: Federal Power Commission, 1973.

developed and the associated commercial uncertainties and financial risk, federal involvement may be necessary "to accelerate progress beyond normal commercial capability" (ERDA, 1975).

The program extended the federal role in supporting RD&D in the electric power industry. The government has traditionally supported basic and applied research. Equipment manufacturers had traditionally completed the remaining stages to implementation with the utilities participating on smaller projects. Under the expanded ERDA role, most large-scale demonstration and commercialization projects were federally funded. The manufacturers' and utilities' RD&D activities were coordinated to support the implementation of the technologies after the demonstration stage.

The many joint projects among the ERDA and DOE, the electric utilities (collectively represented by EPRI), and the manufacturers are evidence of the effort to coordinate RD&D. Over 100 research projects valued at $800 million were conducted jointly under EPRI and ERDA up to 1976. In 1979, EPRI signed an agreement with DOE that formalized and expanded the existing level of cooperation between the two. From 1973 to 1981, the value of joint projects climbed from $106 million to at least $1.2 billion (EPRI, 1981).

This atmosphere of cooperation progressed until 1981 when there was a major change in DOE RD&D policy because of financial constraints faced by the government. Based on the assumption that the technologies that had been moving toward implementation were now far enough along for the commercial sector to assume the risk, DOE withdrew funding. Table IX.2 shows DOE RD&D budgets for 1981, 1982, and estimated 1983, 1984, and 1985. Funding has been cut from almost all DOE areas, but electrically related technologies have been particularly hard hit. The difference between actual funding levels in 1981 and projected funding levels in 1983 is great. The expected annual decreases for electric power related RD&D funding by program are approximately:

Solar	– $420 million
Geothermal	– $145 million
Nuclear fission	– $250 million
Electric energy systems	– $ 20 million
Energy storage	– $ 45 million
Hydropower	– $ 10 million

EPRI and the General Accounting Office (GAO) disagree with the DOE supposition that these cutbacks can be picked up by industry. EPRI's total budget for 1981 was about $300 million, the highest level to date (EPRI, 1981). Total energy RD&D funded by the private sector has been estimated by the Industrial Research Institute Corporation at $2.4 billion in 1979. RD&D funding cutbacks are likely to cause a great deal of change in RD&D planning. EPRI has identified twelve projects not likely to be completed

Table IX.2
National Need: Energy, 1981–1985 (Functional Code 270 in millions of dollars)

Major Missions and Programs	1981	1982	1983	1984	1985
	Actual			*Estimate*	
BUDGET AUTHORITY					
Energy Supply:					
Research and Development:					
Existing law	3,808	2,755	2,010	2,099	2,101
Proposed legislation			185		
Subtotal, Research and Development	3,808	2,755	2,195	2,099	2,101
Direct Production (net):					
Uranium enrichment	422	1		24	16
Petroleum reserves	−655	−908	−848	−806	−781
Power marketing	− 96	1,841	2,071	1,886	1,892
Incentives—Nonconventional Fuel Production	−1,274				
Subtotal, Energy Supply	2,224	3,689	3,418	3,203	3,228
Energy Conservation	728	163	27	19	4
Emergency Energy Preparedness	2,791	191	242	633	235
Energy Information, Policy, and Regulation:					
Existing Law	1,089	871	772	809	764
Proposed Legislation			− 60	− 60	− 60
Subtotal, Energy Information, Policy, and Regulation	1,089	871	712	749	704
Deductions for Offsetting Receipts	− 62	− 69	− 69	− 69	− 69
Total, Budget Authority	6,789	4,846	4,330	4,536	4,102

Source: DOE budget for fiscal year 1983.

because of the DOE policy change (see Table IX.3). Additionally, the GAO has looked at six different cases and does not believe that three of these (atmospheric fluidized combustion, pressurized fluidized-bed combustion, and fuel cells) will make it to market as DOE predicts. Wind and solar RD&D will continue but only if tax incentives stay in effect. The GAO

Table IX.3
Joint Projects with DOE at Risk Because of Changing Federal Funding Policies (in millions of dollars)

Project	EPRI	Other	Total
Wilsonville Solvent Refined Coal (SRC)	27	50	77
Exxon Donor Solvent	47	332	379
H-Coal Pilot Plant	23	518	541
Heber Geothermal Demo	13	113	126
Dry Cooling Pilot Plant	12	4	16
Plume Model Validation	24	7	31
4.8 Megawatt Fuel Cell Demo	19	72	91
Battery Development	37	41	78
Battery Energy Storage Test (BEST) Facility	21	15	36
Molten Carbonate Fuel Cell	13	68	81
Spent Fuel Storage	7	37	44
Hydrogen Water Chemistry for Boiling Water Reactor (BWR)	10	5	15

Source: EPRI, 1981.

believes that municipal solid waste fuel for electricity had reached the commercialization point.

In line with the federal cutbacks, the utilities, collectively and individually, have altered their RD&D strategies. EPRI has cut back on planned RD&D because incoming revenue has been less than projected. Individually, utilities have cut back on their own RD&D programs.

EPRI's allocation of funds by research areas has shifted to a policy that would reduce utilities' capital expenditures and extend existing plant and equipment life. Both of these areas were targeted because of member utilities' reduced financial ability to expand capacity if this became needed. A major shift has occurred in EPRI's plans for the timing of impact of their research projects. Short-term projects have been expanded to 69 percent of the budget (EPRI, 1981).

With the reduced role of the government in energy and specifically electric power RD&D, EPRI and the equipment manufacturers will be expanding their roles in this area. Currently, EPRI is already addressing this issue by funding a few of the many demonstration projects that are in danger (EPRI, 1981).

C. TYPES OF RD&D PROGRAMS

One of the purposes of this chapter is to discuss appropriate roles for the various organizations involved in RD&D under possible future scenarios. The specific role each organization plays depends on its internal goals, the environment in which it operates, and the role other organizations choose to play. Traditionally, the United States depended on the private sector to conduct the bulk of electric power research, both in the fields of supply and utilization. However, one would reasonably expect the free market approach to falter as: (1) the ability of the companies involved in RD&D to reap the profits of their investments is diminished; (2) the size of the risk increases; (3) the probability of successful application decreases; and (4) the time to application is lengthened. Since many of these problems have become increasingly serious since World War II, it is not surprising that the federal government has played an increasingly important role. It is also not surprising that electric utilities have seen fit to combine their research efforts (most dramatically in the formation of EPRI), given the tradition of utilities freely sharing technical developments.

In discussing possible RD&D programs, it is useful to form a typology of RD&D program areas:

A. *Improving present systems.* Programs designed to improve the efficiency, productivity, and operational life of technologies widely used at the present time to generate, transmit, and distribute electric power (e.g., more efficient coal-fired plants, better safety controls on nuclear plants, higher voltage transmission lines).

B. *Permitting increase in capacity.* Programs designed to support major expansion of facilities utilizing present technologies (e.g., improved methods for removing small particulates from coal plant emissions, development of suitable nuclear waste storage facilities, improvements in coal slurry transportation techniques). Programs of this type would probably involve primarily environmental protection and fuel assurance projects.

C. *New capacity.* Programs designed to provide generation, transmission, and distribution technologies with characteristics markedly different from presently utilized techniques (e.g., fusion power, solar satellites, photovoltaic cells). In general, there would appear to be three types of programs included in this category: (a) very large and expensive systems such as the first two examples; (b) relatively small systems such as the last example; and (c) radical new uses of present type systems such as nuclear power parks.

D. *Fundamental science and engineering.* Programs designed to develop a foundation for major breakthroughs in power generation, transmission, and distribution in the future (e.g., organic superconducting material, novel electric power storage techniques, microwave power transmission). This type of program differs from those listed in the previous category in that the latter involves programs presently being conducted specifically to improve electric power generation. The programs included in this category involve more fundamental research that will provide a store of scientific and engineering knowledge for use in the later stages of the three-decade period being considered by this assessment.

E. *Transfer programs.* Programs to monitor RD&D programs in related technical areas and to examine means of integrating successful results into electric power development programs (e.g., use of space program RD&D to expedite solar satellite development, use of biotechnology advances to improve power storage systems). These programs would involve more than increased monitoring efforts but would require programs to conduct RD&D to test the practicality of technology transfer and to determine effective means of accomplishing that transfer.

F. *Improving end use technologies.* Programs to develop and improve practical end use applications of electric power (e.g., wider use of electric heating in the metals industries, use of electric power drying in agriculture, use of electricity to speed chemical and biological reactions). End use application RD&D programs could, of course, also be categorized into the five types listed above. However, in this assessment the emphasis is not on how electric power will be used but on the impact of that use on the supply system. Therefore, all types of end use programs are included in this category.
(KH:IX-1.)

It is essential that any viable national electric power RD&D agenda include elements of each of these six RD&D programs. However, the relative mix of programs and the overall size of the efforts will depend on the way policy makers see the nation's electric power needs developing. For example, if it appears that economic growth in the United States will be extremely limited, RD&D programs might be conducted at a minimal level with efforts concentrated in improving present systems. On the other hand, if it is believed that the nation is headed toward a high-energy, high-electric-power, high-technology future, a large part of the RD&D efforts might be best devoted to programs creating new capacity.

As one examines the list of program types, it becomes evident that the appropriate role of institutions differs between categories. Programs in the first category have traditionally been sponsored by private industry, equipment manufacturers, utilities, and, more recently, EPRI. (Public utilities such as the Tennessee Valley Authority and the Bonneville Power Authority have also contributed to electric power supply RD&D.) It would seem reasonable to expect that this situation will continue, unless the nature of the utility industry itself changes dramatically. Conversely, in recent years the federal government has had a major role in both the finance and conduct of the third category of RD&D programs and it is probable that this will also continue.

D. TYPES OF RD&D APPROPRIATE FOR EACH SCENARIO

Some level of the six RD&D categories discussed in the previous section is appropriate for the six electric power scenarios. It is important to note, however, that different mixes of RD&D programs will be appropriate for each scenario. This section will stress such differences in RD&D policy. A more detailed presentation of the scenarios as they relate to RD&D policy is available in Appendix G of the NSF report.

"Average Future" Scenario. The situation described by "Average Future" represents essentially an extension of past experience. The Gross

National Product grows at a moderate rate, as does energy use, population, and percentage of energy used as electricity. The population ages, decentralizes, and continues to shift toward the Sunbelt. The bulk of new electric power generation requirements are met primarily by the addition of large coal plants. A considerable amount of increased power requirements will be met by the completion of nuclear power plants now being planned. Existing oil and gas power plants are phased out or used primarily for peaking power needs. Additional transmission and distribution facilities are required to accommodate shifts in population both between regions and between centralized areas and more remote areas. Increased power requirements will increase the potential for environmental abuse.

Although the "Average Future" Scenario will provide the nation with problems in the electric power area, most of these problems involve change in scope of technology rather than its nature. That is, technical requirements will lie in improving existing technologies rather than in developing new ones. The federal government may also be motivated to contribute to solutions of some of the problems caused by increased power requirements (e.g., acid rain, atmospheric carbon dioxide buildup, nuclear waste handling, etc.).

Although little increase in per capita power use is postulated in this scenario, the increased movement of energy usage from oil and gas to coal and nuclear power will probably encourage increased use of electric power, particularly in industrial areas. Government and power industry policies might be directed to encourage innovation in these areas.

"Nuclear Resurgence" Scenario. In this scenario, a major increase in electric power usage is considered. Adequate solutions must be found for disposal of radioactive waste materials, both because of the increased dependence on nuclear power postulated in this scenario and because of the implied continued increase in its importance beyond the year 2010. Environmental problems associated with coal usage will also be larger in the "Nuclear Resurgence" Scenario than in the "Average Future" Scenario and, hence, long-term solutions will be even more important. Given the increase in electric power usage, both total and per person, it seems probable that new and improved end use technology will be needed as interest in this area can be considered to be high.

"Mega-plant" Scenario. This scenario envisions both a major increase in electric power usage and the development of new types of generation equipment to supply that need. Comparable improvements will be necessary in transmission capabilities if the new facilities are considerably larger than present plants. Although the percentage of electric power generated in 2010 by the new type of facilities is small, the implications are that the fraction will grow rapidly after that time. The development of these new types of facilities will only be undertaken if they represent major technical advantages over present systems—cheaper and more abundant fuel sources, lower operating costs, markedly increased efficiencies, and so forth. Radical

developments of this type are typically long term, expensive, and fraught with risks.

Because of the radical nature of the technical advances associated with this scenario, it will be important that a very solid scientific foundation be established and that careful monitoring of parallel developments be maintained. As in the "Nuclear Resurgence" Scenario, the high usage of electric power will probably be paralleled by increased interest in new advances in end use technologies.

"Small Coal Plants" Scenario. This scenario envisions a considerably larger amount of electric power being used than in the "Average Future" Scenario, although the percentage of electric power is actually lower. In this case, additional power would be provided primarily by increasing the number of relatively small coal plants. The scenario also calls for considerably greater cogeneration and generation by third parties. The increased dependence on small coal plants will emphasize the development of coal generation technologies. One would assume that small coal plants would have to be developed that would have efficiencies similar to those of existing larger coal plants. Because the scenario calls for less dependence on electric power than most of the other scenarios, it is envisioned that little research would be done for developing generation capabilities of an entirely new type. Moreover, the low percentage of the total energy used to produce electric power would indicate a low priority for research in developing a new electric end use technology.

"Post-industrial" Scenario. In this scenario, a relatively high percentage of dependence on electric power is assumed, but total energy usage is low. Therefore, total use of electricity is somewhat smaller than in the "Average Future" Scenario. Moreover, a large fraction of electric power requirements will be furnished by dispersed generating facilities and a very large amount produced by cogeneration and third party generators. This scenario would seem to indicate a relatively low importance for improvement on present types of generating capabilities. On the other hand, the need for developing new types of generating, storage, and distribution equipment would be very high. Hence, research into new characteristic equipment, improvements in basic science, and a close monitoring of these developments in related technical areas would have a high priority. In this case, the emphasis on equipment with new characteristics would be somewhat different from the "Mega-plant" Scenario. Instead of developing large power plants, more emphasis would be placed on facilities that could generate power locally. In spite of the low level of electric usage in this scenario, one would assume that research in end usage would be important because the nature of the economy would require important new technical advances.

"Economic Malaise" Scenario. Since this scenario postulates neither a large increase in the total amount of energy generated nor in the fraction of

the total energy usage provided by electric power, there would be relatively little need for substantial research activity. Efforts in research would be best directed toward improvements to the present system through increasing product lifetimes and efficiency. To monitor developments closely in other technical and social fields, it would be useful, however, to continue research in the basic sciences in order to lay a foundation for the expanded power needs postulated after 2010.

E. ENVIRONMENTS CONDUCIVE TO PARTICULAR TYPES OF RD&D

This section presents government policies that encourage particular types of RD&D in the electric power industry. The discussion of such policies is not meant to be comprehensive but rather to provide a list of examples demonstrating the influence policy decisions have on RD&D.

Improvements to present technologies. In the United States, improvements to present technologies have been accomplished primarily by the private sector. Hence, the key role of the federal and state governments would be to establish an environment that encourages innovation in the private sector. Probably the most effective way of accomplishing this is to allow utilities to retain additional profits from technical innovation successes. This could be accomplished by including premiums in rate-setting procedures. Since rate-setting procedures typically lie in the domain of state agencies, federal government action would probably be limited to development of guidelines or specific requirements such as those indicated in the Public Utility Regulatory Policy Administration (PURPA) legislation.

Increased capability of present system. In this area, the federal government, private utilities, and equipment manufacturers have all played major roles in recent years. Since solution of these problems does not directly relate to utility profits, RD&D would seem to be primarily motivated by such tools as laws and regulations, tax incentives, and federal support of RD&D, either in government laboratories or at private facilities.

New characteristics. In examining the climate conducive to the development of generating sources with basically new characteristics, two types of programs need to be kept in mind. One is very large plants, such as fusion power, solar satellites, advanced fission systems, and so forth. The size of these projects and the inherent risks involved are too large for private utilities to undertake alone. Hence, a large measure of government support of the basic research involved in such projects is desirable. This can take the form of direct government support of research, of government funding industry research, or of government subsidizing research by private organizations. Experience indicates that large-scale government support is necessary at least through the demonstration plant stage, and subsidies during the construction of early commercial plant facilities also may be desirable. Since even by government standards these projects can be very expensive, the government must select certain systems to fund and must decide for what period support should be continued and to what development stage parallel research should be supported.

The other general types of new characteristics programs are those of a much smaller size, such as photovoltaic systems. In these programs the government's best role may not lie in direct support but rather in providing an environment that encourages the development of new types of technologies. Experience has shown that often new approaches to problems, new types of technologies, and radical technological breakthroughs are developed by people outside the traditional industry structure. In this case, the primary role of government would be to establish circumstances by which new entries into the electric power field are encouraged. Encouragements in this case might include such items as tax incentives, guaranteed buys, relaxation of certain restrictive regulations, guaranteed loans, and so forth (KH:IX-2).

Basic science research. Increasingly over the last two decades, the federal government has undertaken support of basic research projects. The reasons for this appear to be that our modern economic system makes it difficult for innovators to realize the full benefits from their innovations; that the social value of these inventions is often greater than private advantages; and that many of these projects are too large to be undertaken by private concerns, given the length of time of the development and the inherent risks involved. Encouragement by the government of basic science development would seem to be primarily enhanced by direct government subsidies to universities and by support of research projects in government laboratories as well as in industrial research centers (KH:IX-3).

Technology transfer. Continuous monitoring of research developments in related technical areas is done by both industry and government. However, activities must be supported that encourage continuous technology transfer, particularly between government laboratories and universities and industrial groups involved in relevant research.

End uses. Two groups are particularly interested in the development of new end use technology for electric power. First are the power companies that seek both new markets for their product and new profit opportunities through diversification. The second group is producers and users of end use equipment. It appears that the biggest opportunities lie in new industrial uses in electric power, given the relative saturation of home and commercial use of electricity. To encourage end product development, policies might be developed that allow the utilities to gain profit from the innovations themselves, as well as from the increased use of electric power. The development of new end use products in the electric power industry will probably be handled by non-regulated subsidiaries that compete in non-regulated markets. With regard to the development of new end use products by other parties, the general rules for encouragement of innovation listed earlier (that is, permitting recovery of profits to justify the risks taken in RD&D efforts, special consideration for new types of industries beginning to work in the area, and similar government policies) would appear desirable (KH:IX-4).

F. SUMMARY

This chapter reviews the background of the nation's electric power RD&D. It shows how RD&D responsibility has shifted away from utilities to suppliers and the federal government. The government's role in RD&D has increased as concerned citizens and energy officials realized the importance of an effective RD&D program to meet anticipated demand for electric power. In addition, as the RD&D effort has increasingly required enormous financial, human, and technical resources, it has become apparent that individual utilities cannot afford an adequate level of support by themselves.

Each of the six alternate scenarios presented would require a different mix of RD&D programs. However, since no one knows which of the futures, if any, will in fact come about, a proper RD&D program must be broad enough to serve any of them adequately. In final analysis, it is apparent that whatever programs energy officials choose to pursue, RD&D must serve as the foundation for the development of an efficient and effective electric power system.

BIBLIOGRAPHY

Department of Energy (DOE). *Annual Report to Congress.* Washington, D.C.: U.S. Government Printing Office, 1980.

Electric Power Research Institute (EPRI). *EPRI Journal.* Palo Alto, Calif.: EPRI, 1979.

Electric Power Research Institute. *1981 Annual Report.* Palo Alto, Calif.: EPRI, 1981.

Electric Power Research Institute. *1982-1986 Overview and Strategy.* Palo Alto, Calif.: EPRI, 1981.

Electric Research Council (ERC). *Electric Utilities Industry Research and Development Goals through the Year 2000.* In *Report of the R&D Goals Task Force.* New York: ERC Publishing Number 1-71, 1971.

Energy Research and Development Administration (ERDA). *Creating Energy Futures.* Washington, D.C.: U.S. Government Printing Office, 1975.

Federal Power Commission. *National Power Survey.* Washington, D.C.: U.S. Government Printing Office, 1973.

Komanoff, Charles. *Power Plant Cost Escalation.* New York: Van Nostrand Reinhold, 1981.

National Electric Reliability Study (NERS): *Final Report.* DOE/EP-0004. Washington, D.C.: U.S. Department of Energy, April. 1981.

CHAPTER X

Historical Perspective on Electrification

A. OVERVIEW

Although electricity's beginnings can be traced to an accident—when Luigi Galvani, in 1791, in the course of experimenting with frog muscles fortuitously produced the first electric battery—the subsequent development of electricity was deliberately plotted by human effort and ingenuity. Today, giant steam turbines fueled by coal, oil, gas, and atomic fission produce the gigawatts of energy that drive our machines, light our homes, and run our appliances—a tremendous leap, indeed, from the first batteries that were developed soon after Galvani's experiments. Of course, it took many men of genius to think up newer and better ways of generating electric current and putting it to practical use. For example, Michael Faraday's invention of the generator, Hippolyte Pixii's contribution of the commutator (making the generation of direct current possible), Joseph Henry's work with electric motors, Sir Humphrey Davy's demonstration of the first electric light, and Thomas Edison's feat in creating a practical system for incandescent lighting are but a few of the critical inventions that have shaped our present electrotechnical society. This chapter traces the important technical, industrial, and governmental developments that evolved into today's electric power industry. Its purpose is to provide a historical foundation from which to better project future electric industry policy directions.

B. EARLY DEVELOPMENT

The development of generating capability went hand in hand with the invention of practical applications for the new power. Charles Francis Brush, the entrepreneurial pioneer of arc lighting (the earliest form of the electric light to be developed) also formed the first central electrical generation plant in the world. In 1879, Brush and a partner created the California Light Company of San Francisco. The plant had two of the improved generators invented by Brush and supplied 22 arc lamps. A price of $10 per week per light was charged to the affluent owners of this latest symbol of progress. Soon after, the plant began to provide power for street arc lighting. Brush also established power stations for his system in New York, Boston, and Philadelphia (Eisenmann, 1967:53-67).

Edison. Electrical lighting's major competition was gas lighting, which, although hazy and unsteady, was cheap and readily available by way of the existing grid work of pipes supplied by gas utilities. When Thomas Edison set out to displace gas lighting with electricity, he dismissed the adaptability of arc lighting for universal residential and commercial uses. Meeting this challenge, Edison and his staff developed the components of an entire incandescent lighting system for which the light bulb served as the keystone. They created an improved dynamo for constant-voltage, low-current, direct current generation with high-resistance lamps hooked in parallel—a major breakthrough in generating technique and wiring that permitted an equal and independent power supply to each bulb. The group also developed an underground conductor network, safety fuses, new insulating materials, electric meters, and light sockets with on and off switches.

After developing the technology for his electricity system, Edison turned his attention toward its production and implementation. September 4, 1882, marked the opening of Edison's first central electricity generation plant (Pearl Street Station) in New York City. The plant supplied electricity to its customers' lamps in lower Manhattan, an area which included personal residences, factories, the New York Stock Exchange, and the city's most renowned newspaper offices. When the lights went on in this area, the most influential merchants, bankers and editors in the nation were loud in their praise of Edison's new system (Hughes, 1976).

The manufacturing enterprises established by Edison—the Edison Machine Works which built the generators, the Edison Electric Tube Company which made the underground conductors, the Bergmann Company which produced electric fixtures, and the Edison Lamp Works which mass-produced light bulbs—merged in 1890 to become the Edison General Electric Company. Two years later, a further merger with the Thompson-Houston Companies (originally an arc lighting systems group that expanded into electric traction equipment and incandescent lighting systems) created the General Electric Company of today.

Although Edison's direct current, low-voltage central plant system spread throughout the United States and the world within the first decade of its existence, the growth was much slower than Edison had hoped. The two primary reasons for the low rate of adoption were that incandescent lighting still cost almost twice as much as gas lighting and the cost of electricity for the new large and small direct current motors developed for industrial use compared unfavorably with non-electric sources of power (such as steam and hydropower) (Passer, 1953:105-17). High costs for electricity were attributable to the low transmission voltages required for direct current circuit generation. Such voltages severely limited the area that one central station could economically serve to a radius of about two miles. To achieve the full advantages of central station generation, higher transmission voltages were in order. Technically, however, such increased voltages required alternating current instead of direct current. This situation posed several dilemmas. First, alternating current motors had not as yet been developed; second, conversion methods from alternating to direct current did not exist; finally, and most importantly, there was no means for stepping down the high transmitted voltages to the lower values required for incandescent lighting.

Westinghouse. In 1881, George Westinghouse, an already established engineer-entrepreneur in the railroad switching equipment business, became interested in the commercial potential of electricity. Unlike Edison, who was vehemently opposed to alternating current on the grounds of its being too dangerous, Westinghouse almost immediately perceived the vast potential of alternating current systems for lighting as well as power applications. Fledgling alternating current systems had already begun to appear in Europe, and Westinghouse was quick to take advantage of existing breakthroughs in alternating current technology. First, he acquired the British patents on the Gaulard and Gibbs transformer—a device that could convert high transmission voltages to lower end use requirements. Westinghouse then supported William Stanley, another end-of-the-century inventor-entrepreneur, in modifying transformer design and circuitry, improving alternators, and establishing his own alternating current incandescent lighting system using the Stanley version of the electric light bulb. Stanley also oversaw the first successful experimental installation of the Westinghouse system at Great Barrington, Massachusetts, in 1886. That same year, the first commercial alternating current lighting system was in operation in Buffalo, New York, and soon after orders for 25 new installations were taken (Prout, 1921:87-112).

Nevertheless, two elements were still missing to make central station alternating current generation and transmission commercially successful. No metering system for it existed—the Edison chemical meter worked only for direct current—and, more significantly, industrial end use of the alternating current system had to await the invention of alternating current motors. O. B. Schallenberger, a Westinghouse engineer, solved the meter problem almost by accident in 1887. He had dropped a small spring on a magnetic coil connected to alternating current voltage and noticed that the spring was revolving slowly. Here was the principle on which a revolving needle meter could be constructed to measure alternating current and here, too, as Schallenberger and Westinghouse were quick to perceive, was a key concept for the development of an alternating current motor—an alternating current inducing motion by way of rotating magnetic fields (Passer, 1953). But, as happens so often in the history of great ideas, another inventor-engineer, Nikola Tesla, had already begun work on the same principle five years before in Hungary. In 1888, he emigrated to New York and took out patents on alternating current polyphase systems complete with motors, generators, and transformers.

Westinghouse, learning of these patents, negotiated for their purchase and also the talents of Tesla. Still, several years of work by Tesla and the Westinghouse team of engineers were necessary before alternating current power systems became practical. They found that the whole alternating current system then in use for incandescent lighting had to be thoroughly revised to make it applicable to the polyphase power generation necessary for induction motors. The engineers also had to decide on the most suitable frequency and phase for power circuits that would work for both lighting and industrial use. By 1892, however, Westinghouse was ready. In a brilliant display at the Chicago World's Fair, Westinghouse showed the world what a completely flexible alternating current two-phase power system could do. The event marked a turning point for the entire electric power and light industry—a turning point whose force was fully realized three years later with the construction of the Niagara Falls power facility.

Niagara Falls facility. Using Niagara Falls' full 6 million horsepower potential of 270,000 cubic feet of water cascading every second over a 160-foot-high cliff had been a dream for decades, a vision whose realization had to await the developments in electricity generation, transmission, and utilization that were occurring in the latter part of the nineteenth century. The major problem with using the falls' energy potential had always been a superabundance of power relative to the amount of land available to permit local use by belt or rope transmission from water wheels. By the end of the Civil War a few industries, mainly paper and pulp mills, had located near the falls, utilizing a mile-long hydraulic canal that carried water from above the cataract to the river's banks below. These mills, each with its own water wheel, used only a fraction of the available 200-foot head, altogether accounting for no more than 10,000 horsepower. When, in 1885, the New York State Niagara Falls Reservation was established as an international park in cooperation with Canada, it absorbed much of the already limited land suitable for industrial and powerhouse use. Hence, the problem of profitably exploiting the potential of the falls became even more acute. The solution lay in somehow transferring bulk power over distances longer than previously possible and especially in transmitting this power to 20-mile distant Buffalo. In 1889, the Cataract Construction Company was established under the leadership of banker Edward Dean Adams and backed by representatives of the leading investment houses in the country—among them J. P. Morgan and William Vanderbilt—for just that purpose. The Cataract Company took over an older scheme for large-scale Niagara power development first proposed in 1886 by Thomas Evershed, a civil engineer employed on the Erie Canal (Adams, 1927). Evershed envisioned twelve hydraulic canals carrying water to a large number of vertical shafts that in turn dumped into a single discharge tunnel, some two and one-half miles long, and pitched to strike the river below the falls. Each of the shafts, about 250 in number, would house a 500-horsepower turbine. Mills and factories would line the canals some distance above the falls. After months of study, this idea was abandoned by the Cataract Company because of the huge excavation costs involved. The plan was modified to a shorter discharge tunnel linked with a short inlet canal and a single large wheel pit that could handle enough water to generate about 110,000 horsepower. In essence, that meant a single center for water power generation instead of decentralized production. The basic problem, how to transmit this power to local industry and to Buffalo, became even more critical under the new centralized scheme. Experts such as Edison, Westinghouse, and Frank Sprague, the developer of electric traction for trolleys, elevators, and subways, were consulted from the project's beginning to consider the problem of power generation and transmission at the Niagara site. None of their suggestions seemed adequate to Adams. After touring hydropower sites in Switzerland, then at the forefront of this kind of development, Adams decided to sponsor an international design competition based on the idea of a single central power station. A Swiss firm was awarded first prize for hydraulic turbines, but there were no first prizes in the most crucial category of all—transmission. Significant, however, for the rapidly advancing state of the art of electricity, half of the fourteen proposals were now for electric transmission as opposed to other methods. Only two proposals were for alternating current transmission, one of which specified polyphase current. By 1892, although company officials could not agree on any one specific electrical method, central electric power generation and transmission became the choice (*EPRI Journal*, 1979).

Construction was started on the discharge tunnel, the inlet canal and the wheel pit for the powerhouse. Preliminary designs for 5,000-horsepower water turbines were being prepared by several Swiss firms, and that in turn swayed opinion toward alternating current power generation. Construction of reliable direct current generators, sufficiently large to work from such gigantic turbines, would have been very difficult compared to making equivalent alternators. Also, the successful transmission of 30,000-volt alternating current power over 110 miles at 77 percent efficiency in Germany by an Oerlikon engineer named C.E.C. Brown (later the founder of Brown-Boveri) and the Westinghouse alternating current successes at Telluride, Colorado, and later at Portland, Oregon, began to weigh the scales heavily in favor of alternating current. However, no definite commitments were made. Instead, the Cataract Company contacted three Swiss electrical manufacturers along with three American firms: Westinghouse, Edison General Electric, and Thompson-Houston (the last two had not as yet merged to become the General Electric Company) to come up with solutions. By late 1892, the General Electric Company and Westinghouse Company presented alternate plans for electrification of the falls. (The Swiss firms were by now out of the competition, mainly because the Cataract Company preferred to have the work performed by an American company.) Charles P. Steinmetz, a political refugee from Bismarck's Germany, had just joined the General Electric research team in time to contribute his knowledge on alternating current transmission systems, transformers, and motor design to the Niagara project. The Westinghouse engineers, however, were too far in the lead in polyphase transmission and induction for General Electric to be awarded the final contract for building the first three generators. In 1893, Westinghouse received the contract for the generators, while General Electric was awarded contracts for the transformers, the transmission line to Buffalo, and the substation there. The Westinghouse generator, after many problems caused by the design of the 5,000-horsepower turbines, was finally constructed for a 25 cycle per second, two-phase alternating current system, generated at 2,200 volts but stepped up for transmission. The central station at Niagara Falls began to provide power for local use in 1895 and one year later for Buffalo. With the exception of two-phase current, the Westinghouse system of generation, transmission, and induction end use remained the standard for electric power for many years. By the time of the Buffalo installation, it was clear that three-phase current was superior for steady transmission. The invention of the phase transformer soon after permitted relatively easy conversion from one phase to the other.

The fierce competition between Westinghouse and General Electric over the Niagara contracts was but one of many battles engaged in by the two emerging giants. The companies were constantly filing suits against each other over infringement of patent rights, and the customers of both were beginning to get concerned over patent guarantees on their orders. Each company excelled in different areas. Westinghouse, with its Tesla and Stanley patents, was far ahead in alternating current polyphase systems, whereas General Electric, thanks to Edison, dominated in incandescent bulbs and direct current systems. Some of GE's strongest patents were in running electric traction equipment for the then burgeoning trolley, elevated, and subway lines. In 1896, the patent deadlock was resolved by a cross-patent agreement between the two companies. Each company could now provide a full line of quality equipment in all areas of electric use without concern over infringements, and customer unease was put to rest (Passer, 1953).

The Niagara facility continued to expand considerably beyond the 1895 installation. Ironically, it was not caused by demand for more electricity from Buffalo (penetration of that market was never very substantial) but rather growing local use. The electrochemical industries (especially plants for the production of aluminum and carborundum) were just coming into existence. A number of electrometallurgical plants were built near the falls within a year of Niagara's first power output in order to benefit from the low-cost source of electricity. These plants used electricity directly—not just to drive motors—for their refining processes. Not only did the original Cataract Company add generating capacity to the first powerhouse, but the company's successor added a second power station. By 1904, the Niagara facility finally delivered the full 110,000 horsepower originally projected—the two powerhouses running 21 5,000 horsepower turbines. In addition, as early as 1895 and 1896, the company that controlled the original surface canal that delivered hydraulic power to mills generated an additional 34,000 horsepower. This was accomplished from a central station utilizing fifteen water turbines located in pits at the point where the canal joined the river. Most of this power also went to the new electrochemical plants (Adams, 1927).

The Niagara facility had a significant impact on the development of electrical power generation. Large central stations had won out over individual power sources, and alternating current was shown to be superior to direct current. Furthermore, the Niagara installation made it clear that the financial community, rather than individual inventor-entrepreneurs, would be the dominant decision maker for determining large-scale technological innovation for the market place.

By the beginning of the twentieth century, the separation between the purveyors of electricity, the utilities, and the manufacturer-vendors of electrical equipment was firmly established. The Niagara facility had made it clear that the huge capital outlays required for large-scale installations imposed the ultimate choice of suitable equipment on the utility and that the manufacturer's innovations had to be tailored to the marketplace. The utilities' role was to determine and to create the demand for electricity—the manufacturers in turn had to provide the most innovative, cost-efficient equipment their R&D departments could create. The manufacturers, like the utilities, had a segmented market: power and lighting companies supplied the residential, commercial, and the industrial sector, while electrical equipment was required at both ends of the power line.

C. REGIONAL ELECTRICAL DEVELOPMENT

Soon after the Niagara project, the interplay between electrical demand and supply led to several significant technical developments. Three-phase transmission replaced two-phase when it became clear that not only were copper costs lower, but that current transmission was steadier. Frequencies were standardized to just two values (60 cycles per second for lighting and

25 cycles per second for heavy machinery) and eventually to only one (60 cycles per second). This facilitated uniformity in equipment standards that not only permitted mass design and production at significant savings, but also provided ready interconnection of power systems. Phase transformers and rotary alternating to direct current converters could be used to adapt already established systems.

West. Historically, the West Coast holds a place in the forefront of electrical development. The explosive growth of California in the mid-nineteenth century, combined with the great distances over which many resources such as wood, water, coal, and electricity had to be transported, pushed this region into early long-distance electrification. California had the first truly long-distance lines carrying high-voltage power from what were then the earliest extensive hydropower sites in the nation. Also, because of the state's unusual agricultural requirements, California had a well-advanced rural electrical supply system long before most of the rest of the nation.

The California Electric Light Company of San Francisco provides a good example of early utility expansion. This company was the first utility to have a central station—it began in 1879 by supplying 21 arc lights using the Brush system. By 1888, the utility had switched to incandescent light and, after negotiation with the Edison General Electric Company, became the Edison Light and Power Company. Five years later, a merger with the local gas company created the San Francisco Gas and Electric Company. Meanwhile, hydropower was being developed in the Sierra Nevadas. Hydroelectricity on a major scale was first produced in the 1880s and 1890s at Lake Fordyce Dam and the Folsom Powerhouse, and, by the beginning of the twentieth century, dozens of hydroelectric plants had emerged along the mountain gorges. The two leading hydro companies in northern California were the California Central Gas and Electric Company and the Bay Counties Power Company. These merged in 1903 into the California Gas and Electric Corporation. The new utility had extensive power-making capacity as long as rain fell, but because of a lack of steam generators, a drought could cripple the whole system. Also, the presence of the San Francisco Gas and Electric Company prevented penetration of that market despite the low costs of hydro-generated electricity compared to steam-generated power. The San Francisco Gas and Electric Company, in turn, could not lower its high electric rates because it was tied to expensive steam generation facilities. Business logic dictated a merger, and in 1905 today's giant Pacific Gas and Electric Company was born.

Southern California, too, led the way in long-distance transmission and hydropower sites. In fact, many of the advances in transformer capability and transmission know-how used at Niagara in 1895 were actually perfected by Westinghouse and General Electric in earlier California installations. The Southern California Edison Company, the other great utility serving

California today, was the outcome of several small system interties brought about by the drought crisis of 1924.

Northeast. In the Northeast with its older, well-established communities, smaller local utilities prevailed well into the twentieth century. There was not the same impetus for the kinds of trunks and interties that developed in other regions. Yet these municipally oriented utilities continued to pioneer in urban generation and distribution systems and in the 1920s led the way in the development of home appliances.

The Hartford Electric Light Company of Connecticut (HELCO) is an outstanding example of progressive technical and management developments in the Northeast. HELCO owed many of its innovations to its policy of working closely with electrical manufacturers—actually serving as a testing ground for many industry experiments. In 1893, HELCO led the East in the use of three-phase alternating current transmission. The company also originated the employment of large storage batteries in ''load'' management and as protection against temporary breakdowns. They were the first to use aluminum conductors, enclosed-arc street lights, and conduits for underground cable. HELCO's small R&D laboratory developed patents for electric cooking and heating elements, ice-making machines, and space and water heaters. True to the tradition of the industry, however, these patents were then turned over to General Electric for manufacture. The utility was also the first to use a steam turbine in the United States. In 1901, HELCO installed a Parson's turbogenerator delivering 2,000 kilowatts. The rights to this new British invention had been acquired by Westinghouse several years before, but American utilities had been reluctant to buy it (Passer, 1953).

Midwest. The Midwest illustrates another important phase in the growth of American utilities. From 1892 until the Great Depression, the electric destinies of this part of the country were mainly controlled by one individual. Samuel Insull, Edison's financial wizard, took over as president of the Chicago Edison Company in pique at being passed over for a top management position during the Houston-Thompson-Edison General Electric merger. Insull built the small municipal utility into an empire of 65 interconnected utilities operating in 23 states. Although some of his financial methods, especially the foundation of large holding companies, led to his downfall during the Great Depression, his initiative in developing advanced central power generation, his extension of the rural network to cover much of the Midwest, and, perhaps most importantly, his innovations in the rate-charging format revolutionized the utility industry. Insull saw that electricity costs could never compete with those of gas because of the capital investments required to build central stations. His solution lay in developing a business strategy based on expansion. Insull established the concept that utilities needed to be natural monopolies to be financially viable. In return for the legal obligation of furnishing electricity to a given region and its consumers, a given utility had to have licensed rights to be the exclusive supplier in that region. Only in that fashion was it possible to build installations based on predictable demand. Insull then established the demand charge, which meant higher rates at the beginning of a new installation, but lowered rates as electricity use increased and fixed costs became covered (McDonald, 1962:85-90).

By 1907, Insull had consolidated Chicago's electric companies into today's Commonwealth Edison Company. In 1912, he formed Middle West Utilities, a giant holding company that developed the Illinois rural grid and neighboring rural networks. Insull demonstrated that, even without the base of irrigation that existed in California, electric service was feasible and financially profitable for most rural communities and farming areas.

Southeast. Yet the most extensive interconnected system was developed in the Southeast almost a decade before the Midwestern boom in electrification. This surprisingly early development can be attributed to the General Electric salesman named Sidney B. Paine. Paine was in charge of a sales department set up to equip the burgeoning textile industry of the late nineteenth century. This industry had mostly been concentrated in the New England region, but several enterprising manufacturers began locating their plants in southern states. The advantages of the Southeast were that water power was as available there as in New England, the mills were closer to domestic cotton sources, and labor was cheaper. Paine had little success in selling electricity-generating systems to New England factories—these were conservative and located close enough to water power to get along with rope or belt transmission. Paine, however, tapped a veritable gold mine of electric business when he solved what had previously seemed an insoluble problem for a mill under construction in Columbia, South Carolina. Ordinary mechanical and power transmission from water wheels to mill machinery had turned out to be impossible because the distance involved proved to be too long (1,000 feet). The owners then called in several electric companies to bid on power transmission. Paine, because of years of previous experience in textile mills, was the only one to recommend a practical system. It involved alternating current generating equipment and polyphase induction motors. These motors, unlike direct current shunt motors, did not spark and provided constant-speed drive even when the lighting was turned on and off. Because they were not a fire hazard, the motors could be placed directly in the factory, obviating the need for mill shafting and belting—complicated devices that always resulted in great losses in efficiency (Passer, 1953). Paine's concept of the individual and group machine drive, possible only with alternating current equipment, won the day. It also revolutionized the textile industry—and many other industries later on—and led to pioneering accomplishments in interconnected grid for the whole Southeast. This grid was pioneered by the Southern Power Company, which linked together seven major independent networks in North and South Carolina, Georgia, and Tennessee over a 1,000-mile-long circuit. Central power was generated at Rockingham, North Carolina, from hydropower sites on the Cape Fear River. The line was projected for extension from Nashville to Memphis so as to provide power as far as the western banks of the Mississippi. Incidental to this industrial electrification, many rural communities in the South benefitted from the cheap electricity that had been made available (*EPRI Journal*, 1979).

The southern grid existed by the beginning of World War I, but the networks had not yet begun to exchange power on a major scale. However, the grid pointed the way to the potential decentralization of the nation's industrial centers since it was apparent that power could be delivered anywhere as long as there were trunk lines long enough to reach the factories.

By the 1920s, the movement toward interconnections based on socioeconomically and geographically related areas and regions became popularly known as "superpower." A diversity of electric energy sources (mostly coal and hydro in those days) provided greater economies of scale and reliability, and, in addition, surplus power could be sold to neighboring locations. The power industry, always the most capital intensive in the nation, gravitated toward the greater economies of scale brought about by more and larger plants linked over ever-expanding customers for their product. Growth, of course, went hand in hand with the industry's being able to stimulate sufficient demand to pay for the next round of plant construction and interconnection.

TVA. The onset of the Great Depression brought to a temporary halt this rapid escalation of power facilities, interties, and electrical demand.* However, the trend for national electrification was not slowed. For it was at this juncture that Franklin D. Roosevelt set a federal policy of actively promoting cheap electricity for the benefit of all citizens. The most dramatic and successful of all the federally backed power projects was the Tennessee Valley Authority (TVA) experiment.

The TVA was inaugurated in 1933. Electrification of the Tennessee River Basin was not its only or even its main purpose. It was to be a multipurpose R&D installation on a grand scale. Flood control, navigation, agricultural and individual development, reforestation, national defense, and lasting improvement of the social well-being of what was then a depressed region were all part of TVA's program. In its early years, TVA was not noted for any great technological advances in its electrical developments; however, the total scope and quantity of the work undertaken was unprecedented.

From 1933 to 1941, TVA implemented its agenda with extensive dam construction for flood control and the production of hydroelectric power at six

*In the years from 1900 to 1929, yearly electricity generation had grown from 2.5 to 91.9 billion kilowatt hours, and costs had dropped from 25 cents to about 5 cents per kilowatt hour (in 1980 dollars) (*EPRI Journal*, 1979).

major sites. Also built were locks for shipping along some 650 miles of waterway, along with dikes, roads, and bridges. TVA created entire towns, installed and maintained fertilizer plants, and oversaw electric transmission and distribution networks. Recreational facilities, too, were also built; boat marinas and lakes for swimming and fishing were developed and operated (Kyle, 1958).

The following two decades (1941-1961) saw TVA's emergence as a great power utility, spurred initially by the enormous demands for more energy required for military production for World War II. After 1961, the federal utility expanded its electric-generating capacity even further—this time, primarily with coal and nuclear power.

TVA's teams of engineers, architects, agriculturalists, and lawyers made contributions in all of the areas they undertook. The federally backed authority was also able to offer electric rates considerably lower than those of private utilities. This originally controversial project served to prove to the utility industry that rates were flexible, a point Insull had already demonstrated several decades before. The low rates encouraged consumption and increased demand. Even the poorest of the Tennessee Valley rural population were able to have electricity in their homes. Their general standard of living and well-being was improved, and TVA came to be the symbol of the idea that electrification was a social equalizer (McGraw, 1971).

In addition to TVA, the government also became involved with other electrification projects during the 1930s. The Bonneville dam on the Columbia River was constructed by the Corps of Engineers with funds from the newly created Works Progress Administration (WPA) and administered (together with the Coulee Dam later built farther upstream) by the Bonneville Power Administration (BPA) under the Department of the Interior. BPA, although sharing many of the same objectives as TVA, never went as far in its adherence to New Deal power policy. Its retail rates were not as binding as TVA's and it never shared TVA's ideological commitment to economic restructuring of a whole region. Over the years, BPA became more of a broker in the distribution of electric power than a direct supplier, with its main concerns focused on water for irrigation rather than electricity.

Another New Deal power project was initiated by the Rural Electrification Administration (REA). The REA lent money to farmer cooperatives for them to purchase electricity in bulk for rural distribution. Most private utilities in those days were not willing to bear the costs of distributing power for widely dispersed rural use. Thus REA programs were the only means of bringing electricity to many of the non-urban areas of the country. Whenever rural electrification cooperatives could not buy power from federally backed generating sources (such as TVA or other federal hydroelectric installations) and had to turn to private or municipal utilities, they met with great resistance. The government frequently had to step in to force these utilities to sell bulk power to the cooperatives (Davis, 1982:72-85).

During the 1930s, the concept of electricity changed. No longer was it the luxury it had been in its early days. Access to electricity became viewed as an intrinsic right of every American.

D. RECENT DEVELOPMENTS

World War II strained the capacities of the utility industry to the limit. While the nation mobilized for war production, the utilities scrambled frantically to keep up. Although electric capabilities were almost doubled, this feat was accomplished without any major technological advances. Such advances had to await post-war development when revolutionary scientific and technical breakthroughs made during the war were put to practical use (*EPRI Journal*, 1979).

After 1945, American industry, now re-geared for civilian production in a booming, consumer-oriented economy, demanded ever-increasing amounts of power. Also, the public equipped their homes with every sort of electric appliance. Not only were manufacturers busy supplying new products, but American defense strategies called for nuclear fuels, metals, and many other materials requiring high levels of electric power.

The pace of electricity consumption was unrelenting. By 1955 the industry was producing some 550 kilowatt hours $\times$ 10^9 compared to a pre-war total of about 100 kilowatt hours $\times$ 10^9. Hence, the electric system was placed under considerable technological strain in its efforts to supply demand. The utilities relied chiefly on the research and developmental talents of the major electric companies, such as General Electric and Westinghouse, to provide the necessary equipment for their increasing needs.

The dramatic wartime and post-war advances in electronics, microwaves, nuclear physics, automatic controls, operations research, and information theory led to a whole new generation of sophisticated electrical engineers. By now, however, a host of new industries had entered the economic scene. Numerous companies sprang into existence from the new technologies, many of which were based on government-sponsored wartime and post-war R&D efforts. Not only was there great demand for research talent for solid-state physics and electronic applications, these fields were also considered more glamorous than electrical engineering, and the electric power industry could not compete successfully.

Ultimately, however, these new fields and industries, such as computers, semiconductors, nuclear physics, and the aerospace sciences, had profound influences on the growth of electrification. The new electronic and nuclear technologies, in turn, stimulated major advances in the power and light industry.

Nuclear power. The most dramatic, even if not the most pervasive, of these impacts occurred in 1954 with the opening of utility participation in the use of nuclear power. Utilizing the vast resources locked in the atom to generate electricity seemed like a godsend to an overburdened industry seeking solutions for what seemed to be unremitting demand. There was a host of political and psychological reasons for wanting the atom to be economically productive rather than solely a military weapon. The memory of Hiroshima was still close at hand when the Atomic Energy Commission (AEC) was created in 1946. Although much of the future development of atomic power was geared to ensure military strength with an overriding imperative for secrecy in the interests of national security, there were overtones of a "public" policy in the staffing and organization of the AEC. The commissioners were all civilians, with a former TVA head, David Lilienthal, in charge. Furthermore, the agency was responsible to the Joint Committee on Atomic Energy composed of members from the United States Congress. In its first years, however, the scientific elite that formulated agency policy seemed to be primarily interested in nuclear weapons and paid little attention to any non-military development of atomic power. For example, the first successful nuclear reactor built at Argonne in 1951, which paved the way for a civilian reactor development, was also designed as the prototype reactor for a fleet of nuclear-powered submarines.

When, in the late 1940s and early 1950s, it became apparent that the United States had lost its monopoly on nuclear arms development, the new Atoms for Peace program was announced and implemented with an eye not only on international relations, but also on being able to keep abreast of nuclear developments all over the world. Concurrently, on the home front, the AEC's relatively inconsequential Industrial Participation Program was geared up to support the building of the Shippingport reactor.

By 1954, the Atomic Energy Act had been amended so that utilities, private and public, could reap the benefits of the immense federal R&D effort that had gone into the development of atomic reactors. The AEC lost some of its monopolistic control of atomic development, but retained control of the nuclear fuel produced for reactor use—leasing to the utilities the quantities they needed to operate their plants. Also, in order to build and operate the nuclear sites, licensing had to be obtained from the federal agency.

Thus, nuclear generation, unlike other developments in the power industry, has always been a government-sponsored operation—and, the "cost" of nuclear energy, even before environmental considerations, was never a realistic endeavor for the private sector to assume totally. However, even more significantly for the future of atomic power, the initial autocratic attitude of the AEC carried over into civilian nuclear affairs when it came to the disposal of radioactive wastes. Even during its purely nuclear weapons development phase, the AEC showed little foresight in the matter of permanent and safe storage of the wastes produced during the production of nuclear fuels for atomic weaponry. Certainly, the threat of international nuclear proliferation demanded strong security measures to ensure that enriched fuels would not fall into the wrong hands. Such security concerns supported monopolistic policies for the nuclear industry.

System integration. During the 1950s and 1960s, the gradual evolution of increasingly interconnected systems reached a new threshold of large-scale integration. An important factor in this integration was the use of extra-high transmission voltages, allowing for the bulk delivery of large amounts of electricity over great distances from large generating facilities located close to necessary fuel sources. Central control areas, which now number approximately 130, were developed to regulate power transfers in capacities ranging from 300 to 30,000 megawatts. On a second level of integration, power pools were formed by formal and informal agreements between utilities to coordinate and exchange power among the members. There are now about 30 such power pools. Paralleling power pools development, there came into being bulk power, extra-high voltage (up to 800 kilovolts by 1965) transmission lines that soon blanketed the nation. These lines are programmed for instantaneous transferral of electricity from one part of the network to another in time of need or emergency. Today, there are three such systems: the Eastern Interconnected System, covering two-thirds of the eastern United States and most of eastern Canada; the Texas Interconnected Systems, covering most of Texas; and the Western Interconnected System, covering the western United States and western Canada. Other advances in the 1950s and 1960s included impressive growth in the unit size of turbine generators (from 200 megawatts in 1950 to 1,300 megawatts in 1970) coupled with greater thermodynamic efficiencies (*EPRI Journal*, 1979). Also, without the contributions of the systems engineers and the computer scientists in developing automatic controls and regulating devices, these enormous grids could, of course, never have been functionally integrated.

With the ability to move blocks of power flexibly around whole regional systems as needed, the utilities achieved great overall economies of scale. The price of electricity fell throughout this period (the price was around 3 cents per kilowatt hour, in 1980 dollars, by 1970) despite the enormous costs involved in building large-scale facilities.

In 1965, the faith of the nation in the infallability of its power supply, a confidence well founded on many decades of reliable and cheap service, was deeply shaken by the massive Northeast blackout that knocked out almost the entire region. It became apparent that systems integration and pooling had been companies of private utilities to prevent investor fraud and other financial councils and a national council were voluntarily formed by the industry. Their purpose is to improve coordination of bulk power supplies and to plan for a reliable regional and national power future. Seven regional councils comprise the members of the vast Eastern Interconnected Network, one corresponds to the Texas Interconnected Network, and the last covers the same area as the Western Interconnected Network. The National Electric Reliability Council (NERC) serves as a national forum for the regional councils.

Its board comprises representative members from the regional councils and from each major segment of the industry.

Regulation. The last fifteen years also saw increasingly extensive regulation of major energy facilities, with electric power plants as a main target of concern. Electric utilities have been subject to some form of governmental regulation almost from their inception. Because transmission and distribution are so costly, there is little advantage in having competing systems cover the same territory. However, in return for having exclusive rights over a given area, utilities have had to accept regulation of their price structures and other aspects of their operation. Since open competition could not serve as a natural regulatory mechanism, consumers had to be protected against the possibility of high rates and poor service that might result from unregulated monopolistic control of the electricity market.

A brief historical background to government regulation will be helpful at this point. Initially, governmental control, where it existed, was mainly local and municipal in nature. Cities ceased granting multiple franchises to electricity companies when it quickly became apparent that this resulted in many weak companies being bought out by one dominant concern which then became an uncontrolled monopoly. Instead, city councils began to view utilities as natural monopolies and to exert regulatory control. Because of wide-spread corruption in big city governments, however, the first decade of the twentieth century saw a wave of reform movements in many states that replaced local urban control with state-wide public utility commissions. In 1907, New York and Wisconsin led in the creation of Public Utility Commissions (PUCs) and by 1922 state regulation of the electricity market had spread to 47 states and the District of Columbia. The effectiveness of these commissions varied from state to state—with some being highly effective in protecting customers from high prices and poor service, while others fell under the control of the utilities. In some cities, municipal ownership was an alternative to commission regulation, and sometimes city residents, as in Los Angeles and Cleveland, had both private and municipal utilities available to them (David, 1982).

In 1920, the Federal Power Act empowered the Federal Power Commission (FPC) to regulate interstate electricity prices as utilities began to expand and exchange power across state lines. When, in the early 1930s, the government began to take more interest in public power, it also became involved with regulating the financial structure of private utilities. After the 1935 passage of the Public Utility Holding Company Act, the Securities and Exchange Commission supervised the financial structure of the holding companies of private utilities to prevent investor fraud and other financial abuses that had crept into utility financing.

The 1960s saw the emergence of the environmental movement and the creation of the Environmental Protection Agency, which was responsible for the establishment and enforcement of pollution abatement regulation.

This decade also witnessed the 1965 Northeast blackout, and with it a more cautionary attitude on the part of the FPC, which until then had vigorously advocated expanding regional interconnection with the ultimate aim of a national grid.

During the 1970s, the Clean Air Act began to require electric plants to reduce the smoke, ash, sulfur, nitrous oxide, particulates, and other pollutants that billowed from their smoke stacks, and the Energy Supply and Environmental Coordination Act of 1974, following on the OPEC oil embargo, tried to force power plants that were burning oil to switch to coal. When the latter law proved to be ineffective because of the many exemptions it included, the Power Plant and Industrial Fuel Use Act of 1978 was passed. This act forbade plants to burn oil and gas beyond 1990. By 1982, however, this prohibition had also been weakened. The Public Utility Regulatory Policy Act of 1978 attempted to set standards that would conserve electricity and favor individual consumers rather than industry. PURPA was a response to the energy crisis. Soaring electric rates had triggered highly vocal consumer movements that challenged power company pricing and questioned public utility policies in most states.

There was also a drastic reorganization of the nuclear establishment during the 1970s. Most of the early chairmen of the AEC had been administrators—starting with Lilienthal, whose TVA methods were soon considered too lax for security reasons, followed by Lewis Strauss, whose main concerns were hydrogen bomb development along with tight internal security, and, later, John A. McCone, an administrative specialist. In the 1960s, the Kennedy administration named Glenn Seaborg, a nuclear scientist, to head the AEC, and, throughout that decade, nuclear development was pushed through forcefully by the physicists and engineers involved with atomic reactors. This group was generally not aware of many of the social ramifications of nuclear power. In their enthusiasm for inexpensive electricity, they did not pay adequate attention to the environment and, especially, to an ultimately safe, permanent disposal of high-level wastes.

By the end of the sixties, the environmental movement was strong enough to make a change in the AEC, which they accused of callous environmental neglect. Several ecology-minded commissioners followed Seaborg, but by 1974 it was apparent that the AEC needed reorganization. The Energy Reorganization Act split off the licensing functions to the new Nuclear Regulatory Commission (NRC) and to the Energy Research and Development Administration (ERDA), which included the AEC's operating functions with several existing research programs. By 1977, when the Department of Energy was created, ERDA was consolidated with other bureaus to form DOE. Most of the Federal Power Commission and parts of the Interstate Commerce Commission became the Federal Energy Regulatory Commission. The remainder of the FPC became the Economic Regulatory Administration within the DOE.

In addition to the new regulations, utilities were beset by serious concerns over the reliability of their electric supply systems. The industry was in a difficult position. Although manufacturers were still supplying the hardware and giving technical advice on interconnections, it was ultimately the responsibility of utility engineers to make the new expanded systems perform reliably. And not only were reliability standards an issue, new federal and regional regulations called for solutions regarding the ecological and environmental impacts of generating and supplying power. The kind of long-term R&D mission necessary to solve this problem was one no individual utility could even begin to undertake. In 1972, breaking the tradition of letting equipment companies handle all technical research, the utilities banded together to form EPRI, the Electric Power Research Institute (*EPRI Journal*, 1979). This organization, funded by regulatory-approved electric power excise taxes, was now responsible for providing the future options that would enable the electric industry to fulfill its mandated role of supplying safe, clean, and reliable power to the nation.

In 1973, a new and most disconcerting factor further complicated the operations of the utilities. The 1973 oil embargo and its attendant fuel price escalation, coupled with overall inflation, increasing construction costs, rising environmental constraints, and regulatory delays along with the by now dashed hopes for really inexpensive electricity from nuclear plants, reversed a long-standing trend. Electricity prices began to soar. What had been a major point of leverage for maintaining a favorable operating environment, the ability to keep prices to consumers declining or at least stable and thus increasing demand, became eroded. As costs for electricity increased, the "certain" demand projections on which utilities had been operating for decades were found to be in disarray. After 1973, the national demand for electricity declined from a steady 7 percent a year—a rate that requires a doubling of capacity every ten years—to a dramatically low 4.2 percent and, by 1982, to a negative growth rate. What happened was that conservation had become an economically attractive alternative strategy for the whole spectrum of electricity users. Federal programs were initiated to develop technology and new approaches for increased end use efficiency, to encourage industrial cogeneration, and to promote self-generation. Alternate and renewable sources of energy such as solar, wind, geothermal, biomass, and others, were being looked at seriously as a means of providing a buffer against foreign oil dependence. Legislation was passed to curtail, and ultimately phase out, the use of oil and gas for electricity generation, making it obligatory for the power industry to develop cost-efficient alternative technologies. Such technologies will need to do more than just take up the slack left by the phasing out of oil and gas fuels; they will also have to provide greater overall generating capacity if demand projections showing any rise in consumption can be credited.

The late seventies were filled with numerous studies that tried to assess future demands for electricity. Conclusions ranged from a high of some 10×10^{12} kilowatt hours by the year 2010 to below the 2×10^{12} kilowatt hours consumed at present (Stobaugh and Yergin, 1979:110). Obviously, there is a lot of disagreement, much of it related to basic assumptions made about the future of the national economy, the effectiveness of conservation measures, and projected changes in consumer life-styles.

The power and light industry now operates in a risky and challenging business environment. It is confronted by confusing planning circumstances, political polarization of the regulatory process, mandated changes in performance standards, and, finally, difficulties in competing for the massive capital the industry needs for growth. Coping with these uncertainties will require some very broad and wise strategies on the part of all the players—industry, consumers, government—involved with electrification.

E. SUMMARY

The electric power industry in the United States, despite its present image as a stable and prosaic institution, has actually experienced numerous transformations in structure, procedures, leadership, and goals over the last century. These transformations reflect not only the changing nature of the industry itself, but also changes in the technical, economic, social, and regulatory environments in which the industry operates. Finally, the few guiding principles of the past, such as steady increase in the demand for electricity, decreasing cost of producing electricity, and greater power generation concentration, can no longer be assumed to be true when designing electric power policy.

BIBLIOGRAPHY

Adams, Edward Dean. *Niagara Power: History of Niagara Falls Power Co., 1886-1918*. Niagara, New York: Niagara Falls Power Company, 1927.

"Creating the Electric Age." *EPRI Journal* (March 1979).

Davis, David H. *Energy Politics*. New York: St. Martin's Press, 1982.

Eisenmann, H. J. III. *Charles F. Bush: Pioneer Innovator in Electrical Technology*. Case Institute of Technology, 1967.

Hughes, Thomas P. *Thomas Edison: Professional Inventor*. Chatham, England: McKay Ltd., 1976.

Kyle, J. H. *The Beginning of the TVA*. Baton Rouge: Louisiana State University Press, 1958.

McDonald, Forest. *Insull*. Chicago: University of Chicago Press, 1962.

McGraw, Thomas K. *TVA and the Power Fight*. New York: J. B. Lippincott, 1971.

Passer, Harold C. *The Electrical Manufacturers, 1875-1900*. Cambridge, Mass.: Harvard University Press, 1953.

Prout, Henry G. *A Life of George Westinghouse*. New York: American Society of Electrical Engineers, 1921.

Stobaugh, R., and D. Yergin. *Energy Future*. New York: Random House, 1979.

Part Four

PANELISTS' VIEWS

The assessment's panel of experts provided most of the wisdom reflected in the preceding pages. In the great majority of instances, the text of the preceding pages reflects the consensus opinion of the core staff and the panel of experts. This is not, however, always the case. In a few instances, there was no consensus, so the text reflects the opinion of the core staff and some, but not all, of the panelists. In other instances, a panelist would have added to the text an elaboration that the core staff, usually for reasons of style or administrative practicality, elected not to add. This part reproduces dissents and elaborations by individuals on the expert panel to the assessment text as authored by the core staff. Expert panelists' comments are preceded by the commenting panelist's initials, the chapter to which the comment pertains, and a number that orders the comment among others by the panelist within the chapter.

Expert Panelists' Individual Comments: Disagreements and Elaborations

DR. SANFORD BERG

SB:IV-1. CWIP affects returns to investors and the rates paid by various customer classes.

SB:IV-2. I really feel that a question concerning the advent of new pricing mechanisms should be addressed. New rate structures, such as time of use and direct load controls, could have tremendous impact on utilities in the future.

SB:IV-3. Why would we want to influence the evolution of future generation mixes not favored by financial (economic) considerations? Economics is the true benchmark for analysis because not only are the costs to the firm examined, but externalities, such as pollution and strip mining, are included in the analysis. This is an important point. I don't want to sound dogmatic, but it would be totally irrational to influence future technologies in the absence of economic analysis.

SB:IV-4. The cost of pollution control should rest solely on the demander of the electricity and not on shareholders or the general public. The consumer is the one who demands the electricity and must pay for the entire cost of supplying the electricity, including pollution control devices. Pollution control is a cost of doing business and should be treated like any other cost. Investors are not responsible for the costs of doing business and, if they were saddled with pollution control costs, there would be a loss of return and a disincentive to invest in electric utilities. The general public, on the other hand, should only bear the costs of pollution control if they are ratepayers. The idea is to internalize the costs to the firm so that they may be passed to the ultimate consumers. This reflects an efficient allocation and efficient pricing. We don't want people not using a service (electricity) to pay for the costs of the service. A final comment on this question concerns the intertemporal cost of pollution control devices. The costs could be allocated over time rather than a fixed charge when the equipment is installed. This would prevent an uneconomic cut in demand in the first period and an uneconomic overconsumption in later periods.

SB:IV-5. This sentence, as written, implies a chauvinism that I do not endorse. Effective RD&D programs can allow the United States to identify and exploit those electricity technologies for which it has a dynamic comparative advantage.

SB:V-1. It should also be noted that the unanticipated increases of 1973 and 1979 resulted in an *ex post* inefficient generating mix (one too heavily dependent on oil). So, some new capacity came on line to back up oil units resulting in a larger proportion of capacity being unused than in the past.

DR. RICHARD BROWN

RB:II-1. This scenario could more likely come about not by "inappropriate policy decisions" related to electric power, but by a set of unavoidable circumstances relating to a general economic slump.

RB:IV-1. However, a shift to electro-technologies that utilize electricity to perform work more efficiently than thermal technologies can result in higher demand.

RB:IV-2. The general population migration to the Sunbelt is a trend that is likely to continue over the next three decades. It is not highly uncertain, that is, increased Sunbelt demand is fairly certain, and should be addressed by regional electric power policies. Such policies may be divergent, for example, Northeast versus Southwest.

RB:IV-3. One should not discount breakthroughs in energy conservation technologies and energy management systems that could offset the need for increased generation capacity.

RB:IV-4. The reader should note that the present electric utility system is largely (almost 80 percent) investor owned.

RB:IV-5. Given that regulatory jurisdictions likely will be focused more on a geographical level (i.e., responding to marked geographical shifts) and to institutional transformations, the federal role of coordination as well as emergency response with respect to bulk power supply will not likely diminish and may need to be strengthened.

RB:IV-6. Given the foreseen potential for societal disruption discussed in conjunction with this principle, there is need for the development of a mechanism at the federal level to deal with balancing efficiency, equity, and risk issues through federal oversight.

RB:IV-7. As a fallout of this principle, an independent, federally sponsored (with industry participation) Electric Policy Institute could be highly effective in continually evaluating the state of the U.S. electric power industry (i.e., an incremental, ongoing technology assessment; recommending congressional and regulatory policy initiatives to aid the industry to adjust to unforeseen economic, social, or technological changes; and coordinating and guiding federal/industry RD&D).

RB: VII-1. These items have not been discussed, are new to the reader, and are not explained. For example, is there a "large-grained modular structure"? These should not be in the summary unless they are discussed first!

DR. DAVID DAVIS

DD: VII-1. This strikes me as unlikely. Canada is a potential local uranium supplier. Furthermore, we could disassemble nuclear weapons.

DD: VIII-1. I don't think so. Coal transportation is relatively inexpensive.

DR. GORDON ENK

Comments on the last paragraph of the "Overview Methodology" discussion. You should provide some support for the nature of how this insight was obtained. Did the six-member issue review board feel or identify any real new insights that were contained in the report?*

GE:II-1. All of these problems noted have been previously recognized but it is not at all clear why they emerge in the "Nuclear Resurgence" Scenario and not, for example, in the "Average Future" Scenario.

GE:II-2. What is the basis for assuming strong government policies? We are presently going through the transition.

GE:III-1. But you might end up with very different actions and policies depending on who held decision-making power on a particular issue.

GE:III-2. This discussion of the meaning of various scenarios or particular factors is very difficult to follow. It is very abstract and seems arbitrary at best.

*For more methodological detail than is presented in the "Overview," the reader is referred to the methodology appendix (Appendix A). The issue review board critiques, though they emphasize the interests of the individual board members, strongly support the statement that the assessment as a whole reflects "high-quality knowledge and insight." For example, Dr. Allison, though concerned that the assessment report "raises far more questions than it answers," found it to be quite "exhaustive in terms of issues raised; scenarios developed; policies considered; technical, economic, political, and social factors discussed; and methodology." Though Dr. Bauer would change certain of the assessment's points of emphasis, he nonetheless feels assessment participants "are to be congratulated" for an effort that is "forward looking," will "provoke much thoughtful discussion," and will "assist policy makers in assessing the implications of a highly complex set of issues." The critiques of other board members are of a similar vein.

General comments on Chapter IV. My overall reaction is that there is nothing new or insightful presented. The real question that needs to be addressed is, What are the really significant longer-term questions facing this industry? I did not find any identified! It appears to be a sort of rearranging of old ideas and is presented in a manner that is less than engaging.

It is striking that there does not seem to be any real "power" of explanation or understanding that was generated from the scenarios and then discussed in Chapter IV.

It is not that there is a lot to disagree with in Chapter IV. It is written at a level of generality so that there is not much sense in finding fault. However, it lacks insights and any new images of what should be considered important! The "overarching" and "operational" questions are hardly insightful, new, or helpful. I doubt that any decision maker involved in policy making (public or private) would find anything of interest in these questions.

GE:IV-1. I believe the purpose of this assessment, since it was a technology assessment, was also to identify the unexpected, unanticipated, higher-order impacts that would result from pursuing various paths or policies. It fails to do this.

GE:IV-2. There is a disturbing lack of clarity as to whose policies are being discussed—government policies or corporate policies. It is unclear whether you mean government policies toward electric utilities or utility corporation policies.

GE:IV-3. I question your example of WPPSS. What does it add to the TA? I am not clear about its purpose or lessons learned other than the fact that big mistakes are costly.

GE:IV-4. Again, these questions are at a level of generality that are hardly insightful to policy makers!

GE:IV-5. Do Continental Group and Milbrath really project the next 30 years? I do not believe they do.

GE:IV-6. I seriously doubt that RD&D, especially technological RD&D, will lead to the identification and evaluation of trends in human value systems.

GE:IV-7. One should be somewhat cautious about a non-critical endorsement of all RD&D funding. We do not really have a good understanding of what is good RD&D and what is bad RD&D. Blind endorsement could be disastrous!

GE:VI-1. The federal government's role in the allocation of private land is very unclear. Overall, the government has almost no role in this area.

GE:VI-2. The section on land and water is very unclear. Basically, your message—there is plenty of land and water for electricity production, if there were no competing uses or constraints—is hardly an insight or helpful perspective.

GE: VI-3. What is the relationship with fertility rates and occupational health?

GE: VI-4. At this point in the chapter, I am asking, Why is this here? How does it relate to the scenarios and to a technology assessment? It sounds like a "primer" on energy and the environment—with all the discussion of a general and widely known nature!

DR. KIRBY HOLTE

KH:IV-1. No one disputes the need for "formal recognition to the importance of equity and risk, as well as efficiency." Such recognition requires the explicit identification of present and future stakeholder issues and consequences. The mechanism by which this may be achieved is not, however, clear. The use of citizen advisory groups, opinion surveys, and public hearings has had partial success. However, consensus is rarely reached. The use of ombudsmen or guardians *ad litem* is appealing in concept but historically has led to adversarial win-lose contests rather than consensus. The debate on benefits of elected versus appointed regulatory commissions also remains divisive. On the one hand, an elected commission may better represent today's stakeholders. On the other hand, political pressures on elected commissioners makes long-term policy making and representation of future stakeholders more difficult.

KH:IV-2. The considerations are more subtle than the question suggests. "Expensing" is a term used to classify costs passed directly to the ratepayer in the year incurred (as opposed to recovery through the rate base, where the costs plus a return on capital is recovered over the life of the facility). In some (relatively few) cases, it may be possible to "break out" environmental costs (i.e., so that they could be expensed) and in some cases environmental costs are already expensed. Fuel, for example, is primarily expensed. Thus, the additional cost of low sulphur oil or coal is an expense. Technically, one could design a coal plant without scrubbers, baghouses, low NO_x combustors, or covered coal, but no one would build such a plant. Furthermore, it would not make sense to compare the cost of this plant to hydro and others, which are, by their natures, already environmentally acceptable.

KH:IV-3. In formulating or modifying electric power RD&D policies, decision makers at all levels, both in the private sector (utility management) and the public sector (government), must first define the objectives of these policies. The objectives are both time related (short range, long range) and geographic (e.g., arid Southwest, Northeast, neighboring countries, Asia, etc.). Other criteria, such as economic growth, national security, and balance of payments, are appropriate; however, objectives must be technology independent.

KH: V-1. There are three reasons for the decline in plant utilization. First, a substantial increase in peak load versus average load caused by summer air conditioning and winter electric heating; second, the replacement of oil- and gas-fired power plants with coal and nuclear plants; and third, in the 1970s, completion of power plants built in anticipation of high electricity demand growth that did not materialize. Rather than retire oil and natural gas plants no longer required to meet kilowatt hour (energy) demand, these plants were reduced from base load to intermediate and peaking duty. The first and third reasons result in an inefficient allocation of resources, whereas the second results in cost savings and increased efficiency in resource allocation.

KH: V-2. The discussion presented is technically correct only if the plant is "project financed." That is, if the revenues of the plant are used to guarantee repayment of the debt. Although project financing is used in some cases, utility borrowing is normally via corporate bonds guaranteed by corporate revenues, rather than revenues from a specific plant or project. Nevertheless, the general conclusions are correct and those utilities with heavy commitments to large coal and nuclear plants are generally viewed as exposed to higher risk. This has, on many occasions, resulted in a derating of bonds, thus an increase in interest costs.

KH: V-3. Increased plant utilization may result in decreased electricity prices if it reflects efficient use of capital and does not result in increased heat rate or increased use of higher cost fuels.

KH: V-4. I would change the emphasis here. With the exception of the "Post-industrial" and "Economic Malaise" Scenarios, the issue of power plant finance—thus, CWIP/AFUDC—is likely to remain controversial and important. This is especially true in the "Mega-plant" and "Nuclear Resurgence" Scenarios, where long construction delays and large capital expenditures threaten utility earnings.

KH: V-5. Although the California Energy Commission did make the recommendations cited, the question of electric utility entry into "for profit" diversification has not been fully addressed by the California Public Utility Commission (which regulates utility rate of return). Nevertheless, several utilities in other states have formed non-regulated subsidiaries and many other utilities have expressed an interest in diversification into non-regulated fields.

KH: V-6. In those cases where utility management expertise does not exist, it could be brought in on short notice relative to the time frame of this study.

KH: VI-1. I would add a paragraph to read: "Finally, comparison of risks between scenarios or between low electricity demand and high electricity demand is inappropriate without consideration of risks in the non-electric sectors and fuel cycles. We cannot ascertain from the available data which of the scenarios would have the lowest risk or if any of the scenarios would result in a lower or higher risk than exists today."

KH: VI-2. Although recent evidence indicates that the burning of high sulfur coal in the northeastern United States is a contributing factor in acid rain, several technology options exist other than advanced SO_2 scrubbers and coal gasification. Furthermore, coal gasification, when used as a retrofit to convert existing coal-fired power plants, is an extremely inefficient use of capital and natural resources.

There is, however, a more important issue. Should government mandate specific technological fixes, and, if so, should these fixes be specific on a case-by-case and plant-by-plant basis or should they be generic, covering all plants and all cases? Obviously, there is no such thing as a generic fix in that every power plant is different and a technology designed for one plant will not generally work on another. This leaves one with the conclusion that the fix must be plant-by-plant or case-by-case specific. Since we have yet to find someone of such divine wisdom as to develop a reasonable technological fix and be able to apply that fix on a case-by-case and plant-by-plant basis, we must conclude that those people who design, construct, and operate the power plants are in the best position to choose the most appropriate "fix."

KH: VII-1. There is no basis by which we can conclude that the existing electric system or any of the scenarios' systems cannot "withstand unexpected failure of just a few or even a single major generating component." Even the New York blackout, as massive as it was, affected a relatively small part of the United States. Furthermore, the New York blackout has led to major redesign of system protection and control. Furthermore, "fine, modular redundant structure" is a feature of all scenarios, with the possible exception of the "Mega-plant" future.

KH: VII-2. There is no basis by which we can conclude that all of the plants must be "relatively" small, as long as there is a sufficient number of plants.

KH:IX-1. I would also add the following program areas:

G. *Environmental and health.* Programs to understand both the environmental and health consequences of electric power production and to develop cost-effective technologies and/or strategies to meet present and future environmental and health standards.

H. *Energy resources.* Programs to expand the number and availability of energy resources for both electric power production and direct end use. These programs include technologies to utilize more efficiently existing fossil and nuclear fuels as well as technologies to utilize energy resources that are not currently economically or environmentally attractive, including renewable energy resources, wastes, and fusion fuels.

I. *Natural resources.* Programs that are directed toward conserving or expanding the availability of natural resources such as water, land, marine life, and so on.

KH:IX-2. Many of the new electricity technologies fall midway between the very large (e.g., fusion) and the very small (e.g., distributed photovoltaic). Appropriate federal government roles regarding these technologies include elements of those described above. At the state level, the most beneficial policy would be to allow utilities full recovery of costs for demonstration and pre-commercial plants, even though these costs are above costs of conventional mature technologies. In addition, tax incentives, guarantees, and so on, now available to non-utility owners, should be extended to utility ownership. Finally, to the extent that utility shareholders are required to assume additional risks, those risks should be compensated through an incentive rate of return; to the extent that utility ratepayers pay additional costs for research, development, and demonstration, the benefits of that RD&D should flow through to the ratepayer as compensation.

KH:IX-3. Utility-funded fundamental and engineering research should be considered a necessary expense suitable for recovery through the rate setting process.

KH:IX-4. In keeping with my earlier expansion of R&D program categories, I would add the following:

Environment and health. Although environmental and health protection regulation is widely accepted as an appropriate government role *and* the implementation of the technology necessary to meet these regulations is an appropriate utility role, a gap remains when it comes to the R&D necessary to develop the standards and the technology. Past policy, in which increasingly more stringent standards are the driving force for R&D (i.e., reactive R&D), should be replaced with objective-oriented programs. The predominant federal role should be in developing a scientific basis for standards. The standards should be goal- or objective-oriented rather than technology specific. Developments of the technologies to meet these standards (objectives) is a shared responsibility between government and private sectors. Appropriate incentives to develop technology in the private sector include tax incentives and full recovery of costs in rates.

Energy resources. The cost of fuel (energy) is traditionally passed directly to the ratepayers through rate adjustments. Utility shareholders do not normally receive a return on investment for fuel costs; thus, little incentive exists to develop lower-cost new energy resources. Policies that would encourage development of these resources include an easing of barriers preventing utilities from forming unregulated fuel subsidiaries, incentive rate of return for power plants utilizing new energy resources, and allowance of a "profit" on fuel for certain new energy resources.

Natural resources. With few exceptions, current federal and state policies related to electric power are directed to preventative care—that is, the prevention of serious degradation to the environment. There is no incentive for research and development that increases the availability or quality of natural resources such as fresh water, agricultural land, and marine life. Technologies such as mariculture, aquaculture, water purification, and so on can be encouraged through tax incentives and by allowing utilities the opportunity to retain profits from new ventures. Additional encouragement would occur if utilities were allowed to use natural enhancement as an offset in meeting other environmental standards.

MR. ALVIN KAUFMAN

AK:V-1. If one is using cost in a broad fashion, the sentence is correct. That is, the stockholder puts his (or her) capital at risk. He can earn a return, no return, or even lose his investment. If he earns less than an adequate return, then he is subsidizing the ratepayer and in that sense bearing part of the cost. In a more restricted sense, however, costs are only borne by the ratepayer. The stockholder carries risk and is compensated for doing so. If the compensation is inadequate, he sells out and puts his capital elsewhere.

AK:VI-1. I do not believe the trend is toward reliance on federal funds for decommissioning, or for reactor cleanup. In the case of decommissioning, many utilities, with regulatory approval, have set up funds to cover these costs. Whether these are adequate to the task is another question.

Insofar as disaster cleanup is concerned, Three Mile Island showed a federal reluctance to rush in to fill the money gap. In the event of a catastrophe, the federal government would probably be the only entity able to handle the problem. The Three Mile Island cleanup, on the other hand, has been left essentially as a local problem. The various proposals for federal aid have gone nowhere, at least to date. Further, private insurance coverage has been substantially increased.

DR. STEVE MURDOCK

SM:II-1. I would have preferred a more elaborate explanation of the logic underlying each scenario and of the extent to which the assumptions within each scenario constrain other assumptions. Perhaps some form of linear programming or optimization routine should have been used to consider the extent to which the scenarios are internally consistent. For example, I find nothing in the scenario descriptions that supports the projection that the "Post-industrial" society will be a low-population society or that the "Nuclear Resurgence" society would be a high-population society.

SM:IV-1. The rationales for these two principles are insufficiently developed. I wonder whether there is sufficient empirical evidence to support the first, and the second may be true only because it is the few decisions that fit this pattern that are most publicized.

SM:IV-2. Is it justifiable to expect decision makers in the electric power industry really to understand the manifold ramifications of industry-wide decisions? This expectation might be justified for all decision makers in all industries, but no more so for utility and other electric power decision makers than for decision makers in other industries.

SM:IV-3. In the discussion that accompanies this principle, I believe the fallacy of assuming that change will be more rapid in the future and situations more complex than in the past has been made. The bases for this unacknowledged assumption seem too impressionistic. They should be made explicit.

SM:IV-4. The demography chapter draws only loose connections between electricity demand and population migration patterns. Given this and the lack of empirical support for this assertion (i.e., for migration rather than population growth as a whole), this statement is simply too strong.

SM:IV-5. Although the use of citizen advisory boards is laudatory from the standpoint of democratic ideals, there is no empirical evidence that the use of such groups improves the performance of the regulatory process or enhances the quality of decision making.

SM:IV-6. The issues raised clearly require R&D activities if they are to be solved, but the confidence placed in such activities seems to suggest that the answers provided may be more complete than science is likely to produce. In addition, I think the section tends to suggest that R&D is capable of being a direct guide to policy making. I am not sure the history of science supports this. Many of the social science research questions raised are basic ones that have not been answered in over 100 years of extensive study. (What factors are responsible for or can allow us to predict economic growth?) Thus, it may be that the tone, if not the substance of this principle, is too optimistic.

SM:VIII-1. The assumption that heavy dependence on coal will lead to a redistribution of the population back to the Northeast and North Central states appears questionable for several reasons. First, because of the extensive electricity distribution systems in existence, locations close to coal resources may not provide a substantial advantage. Second, because much of the impetus for industrial displacement from the Northeast and North Central states resulted from the lack of modern infrastructure in these areas, it is unlikely that industry and population will relocate back to such areas on the basis of energy costs alone. Finally, unless there were extensive alterations in the tax and regulatory structures of the Northeast and North Central states, it is unlikely that coal development would increase substantially in these areas. It seems much more likely that the West would show the most rapid growth as a result of coal development with other areas being little affected by such development or perhaps experiencing population decline.

APPENDIX A

Research Design and Methodology

A. OVERVIEW

There were two central elements to the methodology used in this assessment. One element was organizational, the combination of the efforts of a small interdisciplinary core staff and a panel of twelve highly skilled professionals in a manner that maximized the effectiveness of each group. The core staff provided the continued efforts required for sustained progress; the experts on the panel provided wisdom that can only come from prolonged experience in a subject matter area. The efforts of the two groups were coordinated through telephone calls, letters, personal visits, and, most importantly, four workshops. The other element was conceptual: the use of six alternate scenarios to test the long-term implications of near-term electric power policy decisions under different assumptions about the nation's long-term future. The six scenarios served two basic purposes in the assessment: as a vehicle for identifying important issues related to future electric power system development in the United States and as an aid in determining methods for adding flexibility to electric power policies (see Annex 1).

The conduct of the assessment involved thirteen subordinate tasks:

a. Selection of research participants
b. Selection of critical factors for scenarios
c. Development of a range of critical factor projections
d. Preparation of preliminary scenario set
e. Conducting of first workshop
f. Preparation of background papers for second workshop
g. Conducting of second workshop
h. Preparation of material for third workshop
i. Conducting of third workshop
j. Preparation of draft analysis chapters

k. Conducting of fourth workshop

l. Preparation and review of final draft report

m. Preparation and review of final report

Explanations about how each of these tasks were accomplished are given in the following sections.

B. SELECTION OF RESEARCH PARTICIPANTS

Core staff. The initial phase of the assessment began on July 1, 1981 when a project staff was assembled. This staff included professionals in resource management, environmental law, energy engineering, economics, social sciences, and demography. This interdisciplinary staff was chosen for its experience and expertise to serve as project coordinators and researchers and to prepare the final assessment report (see Annex 2).

Expert panel. This panel of highly skilled professionals are chosen to work closely with the core staff throughout the conduct of the assessment. The primary criteria for panel member selection were knowledge, experience, and understanding of certain key areas regarding electric power. The key areas were defined by the core staff as electric power technology, finance, environmental studies, sociology, political science, and energy supply. Prospective panelists were to have diversity of background, training, and specialization in these key areas. It was also believed desirable that the panel come from different geographical areas to minimize the possibility of regional bias.

The core staff used a three-step process for selecting a panel of experts. Step one of the selection process was the development of a list of organizations that had current or potential interest in the future of electric power. Once the list was compiled, the staff identified one or two people in executive positions in each organization. Next, the staff contacted each of these people and, after explaining the nature of the assessment, asked him or her to suggest approximately six people whose abilities and knowledge would contribute to the expert panel. From this method, a pool of approximately 60 prospective qualified nominees was established.

The next step of the panel selection process consisted of the core staff choosing groups of people to match the key areas already defined. From each of the respective groups, sets of two people representing somewhat different viewpoints were selected. Once this task was completed, the core staff contacted prospective participants by telephone to discuss the nature of the technology assessment and to secure commitments from them. Twelve people were empanelled for the assessment (see Annex 3).

Advisory board. An advisory board was created as a review committee for the assessment's activities and findings. The board served to ensure that the assessment remained relevant and useful for electric power policy makers. Therefore, board members were chosen not only for their expertise, but also to represent a variety of institutions concerned with electric power. The assessment advisory board consisted of four members (see Annex 4).

C. SELECTION OF CRITICAL FACTORS FOR SCENARIOS

The basic research design for this assessment included the concept that the core vehicle for analyses at the first assessment workshop would be two preliminary scenarios, each a variation of a centralist "normal political response" scenario. These scenarios were to represent projections of how the nation's electric power generation and distribution system might develop under a government electric power policy that is essentially reactive and were formulated against electricity demand backgrounds that are, in one case, "high" and, in the other, "low." The first task of the core staff in developing these scenarios was the identification of those factors that will both drive the demand for electric power and constrain the manner in which the electric power system strives to meet that demand.

Although the total demand for electric power was the basic element of each scenario, simple projections of this demand in the year 2010 would have been both simplistic and naive and would have provided little basis for policy analysis. For the panelists to accept the projections and to use them as vehicles for policy examination, details had to be given about the overall state of the society at that time. A reasonable explanation as to how that state evolved had to be offered. Changes in overall energy use, in size of households, and in transportation, communication, and equipment efficiencies; developments in information industries, international trade, and use of space; new laws and regulations in the fields of environmental protection; handling of wastes and societal equity—all of these factors and a host of others will not only have an impact on the need for electric power, but will also define the means by which those needs are met.

For obvious reasons, the scenarios had to be of reasonable length, and this requirement dictated that only a few elements or factors could be used to describe future society. Thus, the tasks involved in establishing a useful data base included determining those factors that were relevant to the policy issues involved, selecting those factors that were most important to the research effort, establishing the present status of each factor, and projecting how these factors might change during the next 30-year period. Selection of factors to be included in the scenario was conducted in four steps: selection of major areas of interest, identification of relevant factors in each area of interest, selection of most important factors, and identification of quantifiable surrogates where necessary.

The selection of major areas of interest was accomplished by the core staff using a modified version of the nominal group technique (see Annex 5). Twenty-two areas of interest were identified (see Annex 6). Core staff members were assigned specific areas for analysis and a total of approximately 200 relevant issues were identified for possible inclusion in the scenarios. For practical reasons, this number was later reduced by staff analysis to the 70 factors that were felt to be most relevant to the assessment (see Annex 7). Operationally, it was impossible to quantify some of these factors in a reasonable way. In such cases, the staff chose surrogate factors. For example, the strength of the environmental protection movement was felt to be a critical factor for the scenarios. Since "strength" as such is non-quantifiable, an artificial parameter, "citizen support of environmental and conservation efforts" was established with an arbitrary level of support (index number) of 100 set for the year 1980. (Later, experience indicated that assessment participants had little trouble understanding or accepting this concept.) Periodically during the development of the critical factor list, panel members were queried for suggestions and comments. Following surrogate selection, factor combination, and further winnowing by the core staff, 58 factors were chosen for inclusion in the preliminary scenario data base. (These factors are listed in Items 1-58 in Appendix B.)

D. PROJECTION OF A RANGE OF CRITICAL FACTOR VALUES

Once the critical factors to be included in the scenarios were determined, it was necessary to make projections about how those factors might change over the next three decades. In order for the panel to use the scenarios effectively, it was necessary that these projections be acceptable to all of the panel members as a basis for discussion. Given the purposefully diverse nature of the group, developing acceptable projections was no easy task. However, meeting this requirement did not mean that mutually acceptable single values for each factor projection had to be determined. Rather, it involved establishing a range of values that participants could agree would probably encompass true factor values and within which some temporarily acceptable values could be chosen for discussion. It also involved analysis to support both the range and the tentative "interim point" choices.

The method chosen for developing the projections was an expert opinion survey, specifically a modified "Delphi" survey (see Annex 8). In this instance, it was decided to use a three-round survey with the first and third round to be conducted by mail and the second round by telephone. This combination appeared most likely to give the greatest amount of participant input in the time frames available. To assist survey participants, a summary of the present status of each factor was prepared. Data on the present status of factors fell into three general categories. First, there were the data that are reasonably "hard" and on which there was little important disagreement: number of automobiles, number of houses, average family income, total installed power capacity, and so on. Second, there were data about factors that could be quantified but about which there was wide difference of opinion; for example, the number of undocumented aliens in this country is estimated by some experts to be approximately 4 million and by other experts to be approximately 12 million; the range of estimates on the percentage of the total Gross National Product represented by the underground economy varies from 6 to 30 percent. The third category involved those factors to which there are no quantifiable answers: How frightened is the public of nuclear power? How much would the average citizen pay to reduce his or her dependence on centralized electric power? How cost effective is solar energy today?

To establish preliminary data bases for the first category of information, routine data sources were examined. Where practical, data from one source were checked against data from other sources for discrepancies. Where discrepancies were noted, reasons for the discrepancies were examined.

Data for the second group of factors were obtained from sources similar to those listed for the first category. However, greater emphasis was given to personal contact with individuals and organizations active in each specific area. For the second category of information, ranges of values were wider than for the first category, and more examination was necessary to account for the range.

The third category was the most difficult to deal with, and input was largely subjective. Careful examination of the literature was undertaken together with personal contact of people active in each area.

Eighty people, including all panel members, responded to at least one round of the survey. The participants were selected from a large pool that included people first considered for panel membership; people recommended by the core staff and panel and advisory board members; and people identified during literature searches. These survey participants represented a tremendous reservoir of talent, experience, and insight in a number of diverse but related fields. The names and associations of survey participants are listed in Appendix B together with a presentation of survey results.

E. PREPARATION OF PRELIMINARY SCENARIO SET

On the basis of the responses from the modified Delphi study, two preliminary scenarios were developed. To accomplish this, a "centralist" scenario was first developed. (This scenario was not used for panel analysis but rather served as a basis for developing the two preliminary scenarios.)

The defining critical factor values for this scenario were selected from the median values projected by the Delphi. These individual values were then combined with supporting data to develop a scenario that represented, roughly, a future to which our society may develop if there are no major surprises.

The next step in scenario development was the preparation of the "high-energy demand" scenario (henceforth called Scenario I). Each of the critical factors was examined to determine reasonable values that would motivate a relatively high energy demand.

For the most part, the values specified for the critical factors in Scenario I were approximately a standard deviation either side of the mean responses from the modified Delphi survey. (For example, the total energy usage postulated in this scenario, 116 quadrillion BTUs per year in the year 2010, was approximately one standard deviation greater than the mean value of 100 quadrillion BTUs per year.) This guideline occasionally yielded values that were inappropriate in light of the views expressed by survey respondents and parameters commonly used in energy/econometric models such as those of the Department of Energy, the CONAES study, Exxon Corporation, and the Solar Energy Research Institute. Inappropriate values were respecified within the ranges of responses to the expert opinion survey.

After new values had been determined for each factor individually, the values were tested against one another and against the group as a whole to ensure consistency, relevance, and reasonableness. All of these items were then utilized to develop Scenario I. This scenario not only described the status of society in relation to the electric power system in the year 2010, but it also listed the basic assumptions on which the scenario was based and the documentation and rationale supporting the trends and events incorporated in the scenario.

Concurrent with the development of Scenario I, Scenario II was developed in a similar manner, using factor projections in consonance with relatively low demand (e.g., a total energy use of 82 quadrillion BTUs per year in 2010). Preliminary Scenarios I and II are given in Appendix C of the NSF final report.

F. CONDUCT OF FIRST WORKSHOP

The first workshop was conducted on January 4-8, 1982, in Austin, Texas. The objectives of the workshop were to identify issues of importance to electric power development in the United States over the next 30 years, to specify areas for additional staff research, and to suggest themes for a final set of scenarios that could be used to identify further and evaluate electric power issues. Eleven of the twelve panelists plus an observer from the National Science Foundation attended this workshop (David Wood was not able to attend because of illness). The workshop began with brief statements by each panel member about his views on important electric power issues (these statements are presented in Appendix D of the NSF final report). The remainder of the first three days of the workshop were devoted to issue identification and analysis.

For this activity, the panel was broken into two subpanels. During the panel selection process, two members had been chosen from each of the areas considered key to the project (i.e., electric power technology, finance, environmental studies, sociology, political science, and energy supply). In forming the subpanels, one member from each specialty area was chosen by random selection. No attempt was made to form either ideologically balanced or ideologically biased subpanels. This process was followed to prevent the possibility of Type I research errors (see D. Campbell and J. Stanley, *Experimental and Quasi-Experimental Designs for Research* [Chicago: Rand-McNally College, 1963]).

Once subpanels were formed, each subpanel was asked to analyze the policy implications associated with each of the two alternate scenarios; however, the sequence of consideration was carefully structured. At the first analysis session, Subpanel A used Scenario I as a basis for analysis, while Subpanel B used Scenario II. (Each subpanel was aware that the other subpanel was using a different scenario, but they were not told what the differences were.)

At the second analysis session, the subpanels were asked, first, to examine different strategies that electric utilities might use to meet the power needs specified in the scenario and, next, to analyze related policy issues.

At third session, Subpanel A conducted an analysis similar to the first one, but utilizing Scenario II as a basis. Likewise, Subpanel B repeated its initial analysis using Scenario I. The fourth analysis session examined the implications of the previously discussed utility strategies on the second scenario and the fifth session was devoted to a comparison of all previous results and preparation for a joint analysis review.

Following the subpanel sessions described above, a joint session was held at which policy issues identified under each scenario/subpanel combination were compared and discussed.

The analysis scheme described above represented a simple 2 × 2 quasi-experimental counterbalanced design. The scheme can be represented diagrammatically as shown below.

	Time 1 (t1)	*Time 2 (t2)*
Subpanel A (PA)	Scenario I (SI)	Scenario II (SII)
Subpanel B (PB)	Scenario II (SII)	Scenario I (SI)

The initial analysis session can be characterized as PA SI t1 and PB SII t1, while the third session can be designated PA SII t2 and PB SI t2. At the fifth session, the following comparisons were made:

PA SI t1 with PA SII t2
PB SII t1 with PB SI t2
PA SI t1 with PB SI t2
PB SII t1 with PA SII t2

Thus, it was in principle possible to separate those differences in policy perceptions due to differences in scenarios from those due to differences in groups.

This design does not, of course, account for changes in attitudes and perceptions that might occur in each subpanel as a result of previous analysis sessions. For example, members of Subpanel A might react differently to Scenario II than they would have if they had not previously analyzed Scenario I. Although this is a disadvantage, the problem was ameliorated by the fact that each panel member had considerable experience in the subject areas involved, and his ideas and opinions were not likely to be greatly changed by short sessions of the type described.

The primary technique used for issue identification was one developed by the core staff for the project, one called Rashoman (after the Japanese motion picture of that name). The technique is a type of stakeholder analysis (see Annex 9). Each subpanel member was asked to consider himself as a particular person who might be affected by the events and trends listed in the scenario being considered. For example, one panel member might be asked to consider the scenario from the viewpoint of a banker who might be asked to provide capital for an electric power system expansion; another panel member might be asked to represent a labor union executive in the power industry. Specific roles were chosen within subpanels by mutual consent.

Once the "players" had been chosen, each person was asked to list those issues that he believed would be important to him, together with a discussion of the issue's significance and suggestions for issue resolution. In all, almost 200 separate issues were raised in this effort (see Appendix E of the NSF report).

One of the purposes of utilizing subpanels and two scenarios was to determine if the issues raised by the panel members were particular to the specific experts involved (Type II research error). If the two subpanels identified similar sets of issues, it would be probable that a different group of experts would also uncover a similar set. Since each subpanel chose a different set of "players" in the procedure utilized, a particularly difficult test was presented. In actuality, differences in wording, stress, and degree of specificity make it difficult to compare the two lists of issues. However, the core staff divided the issues presented by both panels into the subject categories identified by each group (see Annex 10). The inclusion of issues by both groups in 30 of 34 categories identified argues that the issue sets are similar though not identical. This impression is further strengthened by a reading of the issue lists and by the nature of overall panel discussion at the remainder of the workshop.

The next step after issue identification was to examine themes, events, and trends for the final scenario set. For this task, two morphological analyses were conducted, one each by two reconstructed panels (See Annex 11). The results of this effort served as a starting point for the development of the six final scenarios.

The final session of the workshop was devoted to a review and clarification of the results of the assessment to date and suggestions for further work. Among other activities undertaken at this session was an identification of the criteria that the panel believed should be employed in selecting issues for further analysis. (These criteria are listed in order of panel-specified importance in Annex 12.) The panel also suggested that detailed studies be conducted in the following subject areas prior to the next panel workshop: financial considerations, environmental and health impacts, technological development, power system reliability, demographic influences, and social interactions. intereactions.

G. PREPARATION OF BACKGROUND PAPERS FOR SECOND WORKSHOP

The time between the first and second workshop was devoted primarily to conducting the studies suggested by the panel and to early development of the six scenarios. In the former activity, one or more core staff members were assigned primary responsibility for research in a specific area and with preparation of one chapter of the final report to present the results of that research. Although one chapter was to be devoted to each subject area specified by the panel, the staff well appreciated the interactive nature of the issues and areas involved and frequent (normally, bi-weekly) meetings of core staff members were held to ensure an appropriately interdisciplinary approach to the assessment. Portions of each chapter were circulated among staff members for information, comments, and advice.

During this period, it was also decided that a chapter should be included in the final report briefly reviewing the history of electric power in the United States. It was believed that such a review would be useful to people without electric power industry experience in understanding the purpose, nature, and results of the assessment.

For scenario development, it was tentatively decided by the core staff that the basic nature of each of the five "non-base case" scenarios would be specified by "technical" factors. That is to say, the end point of each of these five scenarios would be basically defined by the level of total energy use in the United States, the fraction of total energy expended that would be utilized as electric power, and by variations in the technological mix of generation sources that would be utilized to provide the needed power. Specifically, it was decided that the following end points would be considered: high total energy, high electric power fraction; high total energy, low electric power fraction; low total energy, high electric power fraction; and low total energy, low electric power fraction. Later, it was decided to consider two variations of the first case.

Once the technical end points had been determined, a generation source mix for each case was postulated. Since the purpose of the scenario approach was to ferret out specific electric power issues, it was decided to choose different mixes for each end point. Obviously, these mixes had to be compatible with other elements of the scenario. It was decided that the high energy, high electricity end point would be associated with two different mixes—one involving a relatively high percentage of nuclear power and one involving the development of new, very large generation approaches. The high energy, low electricity end point was associated with a coal-dominated system; the low energy, high electricity end point, was associated with distributed sources, primarily solar; and the low energy, low electricity end point, was associated with an extension of the present mix. These general characterizations of the scenario set continued throughout the assessment.

The next step in scenario development was to identify feasible intermediate points for each of these factors. Finally, additional information was added to give clarity, credibility, and interest to the scenarios. This information was taken primarily from the data gathered in the modified Delphi study. It should be stressed that the scenarios described are plausible ways by which the end points may be meshed. There are obviously many other feasible paths as well as other end points. Final scenario formulation is described in Chapter II.

The method by which the alternate scenarios were developed was unusual in that it combined elements of both normative and projective forecasting. The alternate scenarios could have been derived by purely projective methods (i.e., defined by choosing trajectories with high probabilities of occurrence). On the other hand, the trajectories could have been chosen normatively to represent the most desired path to a particular outcome. The method used combines elements of each. It recognizes that society must make choices, but that these choices may not lead to the desired effects.

H. CONDUCT OF SECOND WORKSHOP

The second workshop was conducted in Estes Park, Colorado, on July 12-16, 1982. In addition to most of the core staff, the following panel members attended: Berg, Davis, Enk, Holte, Murdock, Wardwell, and Yu. Dr. Frank Huband of the National Science Foundation also attended as an observer. The primary purposes of this workshop were to review the research completed to date and to examine the primary scenario concepts.

The review of the research was conducted in three steps. First, each panel member was asked to consider carefully the chapter outline that most coincided with his interest and experience and to scan outlines for the other chapters. Next, the author(s) of each chapter met with those panel members who had concentrated their efforts on that chapter. During this meeting, detailed comments, suggestions, and questions were discussed. Finally, the author(s) of each chapter presented, to the group as a whole, a brief overview of research results to date; material discussed at the smaller meetings; and plans for future research efforts. As a result of these discussions, a number of suggested improvements to the chapter outlines were presented.

In a similar manner, panel attendees reviewed the outlined scenarios presented and suggested modifications, expansions, and other possible improvements.

I. PREPARATION OF MATERIAL FOR THIRD WORKSHOP

Between the second and third workshop, the core staff worked to complete the background chapters (Chapters V through X), to improve the set of scenarios, and to begin efforts to utilize the background material and scenarios to define policy issues and alternatives. The chapter that had been most seriously criticized at the second workshop had been the one dealing with the relationship between the nature and extent of electrification and the quality of life in the country. Specifically, the panel felt that a correlation between the two had not been adequately demonstrated; the panel recommended that this chapter should not be included in the final report if correlation could not be better shown. During the interim period, the core staff sought to find such demonstration in a number of areas. Particular efforts were made to establish a two-step correlation utilizing research conducted by Dr. Ken Land of the Department of Sociology of the University of Texas at Austin. Dr. Land has shown a positive correlation between industrial productivity and improvement in important measures of quality of life. If a correlation could be shown between electric power usage and industrial productivity, a secondary relationship between electric power usage and quality of life could be inferred. Unfortunately, despite a general acceptance in the electric power industry of the hypothesis that increased electrification results in increased productivity and a plethora of anecdotal material, no acceptable correlative data were found. Hence, the chapter is not included in this report.

During this period, the staff began preliminary analysis of the impact that different electric power policies might have on the nation. As a first step, the staff examined the impact that different technological research, development, and demonstration policies would have on different segments of the population. The tools used in this analysis were impact wheels and stakeholder analysis (see Annex 9 and Annex 13). The results of this analysis are shown in Appendix G of the NSF final report.

J. CONDUCT OF THIRD WORKSHOP

The third workshop was held in Estes Park, Colorado, on August 30-September 1, 1982. The following panel members attended: Kaufman, Thompson, Wood, Yu, and Zentner. Dr. G. Patrick Johnson of the National Science Foundation also attended as an observer. At the start of the session, the attendees reviewed and suggested further improvements to the background chapters, the scenario set, and the initial stakeholder analysis. However, the major efforts of the workshop were directed toward the identification and organization of the principal findings of the assessment. This process was initiated with a nominal group exercise (see Annex 5), which identified 32 major findings. These were then organized into seven overarching principles. Each attendee was then asked to write a short paper presenting his views of the findings to date associated with each of these principles. These papers formed the preliminary foundation for the analyses presented in the "Principles" chapter (Chapter IV) of this report.

K. PREPARATION OF DRAFT ANALYSIS CHAPTERS

After the third workshop, the core staff focused primary efforts on the three analysis chapters (Chapters II, III, IV). Chapter II work included the organization of the scenarios into their final format, while work on Chapter IV involved the organization of all of the data, observations, and insights produced during the assessment into a cohesive, comprehensive format. The production of Chapter III, however, involved new analysis by the core staff of the need for flexibility in electric power policy decisions and of possible means of gaining that flexibility. The methodology used for this analysis was alternate scenario planning (see Annex 1). Specifically, the staff studied the policy action set (Policy Set Alpha) which the scenario development exercises had associated with the "Average Future" Scenario. The staff then projected the impact this policy set would have on each of 20 different groups in the country (stakeholders). The staff next examined the differential impact that following Policy Set Alpha would have on the stakeholders if the "Nuclear Resurgence" Scenario were to come about. Possible modifications to Policy Set Alpha that would make it more responsive to the possibility of the "Nuclear

Resurgence'' Scenario were then suggested. A brief analysis of the implications of each of these modifications was made. This process was followed for each of the other four ''non-base case'' scenarios. Finally, modifications to Policy Set Alpha that would best accommodate all possible scenarios were suggested, again with an analysis of possible consequences. A final version of all of these analyses is given in Chapter III.

L. CONDUCT OF FOURTH WORKSHOP

The fourth workshop was conducted in Washington, D.C., on November 27-29, 1983. The following panel members attended: Brown, Davis, Kaufman, Thompson, and Wood. Drs. Johnson and Huband of the National Science Foundation also attended as observers. The principal activity of this workshop was a review of the analysis chapters, as well as the project as a whole. Improvements for the final reports, including a reordering of chapters, were suggested and methods for promulgating the results of the report were discussed.

M. PREPARATION AND REVIEW OF FINAL REPORT

Immediately after the fourth workshop, preparation of the final draft report began. Portions of the final draft report were sent to panel members and a small number of other people for review and comment. The final draft report was sent to panel members for final review in two parts: (1) the background chapters and (2) the overview and analysis chapters. Comments from panel members were incorporated into the final report in two ways. Where possible and appropriate, the comments were incorporated directly into the text of the report. When the comments did not appear to represent the consensus of the panel and staff, or when for other reasons it did not appear appropriate to incorporate the comments directly, they were attributed to individual panelists and included in Part Four of the report to NSF (and of this book).

Incorporation of comments by the panel into the final draft report completed work by the core staff on the assessment. The final report was then sent to the issue review panel for their comments and observations. These comments and observations were also included in Part Four of the NSF report.

N. ANALYSES OF METHODOLOGIES EMPLOYED

Regarding the two basic elements of this assessment's methodology, the use of alternative scenarios and a core staff/expert panel assessment team, it is the consensus within the core staff that both elements worked well. The alternative scenario approach provided detailed contexts within which to test realistically available policy options under a variety of disparate assumptions about the future. This made it possible to identify non-obvious electric power policy issues and implications. Comparisons among scenarios highlighted the matters on which policy flexibility will be important and helped to focus thinking about the ways in which particular types of flexibility might be obtained.

The continuing interest and involvement of expert panel members over the nearly two-year period of the assessment contributed immensely to the progress of the assessment. These members, by virtue of their regular professional involvement with subject matters pertinent to the assessment, brought to the assessment both factual knowledge and theoretical insights that would have been very difficult, if not impossible, to obtain from literature or other secondary sources. The daily reliance on a core staff and intermittent and, occasionally, intense reliance on the expert panel members made it possible for the assessment to effectively and efficiently draw on high-quality knowledge and insight.

ANNEX 1
USE OF SCENARIOS IN POLICY PLANNING

One method of examining and presenting the interactions between projections of a number of technical and non-technical factors is to combine them into an integrated "story" of the future. Such stories are often referred to as "scenarios" and are quite useful in technology planning. Since a scenario can present a multifaceted portrait of the future, it allows realistic consideration of "real world" situations and adds both breadth and depth to decisions about future operations. Moreover, because of its "story" orientation, it often allows the organization to consider alternative futures in a serious but non-threatening manner. In fact, many organizations in recent years have found it useful to employ a series of alternate futures (usually two to four) to encourage the development of more flexible plans.

One of the most important steps in the development of scenarios is the selection of one or more themes to be considered. Normally, when a single scenario is used, the people preparing the scenario will employ a series of most likely single factor projections to develop a "most likely scenario." If alternate scenarios are developed, themes can be chosen that will cause organizational planners to consider various exogenous developments that might materially affect the organization. (In practice, the most commonly used alternate scenario themes are "optimistic" and "pessimistic" ones. These are often not very useful because they do not require focusing on specific potential problems.)

A number of methods are available for preparing scenarios. However, regardless of the method of generation employed, scenarios should be plausible, self-consistent, relevant, thought provoking, and appropriately comprehensive. When alternate scenarios are used, the scenarios should be similar in format and scope and sufficiently different to evoke different planning considerations.

Although computer programs and specialists in scenario preparation can be useful in preparing scenarios, usually organizational personnel can develop scenarios that can adequately serve planning requirements. The first step is normally to select the factors to be considered in the scenario, to determine the present status of each factor, and to project how each factor is most likely to change in the future. Experience has shown that the selection of factors can often be best accomplished in a series of steps. First broad "areas of interest" that need to be considered in the planning process are determined. The nature of these areas of interest will depend on the particular organization involved. Next, those factors within each area of interest that might be significant to the organization are listed. This list of factors typically will be too large for detailed consideration, and, therefore, the factors list must be reduced to a manageable size. (The number of factors to be included, obviously, is dependent on the circumstances involved; however, 40 to 80 has proven to be a reasonable range for both manageability and comprehensiveness.)

The next step in scenario development is to assemble the factors into related groups. This grouping not only simplifies the examination of factor interrelations, but also makes it easier for scenario readers to absorb the large volume of information contained in most scenarios. Once factors are grouped, the present status of each factor is defined and projections are made of how each factor is most likely to change in the future. When these projections have been completed, the individual projections are compared; inconsistencies, contradictions, and unreasonable developments are identified and resolved. The sum of the individual "most likely" factor projections forms the basis for the most likely scenario. Normally, projections are expressed in quantitative terms, and it is often desirable to translate them into a narrative format to make the scenarios more interesting and understandable to the people who will use them.

If alternate scenarios are to be developed, a theme must be chosen for each. Once themes have been determined, the people preparing the scenarios re-examine the original factor projections to see how each might be changed if the future suggested by each alternate theme were to occur. Each set of new factor projections, once examined for internal consistency, serves as a guide for preparing an alternate scenario based on the chosen theme. When all scenarios are completed, they must be checked for consistency, clarity, and completeness and organized into a format for practical use.

It is obvious that use of alternate scenarios in planning can involve considerable expenditure of time and effort. However, experience has shown that the technique illustrates for planners the importance of flexible planning, serves as an excellent interorganizational communication tool, provides a vehicle for integrating relevant technical and non-technical factors in the planning process, provides a basis for an effective monitoring plan, and identifies important decisions that will have to be made in the future.

(For additional information see: John H. Vanston, Jr., W. Parker Frisbie, Sally Cook Lopreato, and Dudley L. Poston, Jr., "Alternate Scenario Planning," *Technological Forecasting and Social Change*, 10 [1977], 159-80.)

ANNEX 2
CORE STAFF

Dr. John H. Vanston, Jr.
Principal Investigator and
Project Director
Technology Futures, Inc.
411 W. 13th Street, Suite 801
Austin, TX 78701

Dr. Ken Roberts
Co-principal Investigator
Southwestern University
P.O. Box 421, SU Station
Georgetown, TX 78626

Mr. David O. Frederick
Co-principal Investigator and
Deputy Project Director
Scientific Foresight, Inc.
411 W. 13th Street, Suite 801
Austin, TX 78701

Ms. Donna C. L. Prestwood
Research Associate
Technology Futures, Inc.
411 W. 13th Street, Suite 801
Austin, TX 78701

Dr. Parker Frisbie
Population Research Center
University of Texas at Austin
18th Floor, Main Building
Austin, TX 78712

Mr. Rick Lowerre
Attorney at Law
2301 Rio Grande
Austin, TX 78705

Dr. Peter Zandan
Technology Futures, Inc.
411 W. 13th Street, Suite 801
Austin, TX 78701

Mr. Georg Zappler
Technology Futures, Inc.
411 W. 13th Street, Suite 801
Austin, TX 78701

Dr. Dudley Poston, Jr.
Population Research Center
University of Texas at Austin
18th Floor, Main Building
Austin, TX 78712

Ms. Carolyn Vanston
Administrative Director
Technology Futures, Inc.
411 W. 13th Street, Suite 801
Austin, TX 78701

Mr. Patrick Drew
Technology Futures, Inc.
411 W. 13th Street, Suite 801
Austin, TX 78701

ANNEX 3
EXPERT PANEL

Dr. Sanford Berg
Department of Economics
Matherly Hall, Room 106
University of Florida
Gainesville, FL 32607

Dr. Richard D. Brown
The MITRE Corporation
1820 Dolley Madison Blvd.
McLean, VA 22102

Dr. David Howard Davis
International Energy Associates, Ltd.
Now: Gettysburg College
Gettysburg, PA 17325

Dr. Gordon A. Enk
Institute for Man and Science
Now: Gordon A. Enk & Associates, Inc.
Makely House
Medusa, NY 12120

Dr. Kirby Holte
South California Edison Co.
P.O. Box 800
2244 Walnut Grove Ave.
Rosemeade, CA 91770

Dr. Steve Murdock
Dept. of Rural Sociology
Texas A&M University
College Station, TX 77843

Dr. Grant Thompson
Conservation Foundation
1717 Massachusetts Ave., N.W.
Washington, DC 20036

Dr. John Wardwell
Department of Sociology
Washington State University
Pullman, WA 99146

Mr. David Wood
Energy Laboratory
Massachusetts Institute of Technology
Room E40-391
No. 1 Amherst Street
Cambridge, MA 02139

Dr. Oliver Yu
Electric Power Research Institute
3412 Hillview Avenue
Palo Alto, CA 94304

Mr. Alvin Kaufman
Congressional Research Service
Room Lm 423, ENR/S
Library of Congress
Washington, DC 20540

Mr. Rene Zentner
Shell Oil Company
Now: University of Houston Law School
Law Center
4800 Calhoun
Houston, TX 77004

ANNEX 4
ISSUE REVIEW PANEL

Dr. Jack Allison
Department of Electrical Engineering
Oklahoma State University
Stillwater, OK 74078

Dr. Doug Bauer
Economics and Finance
Edison Electric Institute
1111 19th Street, N.W.
Washington, DC 20036

Dr. Raphael Kasper
Environmental Studies Board
National Research Council
2101 Constitution Avenue
Washington, DC 20418

Mr. Blair Ross
Energy Resource Planning
American Electric Power Service Corp.
Columbus, OH 43215

ANNEX 5
NOMINAL GROUP CONFERENCING

Nominal group (NG) conferencing is a technique designed to improve the use of expert opinion. It is most effective when a small panel of experts, five to seven, are available for approximately one-half day. An NG conference is normally conducted in five phases. First, the moderator explains how the conference will be conducted, details its objectives, and presents the panel with a question or problem. Next, the panel silently considers the question or problem for approximately 20 minutes and lists as many feasible solutions as possible. In the third phase, each panelist, in turn, writes one solution on a chalkboard or large sheet of paper. This is done silently and all panelists read each solution as it is posted. After each panel member has written one solution, the process is repeated with each person listing a second solution. This procedure is continued until all solutions have been posted. When this is completed, the discussion phase begins. The moderator reads each solution and encourages explanations, expansion, and discussion. During this discussion, posted solutions may be combined, modified, subdivided, and so on and new solutions may be added. In this phase of the conference, active participation by all panel members is encouraged. When discussions have run their course, the final scoring phase is begun. Each panelist is asked to rate each of the solutions on a formal basis. Possible rating schemes include rank ordering, numerical ordering by importance, or selection of five most important solutions. The exact rating system utilized will depend on how the results will be used.

Experience has shown that NG conferencing can be a very useful tool for eliciting imaginative ideas from a group. The method ensures that all members actively participate in the conference and minimizes many of the social dynamic problems associated with committee meetings.

ANNEX 6
SCENARIO AREAS OF INTEREST

Transportation
Population
Economy
Raw materials/resources
Law and regulations
International affairs
Natural environment
Health and welfare
Life-style preferences
Energy demands
Energy supply—traditional
Energy supply—unconventional
Technology
Urbanization/central city
Labor force
Communication/information
Public fears
Military
Agriculture
Politics
Education
Miscellany

ANNEX 7
SCENARIO CRITICAL FACTORS
(Concept Level—Some Not Yet Reduced to Operational Variables)

1. Defense budget as a percentage of GNP
2. Refiner acquisition cost of crude oil (domestic/imported)
3. Size of civilian labor force
4. Shrimp catches (total fish)
5. Number of working women
6. Percentage of average work week at home
7. Sectoral distribution of population (agriculture, manufacturing, service: private/government)
8. Automation of work
9. Percentage of agriculture sales attributable to large farms
10. Ratio of electric storage to production costs
11. Total agricultural sales
12. Real disposable income
13. U.S. GNP
14. Electricity cost per kilowatt hour: industrial and residential/commercial
15. U.S. inflation rate (1980s, 1990s, 2000s) by major components and average
16. Capital cost/Mw capacity of power plants (coal, nuclear, solar photovoltaic)
17. Measure of technology subsidies
18. Capital formation rate
19. Capital allocation by sector
20. Measure of corporate structure/centralization
21. Energy efficiency of manufacturing sector
22. Size of part-time civilian labor force
23. Water available for electricity production—by region
24. Annual per capita miles private travel
25. Percentage of annual per capita miles private travel intracity and intercity
26. Percentage of annual per capita miles intercity that is by mass transit
27. Percentage of annual per capita miles intracity that is by electric power
28. Average household size
29. Number of households headed by a single person
30. Number of households below poverty line
31. Hours of leisure time per person
32. Number of people belonging to top five environmental groups
33. Private U.S. assets located abroad

34. Population distribution by metropolitan and rural (four regions)
35. Metropolitan population distribution by central city and suburban
36. Population density
37. Population size
38. Immigration measure: legal and illegal
39. Percentage of population 65 years and older
40. Percentage of population that is minority
41. Percentage of population having two or more years education beyond high school
42. Measure of foreign penetration of U.S. markets/foreign purchase of U.S. assets
43. Value of U.S. exports/U.S. ownership of foreign assets
44. World GNP (excluding U.S.)
45. Air pollution index (maybe areas violating SO_x)
46. Water pollution index
47. U.S. park land acreage
48. Measure of actual health risks of power generation
49. Measure of perceived health risks of power generation
50. Measure of relative strength of city, state, and federal governments (maybe number of votes or budget size)
51. Level of international trade
52. Measure of internal political factionalism (number of candidates or votes)
53. Measure of environmental regulation (budgets or workforce size)
54. Total U.S. energy consumption
55. Energy consumption composition:
    - coal (excluding synthetic gas and oil)
    - oil/imported
    - oil/domestic (including shale and synthetic oil)
    - natural gas/imported
    - natural gas/domestic (including synthetic natural gas)
    - nuclear
    - hydropower
    - solar (thermal and photovoltaic, excluding wind, biomass, etc.)
    - other (wind, biomass, ocean power, geothermal, etc.)
56. Annual domestic electricity production (kilowatt hours)
57. Composition of electricity production (kilowatt hours), 2000, 2010, and 2030
    - coal/conventional
    - coal/unconventional (fluidized bed, magnetohydrodynamics)
    - nuclear fission
    - nuclear fusion
    - oil and gas
    - solar photovoltaics
    - solar power tower/satellites
58. Measure of utility institutional structure (number of investor-owned utilities/number of municipals)
59. Measure of industrial cogeneration
60. Electricity transmission distance (by region)
61. Electricity transmission efficiency
62. Average capacity of new coal/nuclear plants
63. Time lag between decision to build power plants (coal, nuclear) and power is available
64. Superconductivity in electricity generation
65. U.S. primary energy consumption by sector
    - electricity production
    - industrial (excluding agricultural)
    - residential and commercial

agricultural
transportation
feed stock

66. Acceptable reserve margins by region
67. World GNP/energy consumption ratio
68. Percentage of U.S. electricity generation from plants of less than ten kilowatts capacity
69. Foreign ownership of U.S. assets
70. Coal consumption composition (first use)

ANNEX 8
THE DELPHI SURVEY METHOD

In any ongoing company or organization, there resides an imposing reservoir of talent, experience, and training—scientists, engineers, salespeople, technicians, and so on. Obviously, a well-formulated planning program will take advantage of the knowledge and wisdom of this collection of experts. Since the contribution of any single expert will be colored by personal biases and limited by the scope of the individual's experience base, it is often desirable to gather input from a group of experts. This can be accomplished by the formation of a committee. Although committees can be very valuable, there are social dynamics embedded in even the best-organized committees, which limit their value. Often, the ideas of people with a great deal of insight and information either are not presented or else are not properly considered. To take advantage of the capabilities available in a collection of experts, while minimizing the limitations imposed by committee dynamics, a number of special expert opinion techniques have been developed. One of the most widely known of these techniques is the Delphi survey, originated and developed by the Rand Corporation. Since the original Rand Delphi study, conducted for the Air Force in the early 1950s, hundreds of government, industrial, and private Delphi surveys have been conducted throughout the world. During this time, many variations of the basic Delphi process have been generated. However, any predictive Delphi should have the following characteristics: original input of opinion by experts, idea feedback procedures, and a standardized display of results. In most Delphi surveys, each participant's estimates and comments remain anonymous to other participants.

One formulation of the Delphi procedure involves four specific rounds. In the first round, the experts are asked to estimate when they expect each of a number of events to occur (or other similar quantitatively oriented questions). When answers are received, they are tabulated, and, in round two, sent to respondents who are requested to reconsider their original projections. Respondents whose round two answers fall in the upper or lower round one quartiles are requested to provide reasons for their estimates. In the third round, respondents are sent the retabulated totals, together with the non-attributed comments gathered in round two. Respondents are requested to once again consider their previous projections and are invited to add comments, if desired. For the final round, the person conducting the survey retabulates and distributes all projections and comments.

Although the conduct of a Delphi survey, on the surface, appears simple, successful use of the technique presents a number of serious challenges: (1) choosing appropriate and useful questions, (2) wording questions so they are neither ambiguous nor leading, (3) selecting and attracting a panel of diverse and knowledgeable experts, (4) maintaining the interest of panelists throughout the survey (which may take several months to complete), and (5) utilizing the results of the survey effectively.

The results of a Delphi survey may be useful to the decision maker in a number of ways. The mean value of estimates gives the decision maker a projection of when a given development may be expected to occur, while the spread of these values can give him or her an indication of the degree of agreement between experts. The survey also may be used to compare the estimates of one group of participants with other groups. Where major differences exist, further study would be advised to uncover differences in perception, information, or analysis.

Often, the most valuable results of a Delphi survey are the identification and explanation of unconventional wisdom. A person knowledgeable in a given field who maintains a non-consensus position throughout a Delphi survey may have unusual insights, ideas, or information. These may be expressed in the written comments in the Delphi survey, or they may be more effectively elicited by personal interview.

(For further information on Delphi surveys, see: H. Linstone and M. Turoff, *The Delphi Method* [Reading, Mass.: Addison-Wesley, 1975].)

ANNEX 9
STAKEHOLDER ANALYSIS

The stakeholder analysis process takes a specific set of policies and analyzes the impacts on the stakeholder and the possible pressures that can result from the policy implementation. This is done against a background, or scenario, to provide a realistic context for analysis. The term "stakeholder" is used in a generic sense to identify the various groups influencing or impinged upon by a policy decision. They are any group whose collective behavior can be directly affected and/or can directly affect the policy's future, but that may or may not be under the policy's direct action (for example, a decision aimed directly at utilities will have indirect effects on customers, suppliers, regulators, etc.). Each group believes that it has a legitimate stake in the policy. The stakeholder perspective forces the explicit recognition of all groups that have a stake in the (to be) implemented policy.

Stakeholder analysis is based on the notion that if policy makers are to gain some measure of understanding of external factors, as they affect and are affected by policy making, two important aspects must exist. First, a framework must be established to integrate explicitly the awareness of the external environment into the overall making of policy. Next, actions must be taken in the policy-making process that minimize impacts and prevent unintended implications.

Two basic principles underlie the stakeholder analysis framework for policy makers. First, the central goal is to minimize adverse impacts and unintended consequences of policy implementation. Second, the most efficient and effective strategy is to keep policy matter simple. Therefore, early awareness of possible issues that might affect several stakeholders simultaneously is advantageous to policy makers. The analysis provides an identification of the stakeholder groups and an assessment of the stakeholders' present behaviors and probable future behaviors within the context of alternate scenarios.

ANNEX 10
RASHOMAN ISSUES

Issues	Subpanel A	Subpanel B
Economic Considerations		
General	5	6
Government financial policies	19	6
Business and industry	15	6
Finance and investment	11	2
Gross National Product	1	1
Liability insurance	3	—
Social services and support	4	—
Subtotal: Economic considerations	58	21
Security Issues		
National	3	4
General	5	1
Subtotal: Security issues	8	5
Social and Demographic Issues		
Attitudes	8	12
Household	1	1
Income	1	1
Education	2	—
Geographical distribution	6	1
Employment structure	2	5
Racial and class distribution	1	4
Political	2	1
Subtotal: Social and demographic issues	23	25
Environment		
Natural environment	11	6
Health and safety	9	2
Subtotal: Environment	20	8
Foreign Policy Issues		
Balance of trade	2	7
Foreign relations	2	6
Subtotal: Foreign policy issues	4	13
Government Control Issues		
Regulatory	20	7
Federal/regional/state/local control	5	4
Standardization	1	—
Public ownership	3	1
Subtotal: Governmental control issues	29	12

Issues	Subpanel A	Subpanel B
Energy Production		
General	3	2
Sources	4	14
Plant construction and location	5	6
Natural resources	1	7
Supply/demand	7	3
Reliability of supply	9	2
Cogeneration	6	3
RD&D	9	1
Innovation and technology	10	8
Subtotal: Energy production	54	46
Totals:		
Economic considerations	58	21
Security issues	8	5
Social and demographic issues	23	25
Environment	20	8
Foreign policy issues	4	13
Governmental control issues	29	12
Energy production	54	46
Combined Totals	196	130

ANNEX 11
MORPHOLOGICAL ANALYSIS

Morphology is the study of form. A matrix of all theoretically conceivable combinations of approaches and configurations is constructed. Known systems are traced on the matrix, which is then inspected to identify untried combinations. Hopefully, some of these may be promising and emerge in the future. Morphological analysis identifies known systems and predicts future systems by displaying possibilities that are not yet in use or even explored. Some industrial practitioners believe that its major value lies in encouraging creativity in technical and managerial thinking.

Originally, the approach was intended for the identification of all the possible answers to a problem or all the different means of achieving a given effect. Today, the method. They may call for consideration by a panel of experts whereby the synergy concepts are identified through new configurations and untried combinations in the matrix.

The first stage in the morphological method requires rigorous analysis to identify the component stages of a given system or a known sequence of events. These events or stages may describe a complex product, such as an airplane or motor car, or may concern a simpler thing, such as Scotch Tape. Alternatively, they may describe a process, such as the fastening of a nut and bolt, the dyeing of a textile fiber, or the manufacture of glass products.

After identification, the component stages or parameters are coded A, B, C, and so on and are entered in the first vertical column of a matrix array of boxes. In the second vertical column, against each of the lettered boxes, are noted the particular means employed in the operation of the system as it is currently known. The second vertical column is headed 1, so that a description of the known system and its elements may be designated A1, B1, C1, and so on.

Alternatives for the achievement of any or all of the individual stages of A, B, C, and so on are then entered in subsequent boxes so that a series of alternates to A1-B1-C1, and so on are displayed. If a full series of alternatives is identified, either from obvious and other known means or from intuitive ideas of possibilities, it then becomes possible to describe a second system as A2-B2-C2, and so on or even a third system as A3-B3-C3, if third alternates to each parameter are available, and so on. Further possibilities arise when new systems are not derived from the same vertical column but are taken as permutations between columns. Similarly, dozens of new possibilities may be suggested.

The next step is to inspect these various combinations (1) to ascribe a meaning or understanding of the factors represented in the new system or subset or to classify it as meaningless or impossible because of an obvious physical or natural constraint and (2) to decide whether the new system is feasible or has gaps in its practical achievement.

These questions are often difficult, but they form an important part of this method. They may call for consideration by a panel of experts whereby the synergy of working on the matrix is exploited toward an understanding of what at first may not be discernible. The response to these challenges is an important part of the creativity in the whole process.

For more information on Morphological Analysis, see James R. Bright, ed., *Technological Forecasting for Industry and Government* (Englewood Cliffs, N.J.: Prentice-Hall, 1968), pp. 190-97. Also, Robert Ayres, *Technological Forecasting* (New York: McGraw-Hill, 1969).

ANNEX 12
CRITERIA FOR ISSUE EVALUATION AND ANALYSIS

Amenability to policy decisions

Pervasiveness in social structure

Electric power relationship

Efficient use of resources, including technology

Equity, that is, public welfare

Amenability to assessment evaluation

Contribution to environment quality

Impact on national security

Contribution to personal freedom

Avoidance of high negative risks or dangers (undesirable consequences); irreversibility

Number of states affected

Timeliness

Economic stability and growth

Integration into social system

Consumer preferences

Lack of previous study

ANNEX 13
IMPACT WHEELS

An impact wheel is an idea-recording tool. With it, one can use a panel of experts to identify higher-order, often non-obvious, impacts and implications of selected decisions or developments. Use of the wheel starts with the specification of an event, trend, technological advance, or societal development that one wishes to analyze. This item is placed in a circle at the center of a large piece of paper. The panel is then asked to identify the direct consequences of the occurrence of this central item. These consequences are recorded in a series of circles surrounding the central circle. Each of these circles are joined to the central circle by a single line. Once five to seven direct consequences are identified, the panel is asked to suggest possible implications that might arise from each of these first order consequences. These suggestions are also recorded in circles that are, in turn, joined to appropriate first order consequence circles by a double line to indicate that they are second order consequences. This process is continued for third and higher-order consequences to the extent that it is useful.

The use of this technique will not give a complete set of impacts. However, experience has shown that it can be a very potent means of identifying unexpected or easily overlooked opportunities, problems, and interrelationships. It is often useful to have more than one group examine the same propositions and compare results. A higher-order implication that appears repetitively may hold particular significance for the organization.

Normally, four to six people plus a moderator make a proper size group for conducting an impact wheel analysis. Analyses usually require 20 to 30 minutes, and a given group can complete about three or four analyses before fatigue decreases the value of the technique. When properly conducted, they can be used to identify both potential future needs and means for meeting those needs.

For additional information on Impact Wheels, contact Mr. Joel Barker, Infinity Limited, Inc., 1301 Cherokee Avenue, West St. Paul, Minnesota 55118.

APPENDIX B

Expert Opinion Survey—Final Results

This appendix contains the results of the expert opinion survey conducted during the winter of 1981. Statistical tabulations and participants' comments on all survey items are presented, though few survey participants were queried on each item.

One hundred and twenty-one authorities agreed to participate. These participants, except for institutional participants, were loosely categorized as knowledgeable in one of four subject areas: demography, economics, environment, or energy. Items to be covered in the survey were divided into the same four subject matter groups and the item groups and participant groups were matched for the survey. Institutional participants received items from all subject matter areas; a few items were sent to more than one group of participants.

The first round of the survey asked for projections on 57 items. It was mailed to the 121 participants during the last week of September 1981; 79 responded. Participants whose responses to one or more items were statistically abnormal were contacted by members of the survey staff to inquire as to the rationales supporting those atypical responses. The rationales received by the survey staff, the written comments returned with round 1 item projections, and the round 1 statistics were mailed to the 79 respondents during the last week in November. With this information, they could reconsider, if they desired, their responses to the initial mailing. Forty-one of the 79 participants responded. Statistics and comments from both the first and second iteration of the survey are presented (separately) in this appendix.

SURVEY PARTICIPANT LIST

Jan Acton	Rand Corporation
Peter Auer	Cornell University
Ben Aguirre	Texas A&M University
Martin Baughman	University of Texas
Sanford Berg	University of Florida
Alan Beringsmith	Pacific Gas and Electric
Alan Blinder	Princeton University
Harley Browning	University of Texas
Eugene Buch	Congressional Research Service
Frank Castellon	Kaiser Aluminum and Chemical Corporation
William Catton	Washington State University
Peter K. Clark	Federal Researve Board of Governors
David Davis	International Energy Associates
T.W. Day	Exxon
Diana DeAre	U.S. Bureau of the Census
Isaac W. Eberstein	Florida State University
Lowell Endahl	National Rural Electric Cooperatives Assoc.
Gordon Enk	Institution of Man and Science
Richard L. Forstall	Consulting Sociologist
David Freeman	Tennessee Valley Authority
William Freudenburg	Washington State University
Bob Fri	Energy Transition Corporation
Omer Galle	University of Texas
Mike Gent	National Electric Reliability Council
John Gray	International Energy Association
Patricia Guseman	Texas A&M University
Sam Hadaway	Public Utilities Commission
Jim Harding	Friends of the Earth
Bob Harris	Clement Associates
Bill Hogan	Kennedy School of Government
Kirby Holte	South California Edison Company
W. R. Johnson	Retired, Pacific Gas and Electric
John D. Kasarda	University of North Carolina
Emil Kasum	Philadelphia Electric Company
Henry Kelly	Office of Technology Management
William Kelly	University of Texas
Dan Kohler	Rand Corporation
Charles Kolstad	Los Alamos National Labs
Charles Komanoff	Komanoff Energy Associates
Bob Kuenne	Princeton University
Ron Kutschev	U.S. Department of Labor

Tom H. Lee	Massachusetts Institute of Technology
Dennis Livingston	Tennessee Valley Authority
Richard Machalek	Trinity University
Rene Males	Electric Power Research Institute
Margaret N. Maxey	Energy Research Institute
Ed McCrackin	South California Edison Company
Larry McCray	National Academy of Science
Alex McIntosh	Texas A&M University
Bridger Mitchell	Rand Corporation
Stephen Murdock	Texas A&M University
Edward Murguia	Trinity University
Kumar Patel	Bell Telephone Labs
Harry Perry	Resources for the Future
Forrest H. Pollard	University of Arkansas
Ronald Rindfuss	University of North Carolina
Lewis H. Roddis, Jr.	Consulting Engineer
Eugene Rosa	Washington State University
Dick Rowberg	Office of Technology Assessment
Jack Schenk	Gulf States Utilities
Robert L. Seale	University of Arizona
William Serow	Florida State University
John W. Simpson	Simpson Business Services
David F. Sly	Florida State University
Chauncy Starr	Electric Power Research Institute
Bill Stitt	I.C.F.
Ian Straun	South California Edison
Theresa Sullivan	University of Texas
Grant Thompson	Conservation Foundation
W.F. Thompson	Philadelphia Electric
Bob Trumbule	Congressional Research Service
Victoria Tschinkel	Department of Environment Regulation
Robert Uhrig	Florida Power and Light
Gregory S. Vassell	American Electric Power
John Wardwell	Washington State University
Alvin Weinberg	Oakridge Associated University
Herbert Woodson	University of Texas
Oliver Yu	Electric Power Research Institute
Rene Zentner	Shell Oil Company
Chuck Zielinski	Wald, Harkrader, and Ross
James Zuiches	National Science Foundation

CAPTIONS FOR ALL SURVEY QUESTIONS

ITEM 1. POPULATION SIZE (MILLIONS OF PERSONS IN THE U.S.)

ITEM 2. METROPOLITAN POPULATION (AS A PERCENTAGE OF U.S. POPULATION)

ITEM 3. POPULATION LIVING IN CENTRAL CITIES (AS A PERCENTAGE OF THE U.S. METROPOLITAN POPULATION)

ITEM 4. U.S. CIVILIAN LABOR FORCE (AS A PERCENTAGE OF U.S. RESIDENT POPULATION)

ITEM 5. WOMEN WORKING FULL TIME (AS A PERCENTAGE OF ALL INDIVIDUALS 16 YEARS OF AGE OR OLDER WHO WORKED FULL TIME DURING THE YEAR)

ITEM 6. U.S. PART-TIME CIVILIAN LABOR FORCE (AS A PERCENTAGE OF TOTAL U.S. CIVILIAN LABOR FORCE)

ITEM 7. SECTORAL DISTRIBUTION OF THE LABOR FORCE (AS A PERCENTAGE OF U.S. CIVILIAN LABOR FORCE)

ITEM 8. MEDIAN FAMILY INCOME (1980 U.S. DOLLARS)

ITEM 9. PERSONS IN POVERTY (AS A PERCENTAGE OF THE U.S. POPULATION)

ITEM 10. ELDERLY POPULATION (U.S. POPULATION 65 YEARS OR OLDER, AS A PERCENTAGE OF U.S. POPULATION)

ITEM 11. U.S. BLACK AND HISPANIC POPULATIONS (AS A PERCENTAGE OF U.S. POPULATION)

ITEM 12. LEGAL IMMIGRATION (MILLIONS OF PERSONS ANNUALLY)

ITEM 13. STOCK OF ILLEGAL IMMIGRANTS (MILLIONS OF PERSONS)

ITEM 14. AVERAGE HOUSEHOLD SIZE (NUMBER OF PERSONS)

ITEM 15. HOUSEHOLDS HEADED BY A PRIMARY INDIVIDUAL (AS A PERCENTAGE OF TOTAL U.S. HOUSEHOLDS)

ITEM 16. EDUCATION (PERCENTAGE OF THE U.S. POPULATION 25 YEARS OR OLDER WHO HAVE COMPLETED AT LEAST 13 YEARS OF SCHOOL)

ITEM 17. PORTION OF AVERAGE WORK WEEK AT HOME (INDEX NUMBER, 1980=100)

ITEM 18. LEISURE ACTIVITY (HOURS PER WEEK)

ITEM 19. U.S. INFLATION RATE (ANNUAL PERCENTAGE CHANGE IN THE CONSUMER PRICE INDEX)

ITEM 20. U.S. GROSS NATIONAL PRODUCT (BILLIONS OF 1980 DOLLARS)

ITEM 21. INDEX OF WORLD GROSS DOMESTIC PRODUCT (INDEX NUMBER, 1975=100)

ITEM 22. AGRICULTURAL SALES (AS A PERCENTAGE OF GNP)

ITEM 23. NATIONAL DEFENSE EXPENDITURES (AS A PERCENTAGE OF GNP)

ITEM 24. FOREIGN PENETRATION OF UNITED STATES MARKETS (U.S. IMPORTS AS A PERCENTAGE OF U.S. GNP)

ITEM 25. VALUE OF U.S. EXPORTS (AS A PERCENTAGE OF U.S. GNP)

ITEM 26. RATE OF CAPITAL FORMATION (CAPITAL INVESTMENT AS A PERCENTAGE OF U.S. GNP)

ITEM 27. CAPITAL ALLOCATION FOR ELECTRICITY GENERATION (AS A PERCENTAGE OF TOTAL CAPITAL AVAILABLE)

ITEM 28. ECONOMIC CENTRALIZATION (PERCENTAGE OF TOTAL U.S. CORPORATE ASSETS CONCENTRATED AMONG THE LARGEST 500 NONFINANCIAL CORPORATIONS)

ITEM 29. WORLD TRADE (BILLIONS OF 1980 U.S. DOLLARS)

ITEM 30. VALUE OF U.S. OWNED ASSETS LOCATED ABROAD (BILLIONS OF 1980 DOLLARS)

ITEM 31. NET VALUE OF FOREIGN OWNED ASSETS IN THE U.S. (BILLIONS OF 1980 DOLLARS)

ITEM 32. CITIZEN SUPPORT OF ENVIRONMENTAL AND CONSERVATION MOVEMENTS (INDEX NUMBER, 1980=100)

ITEM 33. PRESERVATION OF FEDERAL LANDS (PERCENTAGE OF TOTAL FEDERAL LANDS WITHDRAWN FOR PARKS, WILDLIFE REFUGES AND WILDERNESS AREAS)

ITEM 34. AIR QUALITY IN URBAN AREAS (NUMBER OF AREA DAYS DURING WHICH POLLUTANT STANDARD INDEX EXCEEDED 100)

ITEM 35. FEDERAL EXPENDITURES FOR POLLUTION ABATEMENT AND CONTROL (AS A PERCENTAGE OF TOTAL NON-MILITARY FEDERAL EXPENDITURES)

ITEM 36. POLLUTION CONTROL EXPENDITURES BY ELECTRIC UTILITIES (AS A PERCENTAGE OF TOTAL ELECTRIC UTILITY CAPITAL EXPENDITURES)

ITEM 37. INDUSTRIAL HEALTH RISKS ASSOCIATED WITH ENERGY PRODUCTION (INJURIES AND DEATHS PER GIGAWATT YEAR)

ITEM 38. RELATIVE SIZES OF U.S. DOMESTIC GOVERNMENTS (FEDERAL,STATE,AND LOCAL GOVERNMENT CURRENT DOMESTIC EXPENDITURES AS A PERCENTAGE OF TOTAL GOVERNMENT CURRENT DOMESTIC EXPENDITURES)

ITEM 39 ANNUAL INTER-CITY PASSENGER MILES (BILLIONS)

ITEM 40. ANNUAL INTER-CITY MASS TRANSIT PASSENGER MILES (AS A PERCENTAGE OF TOTAL INTER-CITY PASSENGER MILES)

ITEM 41. FEDERAL SUBSIDY FOR ENERGY RESEARCH AND DEVELOPMENT (BILLIONS OF U.S. DOLLARS)

ITEM 42. U.S. ENERGY CONSUMPTION (QUADRILLION BTUs)

ITEM 43. U.S. ENERGY CONSUMPTION BY PRIMARY FUEL (HEAT CONTENT OF FUELS IN QUADRILLION BTU)

ITEM 44. COMPOSITION OF FUTURE U.S. COAL CONSUMPTION

ITEM 45. ENERGY CONSUMPTION BY END-USE SECTOR (EACH SECTOR AS A PERCENTAGE OF TOTAL ENERGY CONSUMPTION)

ITEM 46. NET U.S. ELECTRICITY GENERATION (BILLION KWH)

ITEM 47. NET ELECTRIC UTILITY POWER SUPPLY BY ENERGY SOURCE (EACH SOURCE'S OUTPUT AS A PERCENTAGE OF TOTAL)

ITEM 48. NET ELECTRIC UTILITY POWER SUPPLY FROM UNCONVENTIONAL SOURCES (EACH SOURCE'S NET OUTPUT AS A PERCENTAGE OF TOTAL UNCONVENTIONAL ELECTRICITY SUPPLY)

ITEM 49. INDUSTRY ELECTRICITY GENERATION (AS A PERCENTAGE OF TOTAL GENERATION)

ITEM 50. NATIONAL AVERAGE UTILITY ELECTRICITY END-USER PRICES (1980 CENTS/KWH)

ITEM 51. AVERAGE U.S. REFINER ACQUISITION COST OF CRUDE OIL (1980 DOLLARS/BARREL; ALL SOURCES)

ITEM 52. ENERGY EFFICIENCY IN MANUFACTURING (BTUs PER DOLLAR VALUE ADDED)

ITEM 53. PUBLICLY-OWNED UTILITY ELECTRICITY GENERATION (AS A PERCENTAGE OF TOTAL UTILITY ELECTRICITY GENERATION)

ITEM 54. LEAD TIMES FOR CENTRAL ELECTRIC POWER PLANTS (YEARS)

ITEM 55. CAPITAL COSTS OF COAL AND NUCLEAR ELECTRIC CAPACITY

ITEM 56. SOLAR PHOTOVOLTAIC ARRAY COSTS (1980 DOLLARS/PEAK WATT OUTPUT)

ITEM 57. SUPERCONDUCTING GENERATOR (YEAR OF FIRST COMMERCIAL APPLICATION)

ITEM 58. INTRACITY ELECTRIC-POWERED PERSONAL TRANSPORTATION

ITEM 1. POPULATION SIZE (MILLIONS OF PERSONS RESIDING IN THE U.S.) [1]

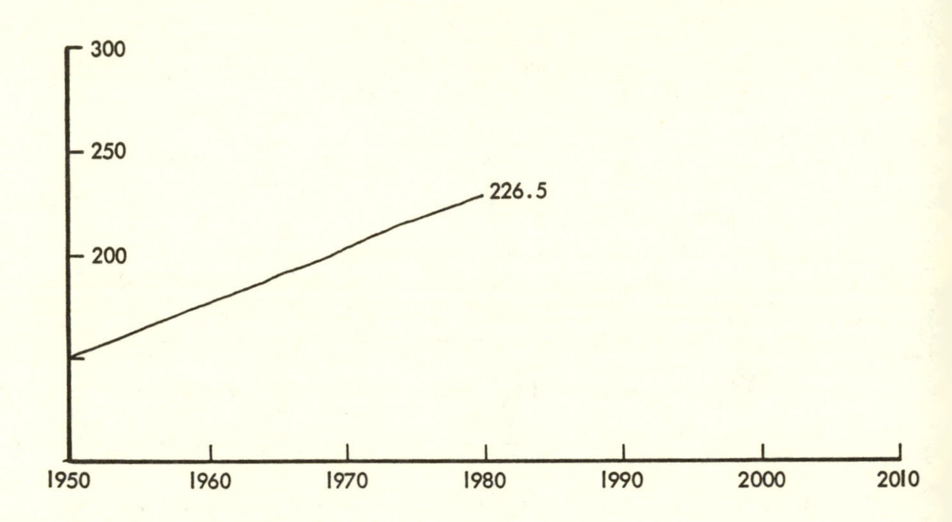

a. Trend Evolution: Please indicate on the above graph how you believe the trend will evolve over the next 30 years.

b. Point Values: Please indicate below your best estimates of the point values (vertical axis values) of the trend in the years 2000 and 2010

2000_____________ 2010_____________

c. Terminology:

d. Sponsor's Comments: The U.S. Bureau of the Census, Current Population Reports, series P-25, No.796, series II U.S. population projection for the year 2000 is 259,869,000.

e. Statistics:

	Round One		Round Three	
Responses:	30		13	
	2000	2010	2000	2010
Low:	220	240	250	260
Mean:	258.1	273.3	262.6	280.3
High:	280	300	267.5	300
Std. Dev.:	11.9	15.7	7.1	13.9

Round One Comments on Item 1 (Population Size)

518. I would expect that smaller average families, associated with low birth rates will produce a slight decline in the overall linear trend. No big surprises expected here.

528. Current census projections are underestimated - witness alleged 1980 census "overcount."

513. Above "best estimates" are simply based on Census Bureau's Series II projections.

525. Census projections leave out stock of "illegals," and they are not fully reflected in undercount.

508. Census Bureau median projections, as of 1976, increased due to higher immigration and a slower fertility rate decline.

529. The key variable to be considered is net international immigration. If it runs about 4 million/decade, Census Bureau will be close. I expect it to be higher, circa 6-8 million/decade.

523. The rate of growth will decrease.

906. I believe we are heading toward a stationary population by about 2040.

908. Systematic inflation, older populations, & expense of children (a "liability" in an economic sense within a family unit) will slow population growth.

511. Next decade will see a slight rise in the birthrate as the "baby boom" cohorts have children. The rate of total growth should continue to slow down and begin to stabilize near zero-growth by 2010.

522. I believe that annual rate of U.S. population increase will continue to decline between 1980 and 2010 relative to the 1970-1980 experience.

901. My estimates were obtained by averaging Series II & III and adjusting by the difference between the expected and actual 1980 population. This assumes that the overcount (or, actually, the underestimate) will not recur in future census years. I think that assumption is unrealistic. If the proportional "error" of 1980 is repeated in successive censuses, the point estimates would become: 2000=269,158 2010=285,645.

Round Three Comments on Item 1 (Population Size)

528. I stick by my estimate (280 million - 2000, 300 million - 2010). The Census Bureau cannot project immigration, and this is ultimately a political issue, not a demographic one. The pressures to admit new immigrants are going to build dramatically.

903. I'll stand by my Stage I forecast (260 million - 2000, 265 million - 2010). I expect population growth to slow in the next century reflecting affluence and deferred child-bearing among women in the U.S.

ITEM 2. METROPOLITAN POPULATION (AS A PERCENTAGE OF U.S. POPULATION) [2]

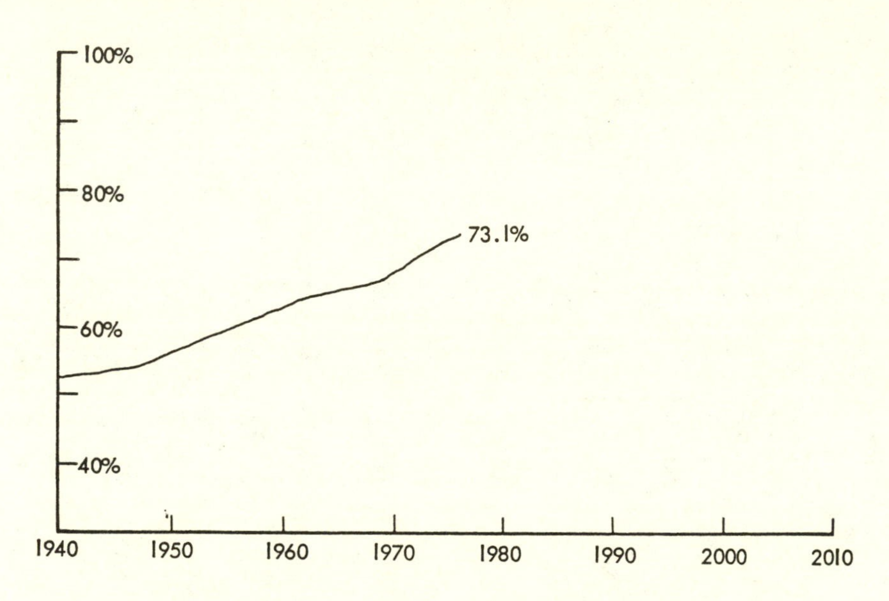

a. Trend Evolution: Please indicate on the above graph how you believe the trend will evolve over the next 30 years.

b. Point Values: Please indicate below your best estimates of the point values (vertical axis values) of the trend in the years 2000 and 201

2000______________ 2010______________

c. Terminology: The metropolitan population is defined as persons living in Standard Metropolitan Statistical Areas (SMSAs). The general concept o an SMSA is one of an integrated economic and social unit with a large popu tion nucleus. SMSAs are defined as the county containing a central city/ cities with a population of at least 50,000 persons, along with the contig counties which are economically and otherwise dependent upon the principal county. SMSAs are defined by the Office of Management and Budget (OMB) an the boundaries may be changed or adjusted periodically by OMB.

d. Sponsor's Comments:

e. Statistics:

	Round One		Round Three	
Responses:	30		13	
	2000	2010	2000	2010
Low:	60%	60%	68%	65%
Mean:	76.6%	77.4%	76.5%	76.8%
High:	87%	90%	87%	85%
Std. Dev.:	6.4%	7.8%	6.2%	7.0%

Round One Comments on Item 2 (Metropolitan Size)

518. Again I would expect a slight leveling off as past trends are modified by rural re-migration. I do not expect, however, that this will alter things much.

525. Much depends on how metro definitions change over time. Many truly rural areas are now counted as metro.

529. Assuming constant boundaries, SMSA proportion is at equilibrium; but if new SMSAs get added, then...

906. I believe we are headed to a mini-decline in metro population but only a small one. The population will stabilize at about 70% urban (about the average for industrial countries).

905. Remember, more SMSAs will be created.

908. I see an increase, but at a slowing rate; 15% non-urban is a "gut feeling" about an irreducible number who don't want city/suburban life.

502. There has to be a ceiling effect on percent of MA and I'm doubtful there can be much gain over 80%.

511. Movement of people out of SMSAs to smaller towns is becoming a major trend in this society, which I expect will intensify in the future.

903. Slowing of the trend of emigration from rural areas into the cities.

901. My estimate is based on the belief that SMSA redefinition will partially (but not fully) offset continuing met losses.

Round Three Comments on Item 2 (Metropolitan Population)

514. My estimate of a slight porportional decline in metropolitan population assumes that SMSA boundaries and definitions do not change and no new SMSAs are created.

903. I'll stand by Stage 1 (2000= 77%, 2010= 80%). I expect urbanization to continue but to reach some sort of saturation level.

ITEM 3. POPULATION LIVING IN CENTRAL CITIES (AS A PERCENTAGE OF THE U.S. METROPOLITAN POPULATION) [3]

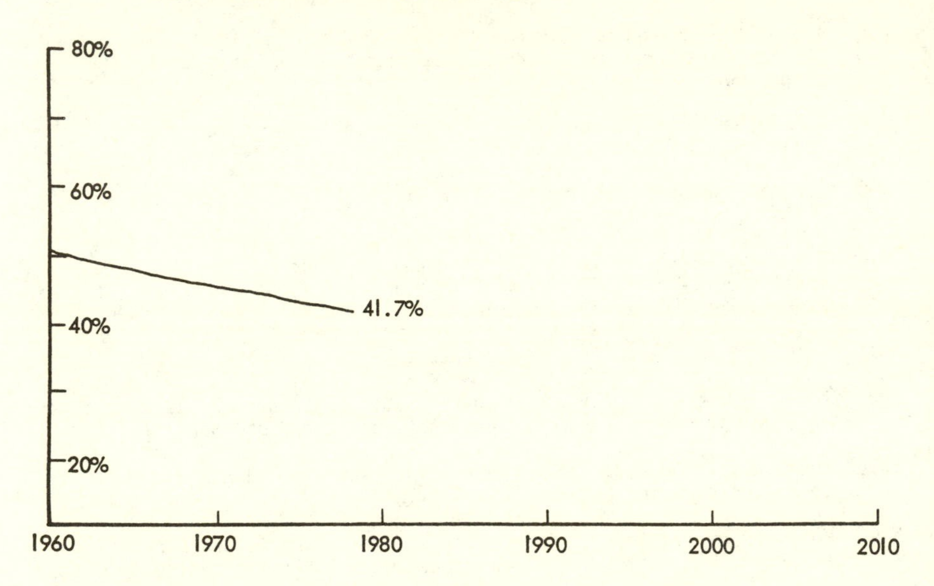

a. Trend Evolution: Please indicate on the above graph how you believe the trend will evolve over the next 30 years.

b. Point Values: Please indicate below your best estimates of the point values (vertical axis values) of the trend in the years 2000 and 201

2000____________ 2010____________

c. Terminology: A central city of an SMSA is the core city and contains a population of at least 50,000.

d. Sponsor's Comments:

e. Statistics:

	Round One		Round Three	
Responses:	30		13	
	2000	2010	2000	2010
Low:	35%	22%	35%	32.5%
Mean:	39.5%	38.6%	40.5%	40.3%
High:	50%	50%	50%	50%
Std. Dev.:	4.3%	6.5%	5.0%	6.8%

Round One Comments on Item 3 (Population in Central Cities)

528. Energy and housing costs may stimulate renovation of new housing in central cities.

508. Continued SMSA (economic) expansion in suburban territory will continue suburbanization of CC population, but increased efforts of CCs to attract jobs, in addition to the higher costs of commuting, will reduce rate of loss. (CC=central city)

104. Fuel/housing costs are likely to increase metropolitan population via reduced commuting and higher urban density.

523. I assume no further expansion of central city limits. I anticipate an increase in multifamily housing in central cities.

905. This definition is rather arbitrary. I expect to see much more urban life styles but not necessarily in the central cities as defined here.

908. Energy costs, commuting time, smaller families and increased investment in inner cities will reverse the decline.

904. I believe the downward trend will flatten out as cities become revitalized and offer socially more conducive settings. Also an input of increasing costs of commuting.

502. Here a kind of floor effect is possible - but the conversion of CC areas to non-residential uses will have an effect.

511. With smaller families, higher commuting costs, higher housing costs, etc., many people who remain in metropolitan areas are likely to prefer the central city over the suburbs. These people should at least halt the population decline in the central cities.

903. Problems of transit and energy cost will encourage return to the core city and "gentification" of center city slum areas into condominium and apartments for young professionals and middle classes.

901. Reconcentrating influences of higher transportation costs about CCs of small SMSAs will partially offset continuing losses to CCs of largest SMSAs. In addition, the CC population of newly-designated met's will slow the apparent rate of decline.

Round Three Comments on Item 3 (Population Living in Central Cities)

514. Most hard evidences to date indicate a) many more people live in central cities than desire to live there; b) large, dense (older) cities are energy inefficient; c) commuting costs and gasoline use are lower in outlying areas; d) gentrification is a drop in the bucket, with the out of the middle class continuing to far exceed immigration of middle-class to cities; e) central cities no longer have locational advantages they once held for many industries, thus more job dispersion.

525. Because of the "aging" of inner suburbs, the distinction between central city-suburb is becoming less important in SMSAs.

903. I see a trend to central city concentration brought on by smaller families, working wives and higher energy costs. Also by non-conventional families whc prefer cities' amusements to lawns and gardening.

ITEM 4. U.S. CIVILIAN LABOR FORCE (AS A PERCENTAGE OF U.S. RESIDENT POPULATION) [4]

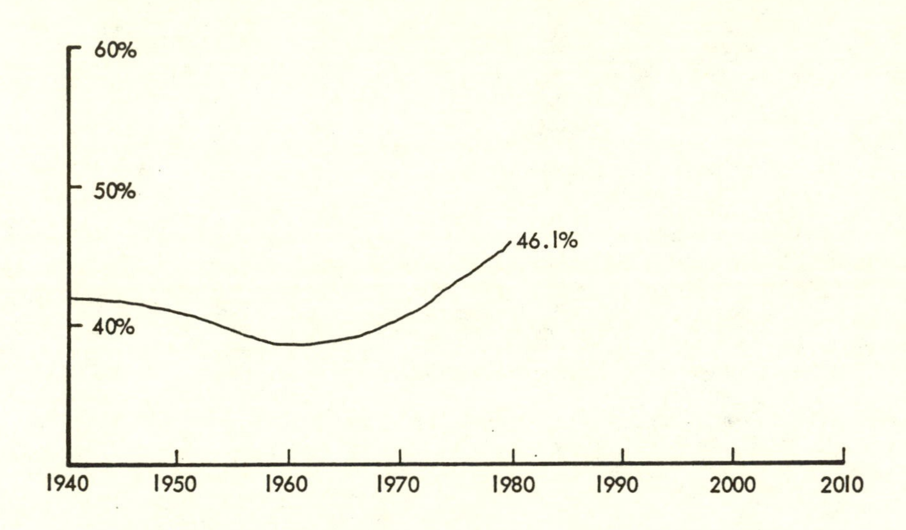

a. Trend Evolution: Please indicate on the above graph how you believe the trend will evolve over the next 30 years.

b. Point Values: Please indicate below your best estimates of the point values (vertical axis values) of the trend in the years 2000 and 201

2000__________ 2010__________

c. Terminology: As defined by the U.S. Census Bureau, the civilian labor force refers to all non-institutionalized resident persons 16 years of age and older, both employed and unemployed, except persons in the Armed Force students, housewives, retired persons, the disabled, or those with no job and not looking for work.

d. Sponsor's Comments:

e. Statistics:

	Round One		Round Three	
Responses:	30		13	
	2000	2010	2000	2010
Low:	40%	38%	40%	38%
Mean:	50.2%	50.4%	50.3%	50.8%
High:	57.5%	60%	55%	57.5%
Std. Dev.:	3.8%	5.1%	3.5%	5.1%

Round One Comments on Item 4 (U.S. Civilian Labor Force)

518. Lower birth rates, coupled with an aging population will cause this sector of the population to decline. The rate of decline will be slowed by an increasing number of persons working beyond age 65.

528. Increased female participation and older age structure are favorable to slightly higher rates.

508. Female Labor Force participation will remain high, labor force participation after age 65 will increase.

523. The growth will stabilize as the small cohorts start entering Labor Force. As large cohorts begin retiring, especially after 2011, it will decline.

906. This should decline as the population ages and small birth cohorts enter workforce ages.

905. Several trends -- women will go up, retired will go up. Also baby boom generation now 30-40. More children in 1980s. More old people about 2010.

904. The dramatic impact of the sixties because of women entering labor force (as well as older work force) will flatten out.

502. The cost of living might encourage even higher rates, up to near 60%.

511. As more and more women are employed -- by choice or by economic necessity -- the labor force should continue to grow steadily until most persons age 25-65 are employed. Many of them may be part-time workers, however. If there is a general collapse, however, the situation will be very different.

903. The increase reflects the gradual increase of working women in the labor force associated with changing work habits, increasing education, and changed family patterns.

901. Continued increase expected because of social and economic pressures to (a) increase female employment participation, (b) reduce minorities unemployment and (c) delay retirement.

Round Three Comments on Item 4 (U.S. Civilian Labor Force)

525. By 2010, most of the baby boom will be in late middle age. Assuming no decrease in retirement age (which would be a serious error), the labor force should be at a historic, if temporary, high in terms of its proportion of U.S. population.

903. I'll stand by Stage I forecasts (2000= 52%, 2010= 55%). Increase in life-expectancy and in proportion of working women, and increase in retirement age, will increase labor force proportion.

ITEM 5. WOMEN WORKING FULL TIME (AS A PERCENTAGE OF ALL INDIVIDUALS 16 YEARS OF AGE OR OLDER WHO WORKED FULL TIME DURING THE YEAR) [5]

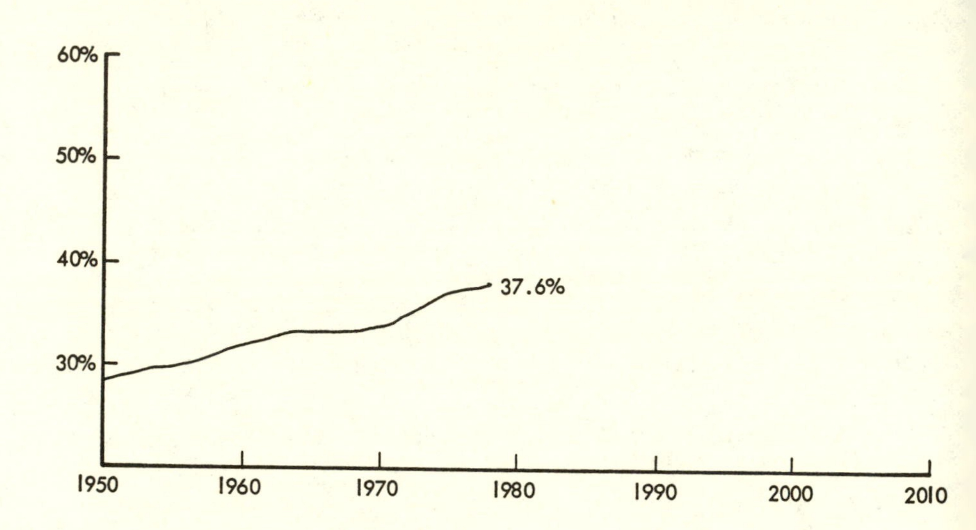

a. Trend Evolution: Please indicate on the above graph how you believe the trend will evolve over the next 30 years.

b. Point Values: Please indicate below your best estimates of the point values (vertical axis values) of the trend in the years 2000 and 20

2000____________ 2010____________

c. Terminology: Full-time workers actually held employment for 35 or mo hours per week during the reference week.

d. Sponsor's Comments:

e. Statistics:

	Round One		Round Three	
Responses:	30		13	
	2000	2010	2000	2010
Low:	39%	39%	40%	39%
Mean:	42.9%	45.4%	42.5%	45.0%
High:	50%	55%	47%	50%
Std. Dev.:	2.9%	4.0%	2.4%	3.7%

Round One Comments on Item 5 (Women Working Fulltime)

518. I don't expect to see change quite as rapid as before, given a shrinking blue collar sector; surely, the service sector can't be expected to expand at a rapid rate given the nature of its growth areas (fast foods, etc.).

508. Will increase slightly (due to economic pressures and more single/divorced women) and maintain high level.

523. It won't see 50%, but it will start approaching it unless we see a sustained increase in fertility.

908. My predictions, which show a fairly dramatic increase in a short time, are driven by overall sense of shifting role of women, economic necessity, increased opportunities, and smaller families with better alternative day care arrangements.

511. I expect this trend to continue to rise steadily -- again, assuming no general economic collapse.

Round Three Comments on Item 5 (Women Working Full Time)

No additions

ITEM 6. U.S. PART-TIME CIVILIAN LABOR FORCE (AS A PERCENTAGE OF TOTAL U.S. CIVILIAN LABOR FORCE) [6]

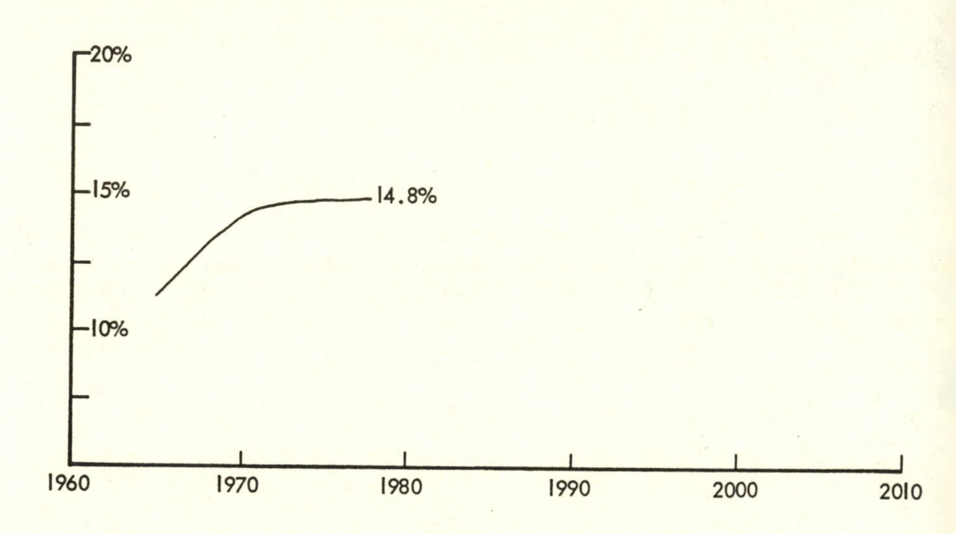

a. Trend Evolution: Please indicate on the above graph how you believe the trend will evolve over the next 30 years.

b. Point Values: Please indicate below your best estimates of the point values (vertical axis values) of the trend in the years 2000 and 201

2000____________ 2010____________

c. Terminology: The part-time civilian labor force actually held employm for fewer than 35 hours per week during the reference week.

d. Sponsor's Comments:

e. Statistics:

	Round One		Round Three	
Responses:	30		13	
	2000	2010	2000	2010
Low:	13%	12%	14%	14%
Mean:	15.9%	16.9%	16.1%	17.4%
High:	20%	25%	18%	20%
Std. Dev.:	1.6%	2.6%	1.3%	1.7%

Round One Comments on Item 6 (U.S. Part-time Civilian Labor Force)

508. Will increase due to economic pressures, especially for housewives, students, and elderly.

304. More +65 workers; part-time.

906. I am very unsure about this one; so much depends on economic conditions.

905. I presume the work week will be shortened.

502. I'm surprised at how flat the curve is on the 70s, but believe it will rise slowly.

511. Despite the recent plateau in this rate, I expect to see it climb rapidly in the future.

Round Three Comments on Item 6 (U.S. Part-time Civilian Labor Force)

903. I agree that the ending of mandatory retirement and increase in life expectancy will drive creation of part-time occupations for older workers. There will be a change in the way "work" is structured, reflecting post-industrialization and the move to an "information" rather than factory-based workplace.

ITEM 7. SECTORAL DISTRIBUTION OF THE LABOR FORCE (AS A PERCENTAGE OF U.S. CIVILIAN LABOR FORCE) [7]

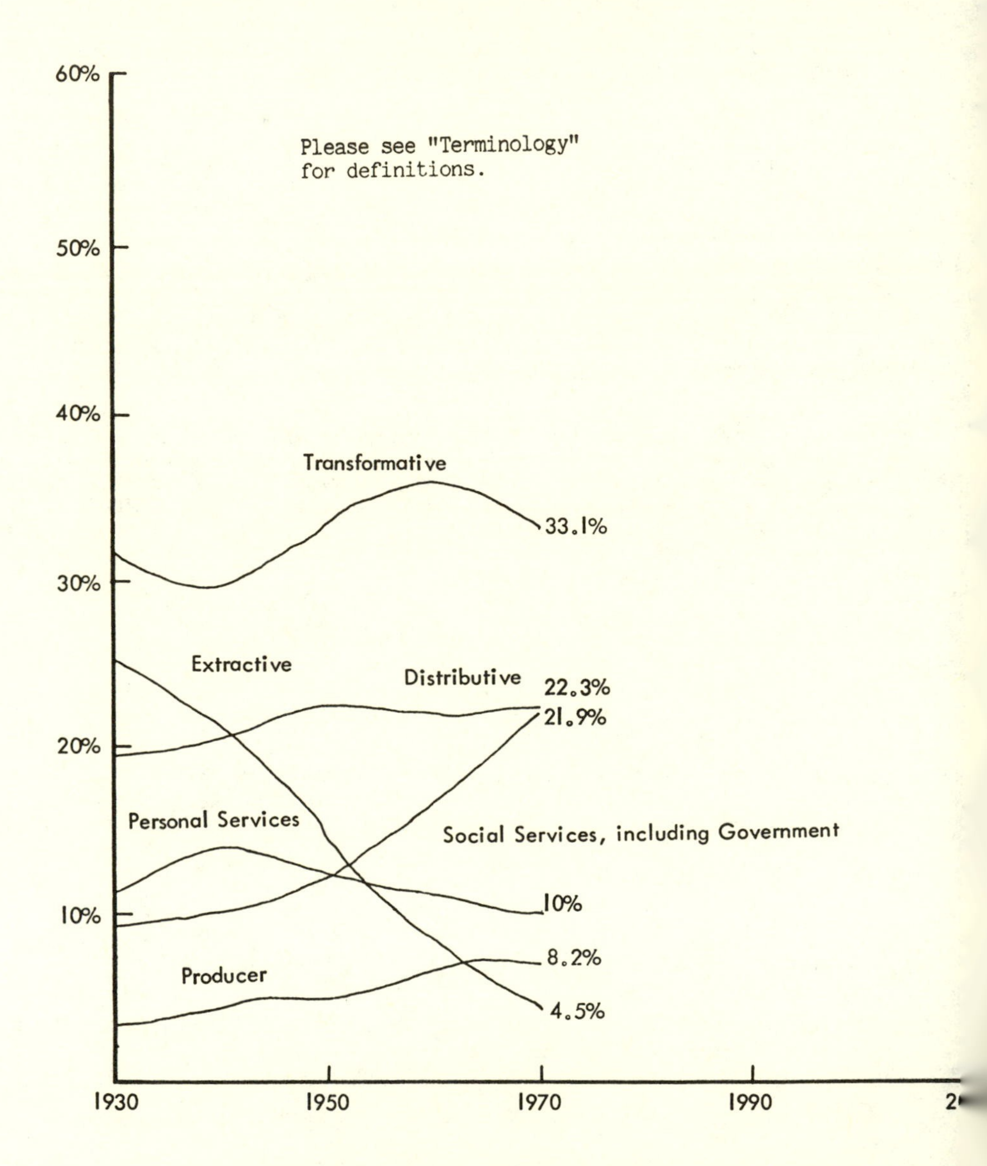

a. Trend Evolution: Please indicate on the above graph how you believe the trend will evolve over the next 30 years.

b. Point Values: Please indicate below your best estimates of the point values (vertical axis values) of the trend for the following years:

	1990	2000	2010
Extractive:	________	________	________
Transformative:	________	________	________
Services:			
Distributive:	________	________	________
Producer:	________	________	________
Social: (Including Government)	________	________	________
Personal:	________	________	________

c. Terminology: The sectoral distribution of the labor force refers to the distribution of the labor force across industry categories. In the past few years, it has become apparent that the well-known three-part typology (extractive, transformative, services) suggested by A.G.B. Fisher and others should be expanded to take into account the wide and growing diversity in the service category. The temporal trend shown relies on the expanded schema developed by Browning and Singelmann. Among the more striking trends are the sharp decline in the extractive sector, the two inflections in the trend in the transformative sector, and the sharp increase in social services.

The sectors are defined as follows:

Extractive (Primary)--Agriculture, forestry, fishing, mining
Transformative (Secondary)--Manufacturing, construction, utilities
Services (Tertiary)

Distributive--Services related to distributing goods and services. Includes transportation, communications, wholesale and retail trade.

Producer--Industries providing services "to producers of goods or that are concerned with various forms of property" (Singelmann, 1978:31). Includes financial, real estate, legal.

Social--Industries providing services in response to collective demand. Includes health, education, welfare, and government.

Personal--Industries providing services in response to individual consumer demand. Includes domestic, hotels, eating and drinking establishments, repair, entertainment, recreation, and miscellaneous services.

Round One Comments on Item 7 (Sectoral Distribution of Labor Force)

104. Most raw materials labor will be done overseas.

Round Three Comments on Item 7 (Sectoral Distribution of Labor Force)

104. The shift from goods to services is slackening and taking its place is a more important shift, now noted in FRG/Swedish govt. studies, from energy intensive industries to information intensive (engineering, chemical, and communications).

906. No change -- I remain high on distributive but believe that micro-processors and other items will lead to significant growth in communication related industries.

Statistics

	Round One			Round Three		
Responses	30			13		
Extractive	1990	2000	2010	1990	2000	2010
Low:	3.0%	2.5%	2.0%	3.0%	2.5%	2.2%
Mean:	4.3%	4.5%	4.1%	4.5%	4.8%	5.1%
High:	7.0%	8.5%	11.0%	7.0%	8.5%	11.0%
Std. Dev.:	1.1%	1.5%	2.2%	1.2%	1.9%	2.7%
Transformative	1990	2000	2010	1990	2000	2010
Low:	21.0%	22.0%	20.0%	25.0%	22.0%	20.0%
Mean:	30.0%	29.5%	29.1%	29.6%	28.8%	28.0%
High:	37.0%	39.0%	41.0%	34.0%	35.0%	35.0%
Std. Dev.:	3.7%	4.7%	5.6%	2.7%	4.5%	4.6%
Distributive	1990	2000	2010	1990	2000	2010
Low:	19.0%	18.0%	15.0%	21.0%	22.0%	22.0%
Mean:	22.7%	23.4%	23.8%	23.0%	24.3%	25.5%
High:	25.0%	30.0%	35.0%	25.0%	30.0%	35.0%
Std. Dev.:	1.4%	2.4%	3.5%	1.2%	2.1%	3.6%
Producer	1990	2000	2010	1990	2000	2010
Low:	7.0%	6.0%	4.5%	7.0%	6.0%	4.5%
Mean:	9.4%	9.3%	9.5%	8.6%	8.9%	9.1%
High:	20.0%	21.0%	23.0%	10.0%	11.0%	11.0%
Std. Dev.:	3.2%	2.8%	3.3%	0.9%	1.3%	2.1%
Social	1990	2000	2010	1990	2000	2010
Low:	10.0%	10.0%	10.0%	10.0%	10.0%	10.0%
Mean:	22.9%	23.3%	23.7%	22.2%	21.4%	20.7%
High:	30.0%	30.3%	32.7%	30.0%	26.0%	26.0%
Std. Dev.:	3.9%	4.5%	5.4%	5.1%	4.7%	5.0%
Personal	1990	2000	2010	1990	2000	2010
Low:	3.0%	1.0%	1.0%	8.0%	8.0%	8.0%
Mean:	10.7%	10.4%	10.8%	12.6%	13.5%	14.8%
High:	25.0%	25.0%	27.0%	25.0%	25.0%	27.0%
Std. Dev.:	4.3%	5.4%	6.3%	4.8%	4.9%	5.7%

ITEM 8. MEDIAN FAMILY INCOME (1980 U.S. DOLLARS) [8]

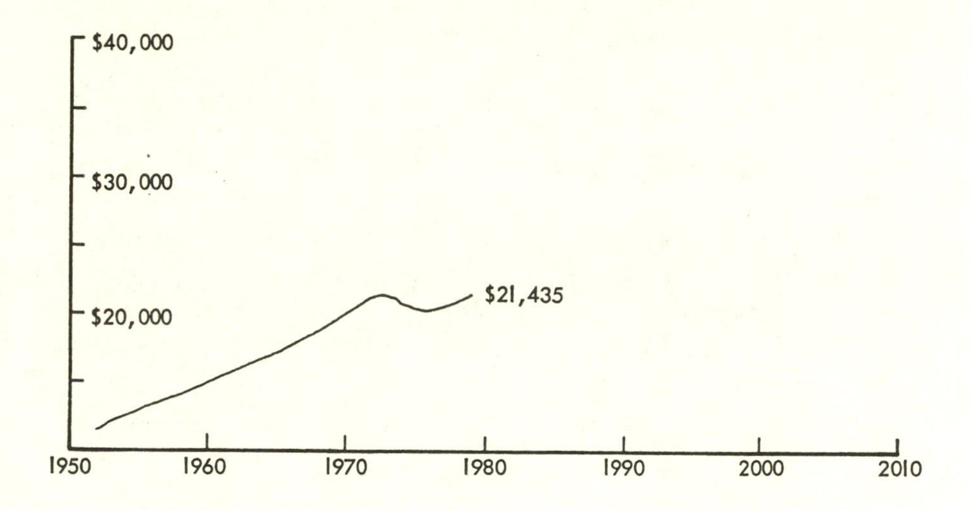

a. Trend Evolution: Please indicate on the above graph how you believe the trend will evolve over the next 30 years.

b. Point Values: Please indicate below your best estimates of the point values (vertical axis values) of the trend in the years 2000 and 2010.

2000______________ 2010______________

c. Terminology: The median family income as graphed is adjusted for inflation. A family is defined as a group of two or more related individuals related by marriage, blood, or adoption and sharing the same dwelling unit.

d. Sponsor's Comments:

e. Statistics:

	Round One		Round Three	
Responses:	29		12	
	2000	2010	2000	2010
Low:	$18,000	$15,000	$20,000	$17,500
Mean:	$25,763	$28,606	$25,700	$28,400
High:	$34,500	$40,000	$29,000	$35,000
Std. Dev.:	$3,750	$6,040	$3,000	$4,800

Round One Comments on Item 8 (Median Family Income)

518. Given current economic trends, I see no basic change in the overall upward climb of median income. I expect slower rates of growth in the social service areas as taxes are reduced and as governmental programs are cut.

508. Increasing economic pressure through lessened, although persistent, inflation will increase the number of workers per family. Also, real money growth will occur through some economic expansion.

523. Family sizes will decline; but individual incomes will rise and female labor force participation will increase. Also "baby boom" earners will move up the age-income profile.

905. Project in a straight line with a pre-1970 slope, because I do not project war or other crisis, such as the oil crisis, to affect a steady rise in income. (2010 value of $36,000)

904. Key point is that this is in terms of real income.

502. This is extrapolation, but at a somewhat lower rate than 1950-1970. There may be a recession in 2000 or 2010 but I can't know whether there will be at that time or not.

511. One possibility (A) is a gradual leveling of this trend for median income, although these could be redistribution around the median. An alternative real possibility (B) would be a moderate to sharp decline in real family income.

903. The discontinuites of the period 1975-1985 are solved with restructuring of the role of government in society, and traditional real growth returns to the U.S. economy, though at slower than historic rates.

901. Continuing decline in 1980s; slower rate of growth in ensuing decades, but even that will be more rapid than real income growth per employed person; median family income growth will reflect increasing labor force participation of married women.

Round Three Comments on Item 8 (Median Family Income)

528. I stand by my previous somewhat pessimistic view. I do not think there will be an increase in real dollars, although there might well be an increase in nominal dollars (and in non-monetary or deferred monetary compensation). Also, the redistributive forces at work in the 60s seem to be at an end.

903. I expect a reduction of unnecessary federal intervention in business and an increase in working wives will combine to raise real family income.

ITEM 9. PERSONS IN POVERTY (AS A PERCENTAGE OF THE U.S. POPULATION) [9]

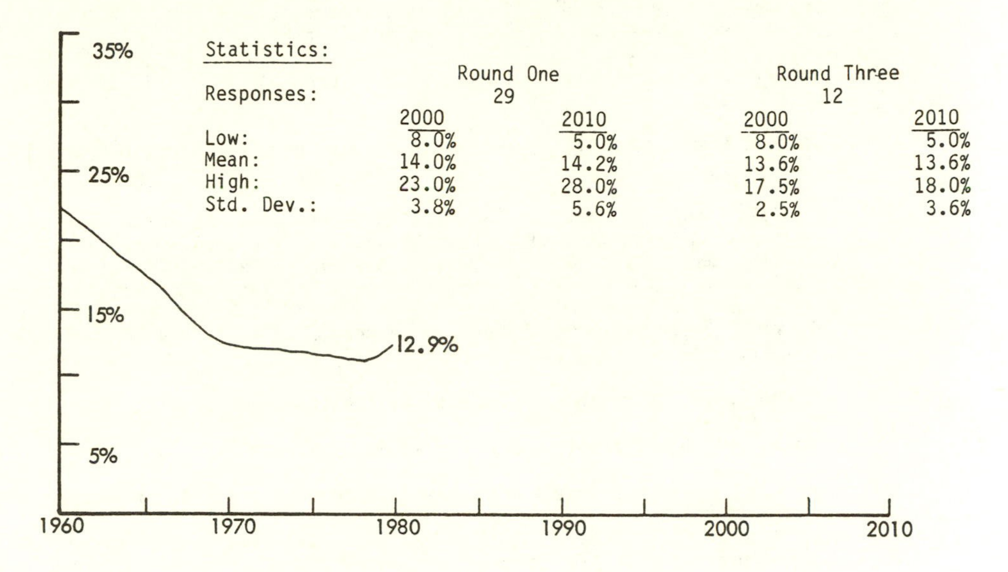

Statistics:

	Round One		Round Three	
Responses:	29		12	
	2000	2010	2000	2010
Low:	8.0%	5.0%	8.0%	5.0%
Mean:	14.0%	14.2%	13.6%	13.6%
High:	23.0%	28.0%	17.5%	18.0%
Std. Dev.:	3.8%	5.6%	2.5%	3.6%

a. Trend Evolution: Please indicate on the above graph how you believe the trend will evolve over the next 30 years.

b. Point Values: Please indicate below your best estimates of the point values (vertical axis values) of the trend in the years 2000 and 2010.

2000____________ 2010____________

c. Terminology: Persons are classified as being above or below the poverty level by use of the poverty index adopted by a Federal inter-agency committee in 1969; the index reflects the different consumption requirements of families, based on their sizes and compositions, sex and age of the family householder, and farm or nonfarm residence. It was determined that a family of three or more persons spends approximately one-third of its income on food. The poverty level for such a family was therefore three times the cost of an economy food plan. For smaller families and unrelated individuals, the cost of the economy food plan was multiplied by factors slightly higher in order to compensate for the relatively larger fixed expenses of these smaller families. The poverty thresholds are updated every year to reflect changes in the Consumer Price Index.

d. Sponsor's Comments: In 1979, the poverty threshold for a family of four was $8,078 (1980 dollars) [10]. In 1980, 12.9% of the U.S. resident population were below the poverty threshold [11].

Round One Comments on Item 9 (Persons in Poverty)

518. Changes in governmental policy along with budget cuts will cause a rapid rise in the numbers of persons in poverty.

525. Some increases are likely with aging-more probably by sex (older women more likely to be poor).

508. Reduced income transfer payments plus larger number of female-headed families.

102. This data does not seem consistent with other data. Attached data on households by income class suggest 14.1% of households have an income under 5000 1978 $ or 5900 1980 $. Since I suspect lower income groups have higher persons per household, something looks amiss.

523. "Poverty" is politically defined. As a result we're likely to see a long range decline.

906. Given present conditions, I expect poverty to increase.

905. Reagan will bring poverty back for a few years.

908. The rise reflects a more technologically complex society, with inflation reducing choices.

904. As we increase the threshold, which we must, we keep redefining poverty. However, this will decline as the society adjusts to the development and enhancement of our ability to make the economy respond as desired.

502. I expect a floor to be reached below which it will be hard to penetrate.

511. This totally artificial measure may indicate declining poverty in the future, but that is no indication at all of actual levels of need or comfort or increased distribution.

903. Attempts to reduce poverty will become increasingly successful through more effective income transfer programs.

901. Depends a lot on the jiggery-pokery now going on with the definition of of the CPI. If that is redefined as planned, the rate of recent growth in "poverty" will diminish as an artifact of the changes in definition, while food prices soar on upward.

Round Three Comments on Item 9 (Persons in Poverty)

514. I am working under the assumption that economic recovery of the private sector during the 1980s reduced inflation, and a bit less government muddling into the affairs of businesses and people will spur further declines in poverty in the future. These declines will not be nearly as large as occupied during the 1960s, however, because of floor effects.

903. Improvements in economy plus more equity in distribution will reduce the proportion living in poverty in the future.

ITEM 10. ELDERLY POPULATION (U.S. RESIDENT POPULATION 65 YEARS OR OLDER, AS A PERCENTAGE OF U.S. RESIDENT POPULATION) [12]

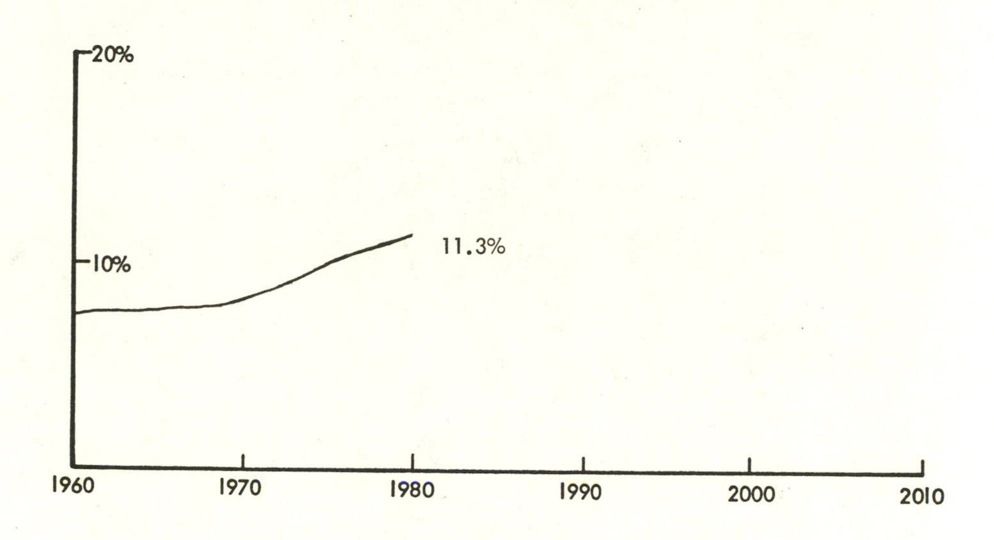

a. Trend Evolution: Please indicate on the above graph how you believe the trend will evolve over the next 30 years.

b. Point Values: Please indicate below your best estimates of the point values (vertical axis values) of the trend in the years 2000 and 2010.

2000____________ 2010____________

c. Terminology:

d. Sponsor's Comments:

e. Statistics:

	Round One		Round Three	
Responses:	29		12	
	2000	2010	2000	2010
Low:	10.0%	10.0%	12.0%	13.0%
Mean:	12.9%	14.6%	13.3%	15.3%
High:	17.0%	20.0%	17.0%	18.0%
Std. Dev.:	1.8%	2.7%	1.6%	1.8%

Round One Comments on Item 10 (Elderly Population)

525. Inevitable unless there is a dramatic increase in fertility.

523. No big change will occur immediately after 2010 as the "baby boom" cohorts reach age 65. Mortality (rate) will continue to decline.

906. The age structure of the nation should lead to increased percentage of the elderly well into the next century.

904. We are just on the threshold of some dramatic improvements in our ability to anticipate and detect disease. Social policies often respond to information on how to live healthier lives.

903. Increase will arise from aging of "baby boom" cohort, longer working life with associated longevity, and improving geriatric medication and health care.

Round Three Comments on Item 10 (Elderly Population)

528. Back to my old theme, immigration. There was considerable pressure on the Select Commission on Immigration and Refugee Policy to admit grandparents and other aged relatives in the "exempt" category. This is another potential source (though a minor one) of increase in the 65+ population.

903. Increase in life expectancy from improved diet, exercise and longer working life will increase proportion of elderly.

ITEM 11. U.S. BLACK AND HISPANIC POPULATIONS (AS A PERCENTAGE OF U.S. POPULATION [13]

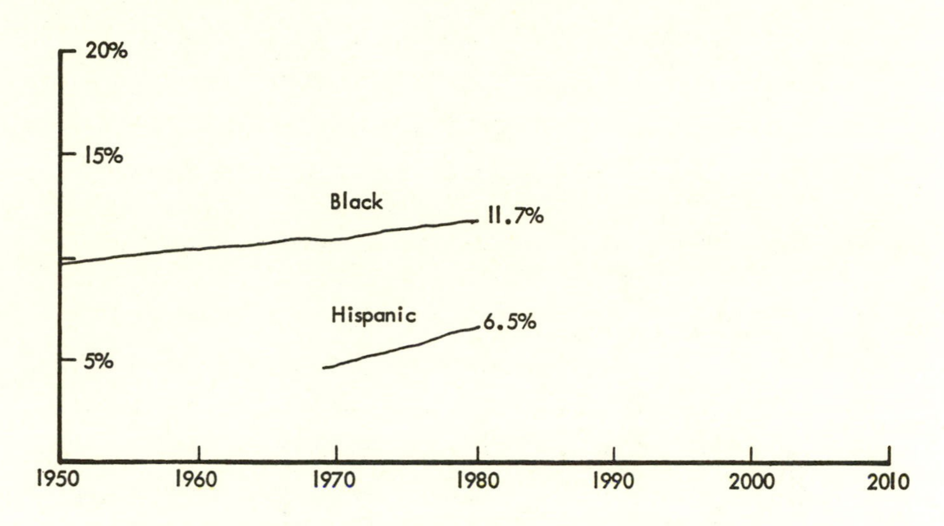

a. Trend Evolution: Please indicate on the above graph how you believe the trend will evolve over the next 30 years.

b. Point Values: Please indicate below your best estimates of the point values (vertical axis values) of the trend in the years 2000 and 2010.

2000: Blacks: ____________ 2010: Blacks: ____________
Hispanics: ____________ Hispanics: ____________

c. Terminology: For census purposes, a Black is a person who classifies himself (herself) as a Black or Negro on the census question dealing with race. A Hispanic is a person who self-identifies his (her) ethnic origin as Spanish (Mexico, Puerto Rico, Cuba, etc.). Persons of Spanish origin may be of any race, although those of Mexican and Puerto Rican origin are usually classified as white.

d. Sponsor's Comments:

e. Your Comments:

Round One Comments on Item 11 (U.S. Black and Hispanic Population)

518. No change in trend for Blacks; Hispanics may make wider use of birth control, reducing their rate of increase to a slight extent.

513. Hispanic population size is so dependent on federal policy and legal implementation of policy. However, the birth rate alone would justify high increase in proportion of Hispanics.

523. I anticipate continued Hispanic immigration.

906. I expect both to increase as a percent of population but Spanish more than Blacks because of heavy immigration.

502. Unquestionably the gap between Black and Hispanic will be narrowed both by higher natural increase and by net immigration.

903. Black rate will slow with increasing affluence. Hispanic rate will also slow but not until after 2000.

Round Three Comments on Item 11 (U.S. Black and Hispanic Populations)

514. While Black population will increase in absolute numbers, it will decline as a percentage of the total U.S. population because of faster increases of other subpopulations (esp. Hispanics, Asians etc.) due largely to their increased immigration.

525. I find it hard to believe that means for the proportion of each group is lower in 2010 than in 2000.

903. Hispanic increase will come from immigration plus faster birth rate in lower SES.

Statistics:

	Round One		Round Three	
Responses:	29		12	
Blacks				
	2000	2010	2000	2010
Low:	10.0%	7.0%	11.0%	9.2%
Mean:	12.5%	10.4%	12.6%	12.5%
High:	14.0%	15.0%	13.0%	15.0%
Std. Dev.:	0.9%	2.4%	0.7%	2.2%
Hispanics				
	2000	2010	2000	2010
Low:	7.5%	8.0%	9.5%	9.5%
Mean:	12.2%	11.3%	10.3%	11.9%
High:	16.0%	17.0%	13.0%	14.0%
Std. Dev.:	2.3%	2.2%	1.9%	1.2%

ITEM 12. LEGAL IMMIGRATION (MILLIONS OF PERSONS ANNUALLY) [14]

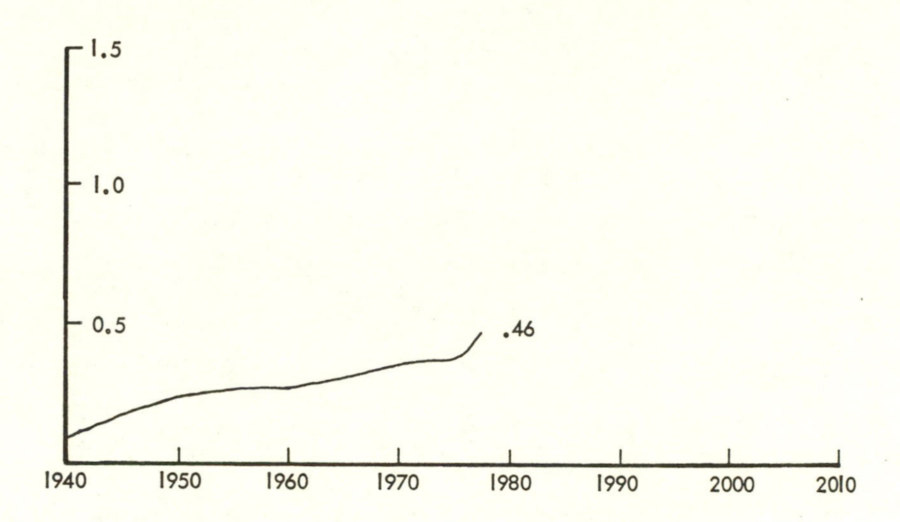

a. Trend Evolution: Please indicate on the above graph how you believe the trend will evolve over the next 30 years.

b. Point Values: Please indicate below your best estimates of the point values (vertical axis values) of the trend in the years 2000 and 2010.

2000 ______________ 2010 ______________

c. Terminology: A legal immigrant is defined as a nonresident alien admitted to the U.S. for permanent residence (one year or longer). The category includes persons who may have entered the U.S. as nonimmigrants or refugees, but who subsequently changed their status to that of a permanent resident. Usually, a 2-year lag exists between the time persons are shown in data as nonimmigrants and the time they may be included as immigrants. A large increase in the number of immigrants to the U.S. in one year may have been due to an influx of refugees two or three years earlier. Also, changes in immigration law (after July, 1968) may influence the volume of legal immigration.

d. Sponsor's Comments:

e. Statistics:

	Round One		Round Three	
Responses:	29		12	
	2000	2010	2000	2010
Low:	.25	.20	.40	.38
Mean:	.56	.62	.58	.68
High:	.80	1.00	.80	1.00
Std. Dev.:	.14	.19	.12	.22

Round One Comments on Item 12 (Legal Immigration)

518. The curve is likely to level out with continued economic woes and a reduction in Indochinese immigration.

513. Again, the legal immigration "quota" is so affected by Congressional mandate that it is difficult to project anything but a series of scenarios.

525. I anticipate liberalization of immigration laws in response to problems of labor scarcity after the turn of the century.

521. We're going to need labor, and immigrants' work hours.

906. It should increase with increasing world population and increasing poverty

905. We are now seeing a backlash against immigration.

904. I believe the U.S. will move to reduce the rate of growth of immigration and stabilize it.

502. I put it flat - it may go down a bit, but I'm doubtful it will rise. (Legalization of present undocumented Mexican population in U.S.A. wouldn't come under this category.)

903. Until recently, the historic postwar rate was around 400,000/yr. I can see this increasing to 500,000/year, but not more.

901. I am assuming periodic regional wars that will generate refugees admitted to U.S.

Round Three Comments on Item 12 (Legal Immigration)

906. I still believe the world economic situation will lead U.S. to accept more and more immigrants.

903. Increasing disparity in economic conditions between U.S. and lesser developed nations in Asia, Africa and Southeast Asia will keep immigration at a rate which increases and legalizes an increasing proportion.

ITEM 13. STOCK OF ILLEGAL IMMIGRANTS (MILLIONS OF PERSONS)

This Item has been eliminated from Round 3 because of the very low response rate to it in Round 1.

ITEM 14. AVERAGE HOUSEHOLD SIZE (NUMBER OF PERSONS) [16]

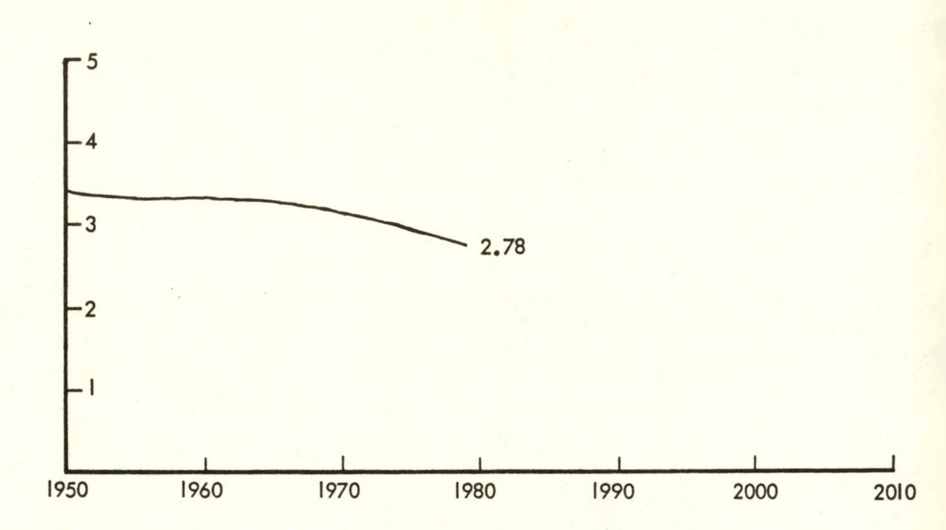

a. Trend Evolution: Please indicate on the above graph how you believe the trend will evolve over the next 30 years.

b. Point Values: Please indicate below your best estimates of the point values (vertical axis values) of the trend in the years 2000 and 2010.

2000__________ 2010__________

c. Terminology: As defined by the U.S. Bureau of the Census, persons shar the same dwelling unit constitute a household. (Thus, while all families a households, not all households contain families.) Average household size i the number of persons in households divided by the number of households.

d. Sponsor's Comments:

e. Statistics:

	Round One		Round Three	
Responses:	29		12	
	2000	2010	2000	2010
Low:	1.90	1.85	2.30	2.00
Mean:	2.52	2.45	2.60	2.50
High:	3.20	3.30	3.00	3.20
Std. Dev.:	0.27	0.32	0.20	0.31

Round One Comments on Item 14 (Average Household Size)

518. I expect a slight dampening of the downward trend, but generally families will get smaller.

528. I expect that housing costs and the raising of children by the baby boom will lead to larger average household size.

521. It's never going to go below 2.

906. I believe the average household size will continue to decline as a result of more childless couples, more single adults living alone and smaller family sizes.

905. Higher incomes mean less need to share. Obviously the minimum is 1.

904. I believe there will be a return to the "preferredness" of having children. In fact, it has begun already.

502. I foresee a decline, but a leveling off. Economic factors may restrict single-person households, but smaller family size will more than out-balance it.

511. I expect this downward trend to continue until it reaches about 2.0.

903. I can see a fundamental change through the 70s as women return to the labor force, stabilizing toward the end of the century as the household remains at a stable level because of economic pressures.

901. All present trends contribute to this continuing decline, and several of those trends have not yet peaked (or bottomed): fertility decline, postponement of marriage, people in HHDs of unrelated persons without related children, single-person HHDs, etc.

Round Three Comments on Item 14 (Average Household Size)

No additions

ITEM 15. HOUSEHOLDS HEADED BY A PRIMARY INDIVIDUAL (AS A PERCENTAGE OF TOTAL U.S. HOUSEHOLDS) [17]

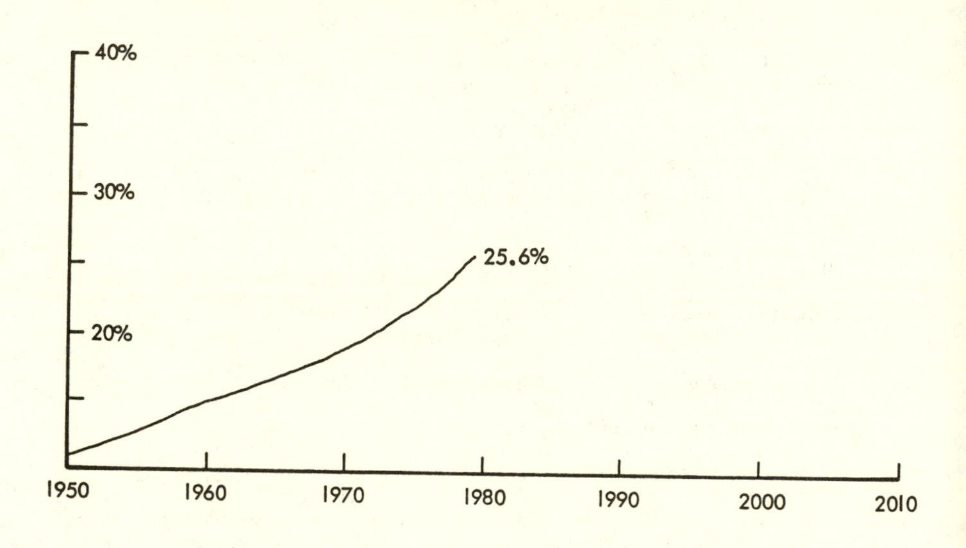

a. Trend Evolution: Please indicate on the above graph how you believe the trend will evolve over the next 30 years.

b. Point Values: Please indicate below your best estimates of the point values (vertical axis values) of the trend in the years 2000 and 2010.

2000______________ 2010______________

c. Terminology: As defined and enumerated by the U.S. Bureau of the Census a primary individual is a household head living alone, or with a person or persons not related to him/her, or living in group quarters (excepting inmates of institutions).

d. Sponsor's Comments:

e. Statistics:

	Round One		Round Three	
Responses:	29		10	
	2000	2010	2000	2010
Low:	24.0%	27.0%	24.0%	27.0%
Mean:	31.4%	33.9%	29.9%	31.8%
High:	45.0%	50.0%	35.0%	40.0%
Std. Dev.:	4.3%	5.9%	2.9%	3.6%

Round One Comments on Item 15 (Households headed by a Primary Individual)

525. Will probably go down on short run due to maturity of baby boom cohort and eventual marriage.

511. Trend due mainly to more and more people living in non-maritial arrangements.

903. The present trend toward non-family households will continue, though at slower rates than in the recent past.

901. This will continue to accelerate until bulk of baby boom is in mid-thirties (at which point they'll give way to an end of postponing marriage and hence be reclassified). Thereafter, the rise in the curve will decelerate (but remain greater than 0).

Round Three Comments on Item 15 (Households Headed by Primary Individual)

903. Increasing financial independence of women will tend to destabilize traditional family structure, as well as lead to deferred marriages.

ITEM 16. EDUCATION (PERCENTAGE OF THE U.S. POPULATION 25 YEARS OR OLDER WHO HAVE COMPLETED AT LEAST 13 YEARS OF SCHOOL) [18]

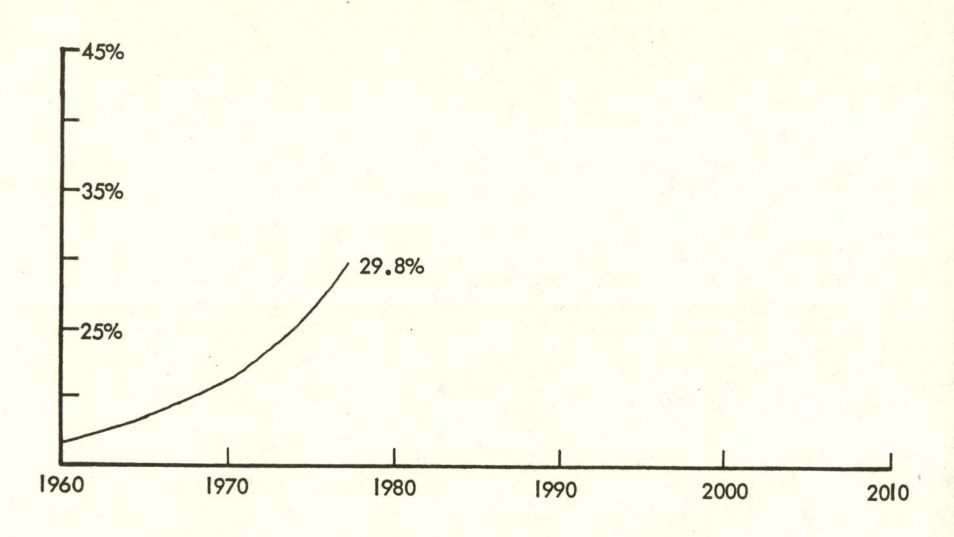

a. Trend Evolution: Please indicate on the above graph how you believe the trend will evolve over the next 30 years.

b. Point Values: Please indicate below your best estimates of the point values (vertical axis values) of the trend in the years 2000 and 2010.

2000____________ 2010____________

c. Terminology:

d. Sponsor's Comments: This question is asked to ascertain your opinion o the percentage of people who will have at least one year's college educatio

e. Statistics:

	Round One		Round Three	
Responses:	29		12	
	2000	2010	2000	2010
Low:	30.0%	30.0%	32.0%	35.0%
Mean:	40.7%	43.6%	38.5%	41.3%
High:	80.0%	85.0%	45.0%	50.0%
Std. Dev.:	10.2%	11.9%	3.4%	4.4%

Round One Comments on Item 16 (Education)

518. This may be overly optimistic but the trend seems to suggest geometric increase. Changes in the structure of the U.S. economy may alter this.

102. Data for 1980 Statistical Abstract indicates that for the year 1979, 36.6% of 25 and over completed high school, 14.7% completed 1-3 years of college, and 16.4% completed 4 years +, yielding a total of 67.7%. On this about 50% of people aged 18-21 are in college. I suggest this (50%) represents an inherent top.

908. Answer depends a lot on economic forecasts and on changing vision of what "13th year" education might accomplish.

502. The magic and economic value of post-high school education has gone, but for social status there will be no big decline.

511. There is an IQ distribution which will soon begin to level off this trend -- unless we continue to water down education to the point where 13 years of school is the equivalent of primary education.

903. Move toward service economy. Expansion of community and other local college systems. Continuing education through life, and college-educated mature or retired persons will increase proportion of society with at least one year of college.

901. Declining male enrollment ratios will depress the rate of increase. When female enrollment ratios also peak out, the curve will flatten (and perhaps even begin to decline). Note: As you know, this state will be strongly influenced by age composition as a function of recent wide fertility fluctuations, which I have not worked in.

Round Three Comments on Item 16 (Education)

528. Notice high drop-out rates currently characteristic of Hispanics.

903. I see increasing penetration of community and extension college programs in communities, as well as acceptance of education all through life and career.

ITEM 17. PORTION OF AVERAGE WORK WEEK WORKED AT HOME (INDEX NUMBER, 1980=100)

a. Point Values: Based on an index number standardized to 100 in 1980, please indicate your best estimates of the index values representing the portion of the average work week worked at home by non-agricultural workers for the following years:

1990 ____ 2000 __________ 2010__________

b. Sponsor's Comments: With the increase of use of home computers and other sophisticated means of storing, manipulating, retrieving, and analyzing information and continued improvements in instantaneous communications, it seems reasonable to expect that more people will be able to perform all or part of their occupational tasks in their own homes. As an example of the sort of response being asked for above, if you believe that by 1990 there will be an increase of 10% in the average amount of time persons spend at home while in pursuit of their occupational tasks, the index value for 1990 would be 110.

c. Statistics

	Round One			Round Three		
Responses	29			12		
	1990	2000	2010	1990	2000	2010
Low:	101.0	102.0	102.0	102.0	105.0	110.0
Mean:	107.8	115.6	126.6	106.7	113.8	121.6
High:	130.0	160.0	225.0	115.0	125.0	145.0
Std. Dev.:	6.3	12.4	24.9	3.7	7.2	10.3

Round One Comments on Item 17 (Portion of Average Work Week Worked at Home)

521. Is this Toffler's new thesis? I really don't see much of it happening.

523. Given the expense of this equipment, I'd be surprised if it affected more than 10% of the work force.

906. I concur with the sponsor's comment and expect a rapid increase in future years.

908. I have always believed there is less in this trend than meets the eye. Offices provide important benefits to employer (control & supervision, ability to evaluate performance, versatility of assignment of tasks); for the worker, there is an important social dimension of the work place; also, many value the discipline that a place of work provides them. Thus, I do not think technology will change this trend dramatically.

904. I think this is very much dependent on the emerging "sociology" of the workplace. I believe the need to "go to work" may offset some of the technological advances.

511. I think Toffler is correct in his projection on this trend.

903. I can't see a major increase in work at home for remainder of this century or even for start of next. Principal reason is need for social contact in work setting. If that is removed, need to develop alternative forum for social contact, which is not yet very visible.

901. I'm skeptical of this trend, even though I do make use of the same set of causes for other purposes in my papers. The skepticism is only a hunch, however: I know of no current data on the question, and I don't fault your logic. But something holds me back from accepting the conclusion.

Round Three Comments on Item 17 (Portion of Average Work Week Worked at Home)

903. I expect people will want to get away from home to work, even if there are changes in the nature and location of the workplace.

ITEM 18. LEISURE ACTIVITY (HOURS PER WEEK)

a. Point Values: Please indicate your best estimates of the average number of hours per week spent by a member of the U.S. urban population in leisure time activities in:

1990__________ 2000__________ 2010__________.

b. Terminology: Hours spent in leisure time activity per week are based on a maximum of 168 hours in a week (7 days x 24 hours daily); other activities in which a person may spend time during the week are sleep, work for pay, family care and personal care.

c. Sponsors Comments: In 1965, this figure was 34.8 hours per week. In 1975, it was 38.5 hours per week. [19]

d. Statistics

	Round One			Round Three		
Responses	29			12		
	1990	2000	2010	1990	2000	2010
Low:	26.0	32.0	35.0	26.0	32.0	33.0
Mean:	40.6	43.6	46.3	39.4	42.2	45.5
High:	50.0	50.0	70.0	45.0	50.0	55.0
Std. Dev.:	4.5	5.8	7.7	5.7	5.5	7.3

Round One Comments on Item 18 (Leisure Activity)

513. These point values are based on anticipated increase in proportion of older, retired population.

906. I expect this to increase because Americans seem more and more leisure oriented.

908. The place that additional leisure time will come from is work-for-pay time. Trend will continue, but slow down somewhat.

502. Flat, essentially, because I believe the long-term secular decrease in leisure - made possible by declines in length of work week -- will level off.

901. I'm reminded of Truman's complaint about economists. I expect that the average may go down, not up, because of declining labor productivity, the perceived need for growth in GNP, the shortage of laborers entering the workforce in the 1990s and the trend [emergent as a function of SSA problems] toward later retirements. But, on the other hand, there'll be all those excess baby boomers moving up the corporate pyramids... So I've equivocated on this one, guessing that two countervailing trends will approximately cancel each other out.

Round Three Comments on Item 18 (Leisure Activity)

No additions

ITEM 19. U.S. INFLATION RATE (ANNUAL PERCENTAGE CHANGE IN THE CONSUMER PRICE INDEX) [20]

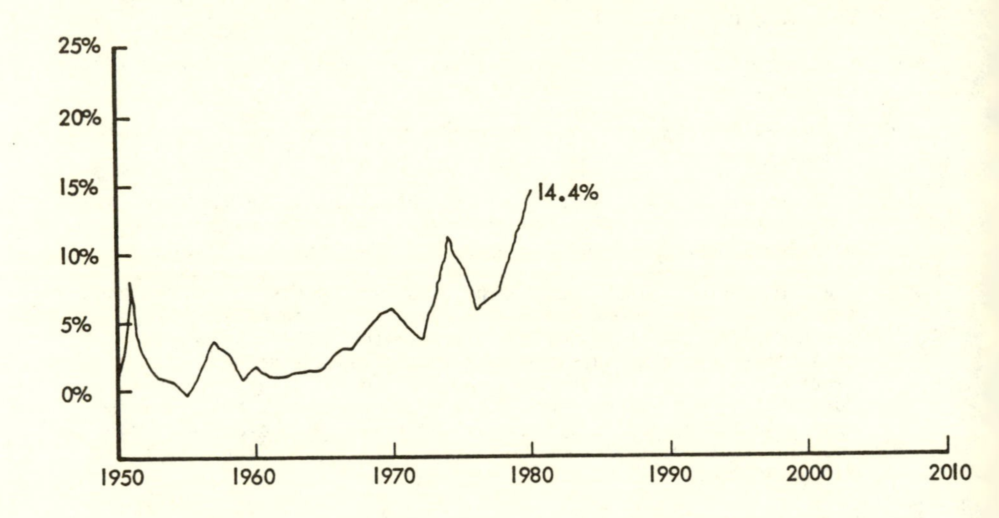

a. Trend Evolution: Please indicate on the above graph how you believe the trend will evolve over the next 30 years.

b. Point Values: Please indicate below your best estimates of the average values (vertical axis values) of the trend during each of the next three decades.

1980's__________ 1990's__________ 2000's__________

c. Terminology:

d. Sponsor's Comments:

e. Statistics

	Round One			Round Three		
Responses	43			20		
	1990's	2000's	2010's	1990's	2000's	2010's
Low:	5.0%	2.5%	2.5%	5.0%	3.0%	2.0%
Mean:	10.7%	10.4%	11.2%	9.7%	7.8%	7.1%
High:	16.0%	21.0%	30.0%	14.5%	14.5%	14.0%
Std. Dev.:	3.4%	5.6%	8.0%	2.5%	2.6%	2.5%

Comments on Item 19 (U.S. Inflation Rate)

304. General flattening in 12-18% range remaining cyclic; probable large increase in mid-late 1980s.

521. There is a "something-for-nothing" attitude in the U.S. that will drive this up.

529. Secular trend will remain upward, unless a major depression occurs.

202. I anticipate years will range between 7 and 14% but the average for the decade will be as indicated above.

906. Given the world economic situation, I expect inflation to be a long term problem.

212. I see Reagan's policies as contributing to a continued rise in inflation (budget deficits, high military spending, productivity lags), followed by a slight dip in the 1990s as a result of continued improvements in energy efficiency, and another rise stemming from the impact of other resource constraints (water, soil).

215 Conditions that existed in the 50s will exist in the late 80s - 2010 population will have a high propensity to consume - they will be coming out of a period of instability into a stable phase. Increased savings will aid investment which will yield increases in productivity - this cycle effect will dampen down inflation.

209. Depends on whether the definition of CPI will be adjusted for changes in market bundle.

201. Some downward move apparent due to excluding housing.

207. Periodic oil shocks repeated throughout the rest of century, causing temporary upturns in inflation, but a general trend down as we learn how to manage this problem.

903. I can see inflation slowly being wrung out of the economy over next decade, with return to lower, but not historic, rates by end of century.

Round Three Comments on Item 19 (U.S. Inflation Rate)

211. Comments 212 and 215 are contradictory. I agree most closely with comments 207 and 903. Comments 201 and 209 assume that measurement changes will affect the inflation rate, while in the short run, the long run rate would be little changed.

903. I see implicit deflator returning to lower levels, though not to historic 3-4%.

ITEM 20. U.S. GROSS NATIONAL PRODUCT (BILLIONS OF 1980 DOLLARS) [22]

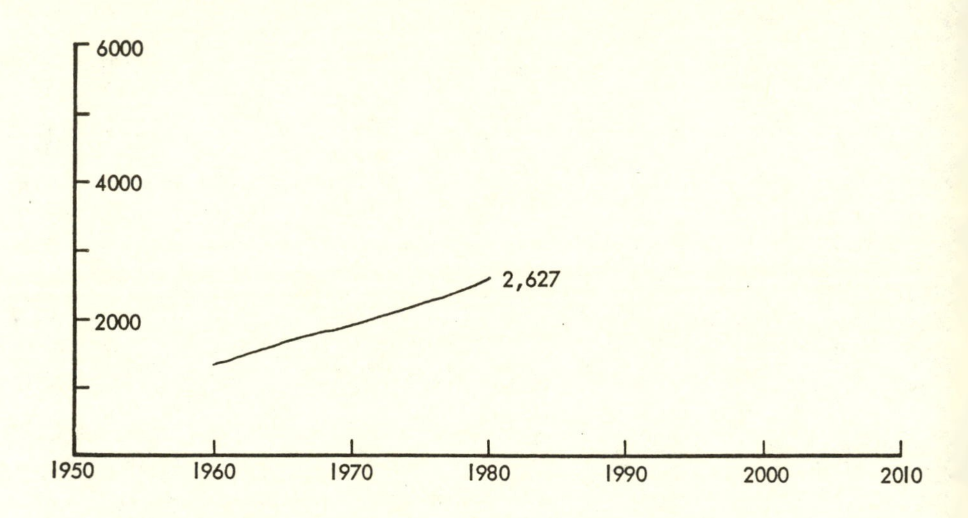

a. Trend Evolution: Please indicate on the above graph how you believe the trend will evolve over the next 30 years.

b. Point Values: Please indicate below your best estimates of the point values (vertical axis values) of the trend in the years 2000 and 2010

2000____________ 2010____________

c. Terminology: Gross National Product is the total national output of goods and services, valued at market prices and adjusted to eliminate double count.

d. Sponsor's Comments:

e. Statistics:

	Round One		Round Three	
Responses:	23		13	
	2000	2010	2000	2010
Low:	2800	3000	3100	3700
Mean:	4140	5070	3998	4850
High:	5290	6100	5290	7500
Std. Dev.:	730	1160	630	1040

Round One Comments on Item 20 (U.S. Gross National Product)

223. Growth Rates: 1980-90: 3.5% /year, 1990-00: 2.7%/year, 2000-10: 2.3%/ year.

206. 3% a year is a reasonable estimate of potential growth.

210. Steady growth trend at historic level of 3.5%.

906. Increase should continue but at a slower rate.

221. Annual Rates of growth: 1980-1990: 2.5% p.a., 1990-2000: 2.5% p.a., 2000-2010: 2.0% p.a. Lower than 1950-1980 averages because of decrease in population growth.

212. My bias is that the U.S. will approach a steady-state economy, with resource conservation/recycling leading to a gradual slow-down in growth as measured by GNP. You should also take into account some estimates of a vast and growing "underground" economy not measurable by GNP (barter, do-it-yourself).

209. Our emphasis on inflation fighting will hurt our chances for growth in the next 15-20 years.

908. I believe in about a 2 1/2 - 3 1/4% growth rate. However, growth will be slower in 1980s and will only speed up in the 90s and beyond.

903. Continued GNP real growth but at slower rates than historically.

104. Low GNP through 1990 because military spending and consumer spending will be big, hurting GNP growth.

Round Three Comments on Item 20 (U.S. GNP)

211. I do not believe the economy can achieve a steady state as comment 212 implies. I would estimate growth rates about like comment 223. Comment 104 implies little change in investment ratios in the 1980s. I believe tax law changes will increase this ratio in 1980s.

903. I see continued GNP growth as competition for domestic and world markets continues and U.S. endeavors to retain its leadership role.

ITEM 21. INDEX OF WORLD GROSS DOMESTIC PRODUCT (INDEX NUMBER, 1975=100) [21]

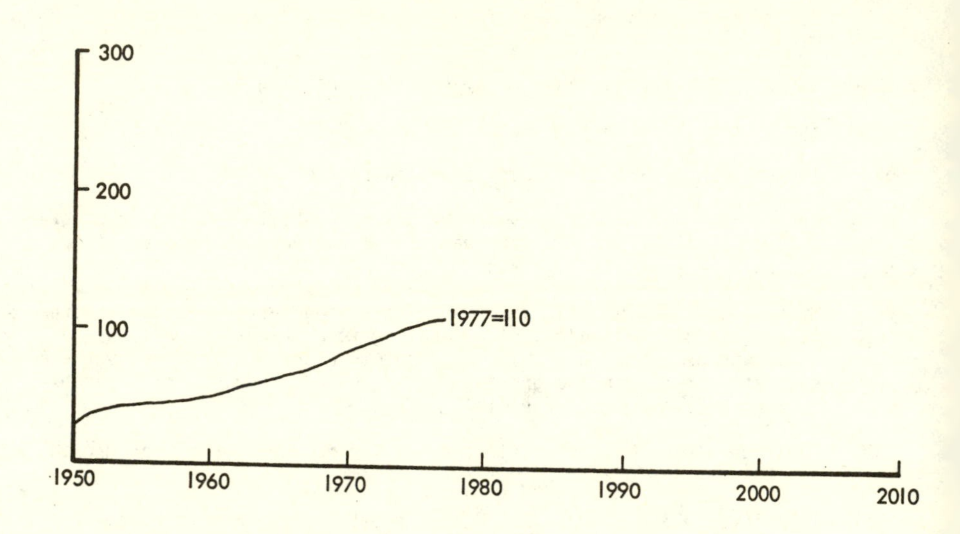

a. Trend Evolution: Please indicate on the above graph how you believe the trend will evolve over the next 30 years.

b. Point Values: Please indicate below your best estimates of the point values (vertical axis values) of the trend in the years 2000 and 201

2000____________ 2010____________

c. Terminology: Gross Domestic Product, as measured by the United Natio Monthly Bulletin of Statistics, does not include the market value of gove ment or private services. With this important exception, Gross Domestic duct is the same as Gross National Product.

d. Sponsor's Comments: With the advent of large government and private s ice sectors in developed countries, GDP may become a progressively less reliable gauge of economic growth.

e. Statistics:

	Round One		Round Three	
Responses:	23		12	
	2000	2010	2000	2010
Low:	115	117	130	140
Mean:	171	214	177	226
High:	242	360	225	360
Std. Dev.:	40	70	34	69

Round One Comments on Item 21 (Index of World Gross Domestic Product)

223. Growth Rate: 1977-2000: 3.5% /year, 2000-2010: 3.0%/year.

205. Dip in 1979-81, thereafter, 4% real growth.

905. Bringing non-cash nations into a cash economy will cause faster growth in GDP. There are plenty of LDCs to bring in.

212. I see continued gradual rise in GDP -- no big breakthroughs in the production capacity of developing countries.

209. See comments to Item 20. After 2000, LDCs will start contributing more to world GDP.

201. Continued energy price squeeze of LDCs.

904. The growth will be fueled by the elevation of the LDCs.

903. Continued world GNP growth. The slower rate of developed countries being offset somewhat by faster rates in the lesser developed nations.

104. Oil prices will keep Third World economics bleak (80% of their import earnings now go for oil).

Round Three Comments on Item 21 (Index of World GNP)

215. World growth will accelerate in the 2000-2010 period as LDCs enter development phase comparable to Western Europe in 1950s and 60s.

104. I'd go with the same conclusions. Political unrest via rough economic conditions (national bankruptcies, firewood shortage, oil deficits) will prevent substantial economic gains in Third World economies.

903. I think growth in world GNP will resume as inflation is contained and real growth is spurred by technology and social organization in LDCs.

ITEM 22. AGRICULTURAL SALES (AS A PERCENTAGE OF GNP) [23]

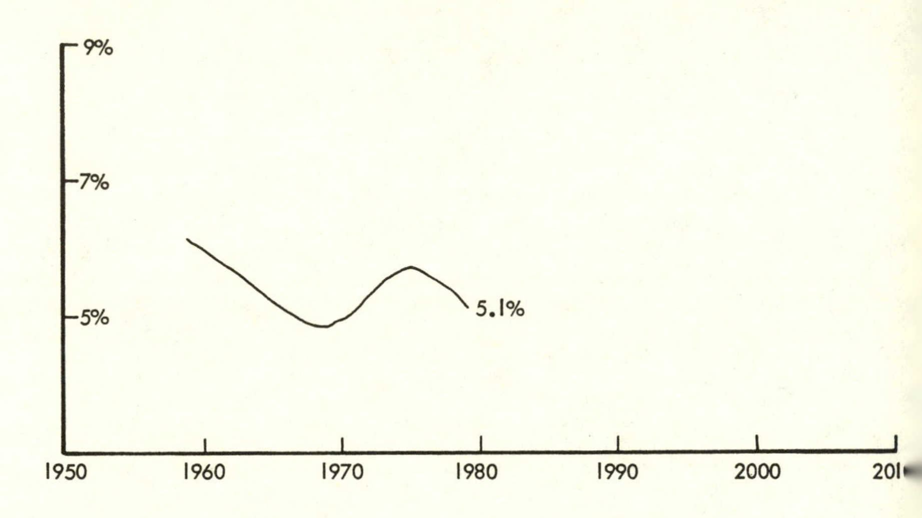

a. Trend Evolution: Please indicate on the above graph how you believe the trend will evolve over the next 30 years.

b. Point Values: Please indicate below your best estimates of the point values (vertical axis values) of the trend in the years 2000 and 201

2000____________ 2010____________

c. Terminology: In order to improve comparability of definitions and enumerations by the Census of Agriculture over time, the term farm refers to all land devoted to agricultural operations which had yearly nominal dollar sales of $2500 or more. Currently included as farms are such divers enterprises as nurseries, greenhouses, sod farms, mushroom operations, cra berry bogs, feedlots, fish farms, and hatcheries; excluded are business en terprises exclusively engaged in forest production and in production of fish, oysters, fowl, etc., from the ocean, game preserves, parks, and the like, when not grown in captivity.

d. Sponsor's Comments: Agricultural census data are available at approxim 5-year intervals. The graph above may therefore be deceptive as to years than 1959, 1964, 1969, 1974, and 1978.

e. Statistics:

	Round One		Round Three	
Responses:	42		20	
	2000	2010	2000	2010
Low:	3.0%	2.5%	3.0%	2.5%
Mean:	5.3%	5.5%	5.4%	5.5%
High:	7.0%	8.0%	7.0%	8.0%
Std. Dev.:	1.0%	1.5%	1.0%	1.4%

Round One Comments on Item 22 (Agricultural Sales)

518. With slow-down in agricultural technology and declining contribution to GNP, the downward trend will continue.

215. Agriculture will become a significant primary energy supplier by the end of the century.

304. An increase in foreign sales anticipated.

513. Should not decline, but remain stable, because of increase in part-time "gentlemen" farmers.

205. Fluctuations of around 5% (I've assumed graph refers to USA).

906. I believe they will increase because of the increase in new genetic varieties and increasing U.S. awareness of world markets.

905. I expect U.S. farm exports to balance its declining share of the domestic economy.

212. Slower growing GNP and higher costs of commodities influence the chart.

223. Will be stimulated by foreign demand not (biomass - methanol - energy) - LDCs will not have had time to develop agriculture. Russia still will have the problem of not being able to feed its population. The United States will need something to help balance its trade. Also some of these LDCs will be starving - world stability will also be one of our motives.

502. With the U.S. comparative advantage in agriculture, agriculture should increase as other world regions will need our produce.

903. Increase in relative value as a result of emerging biotechnology and mariculture -- agriculture becomes even more high tech.

Round Three Comments on Item 22 (Agricultural Sales)

215. Growth of agriculture as primary energy supplier and increasing comparative advantage of U.S. as an agricultural producer will increase GNP share.

213. Continued downward trend will reflect a high productivity in agricultural relative to services and an increased share of services in GNP due to the income elasticity of services.

906. I'll stand by my high estimates (2010=7%) due to the world food situation and our growing technology.

903. Agriculture is an increasingly hi-tech area. I expect that bio-engineering and continued technology improvements will increase its share of GNP.

ITEM 23. NATIONAL DEFENSE EXPENDITURES (AS A PERCENTAGE OF GNP) [24]

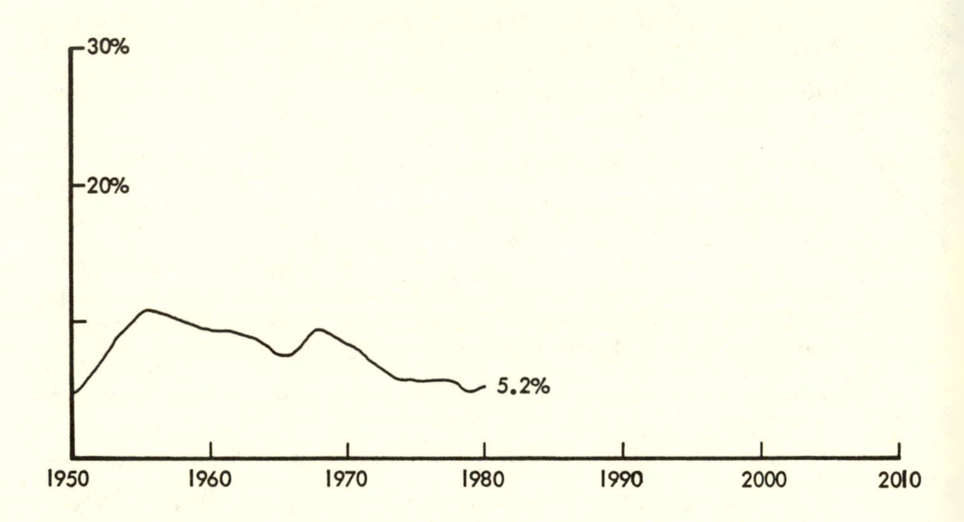

a. Trend Evolution: Please indicate on the above graph how you believe the trend will evolve over the next 30 years.

b. Point Values: Please indicate below your best estimates of the point values (vertical axis values) of the trend in the years 2000 and 20

2000____________ 2010____________

c. Terminology: Defense expenditures include veterans outlays.

d. Sponsor's Comments:

e. Statistics:

	Round One		Round Three	
Responses:	22		13	
	2000	2010	2000	2010
Low:	4.0%	3.0%	5.0%	4.0%
Mean:	5.7%	5.3%	5.9%	5.7%
High:	9.0%	7.5%	10.0%	12.0%
Std. Dev.:	1.2%	1.1%	1.5%	2.1%

Round One Comments on Item 23 (National Defense Expenditures)

223. Sometime in the 1990s one of two outcomes will be clearly apparent: (1) we will have reached an accommodation with Russia which permits both sides to reduce arms expenditures, or (2) we will acknowledge a lack of will to compete with them.

906. I believe this facet will remain about where it is, the economy cannot afford much more.

905. The Reagan administration will cause an aberration.

212. A steady rise in the 80s, then a drop as weapons come on line, but it will not drop below 70s low. Big unknowns here, of course, re cold war.

201. Temporary increase due to Reagan's national defense policy, and accumulated Vets' benefits.

908. Obviously this is highly dependent on war/peace. Will undoubtedly rise, but growth of total GNP should begin to flatten and reduce percentage by after the turn of the century.

904. We will remain confused about the role of national defense but the defense industry lobby is dramatically strong and will continue to fight back.

903. An increase in percentage of GNP as: (1) weapon systems become more expensive; (2) manpower costs increase; and (3) the need to offset Soviet Union, Japan and probably China continues.

Round Three Comments on Item 23 (National Defense)

903. I expect that in view of the growth of China and the increasing dependence of Japan, we will continue to maintain defense readiness both for our own defense and for foreign operations, e.g., Persian Gulf, Africa.

ITEM 24. FOREIGN PENETRATION OF UNITED STATES MARKETS (U.S. IMPORTS AS A PERCENTAGE OF U.S. GNP) [25]

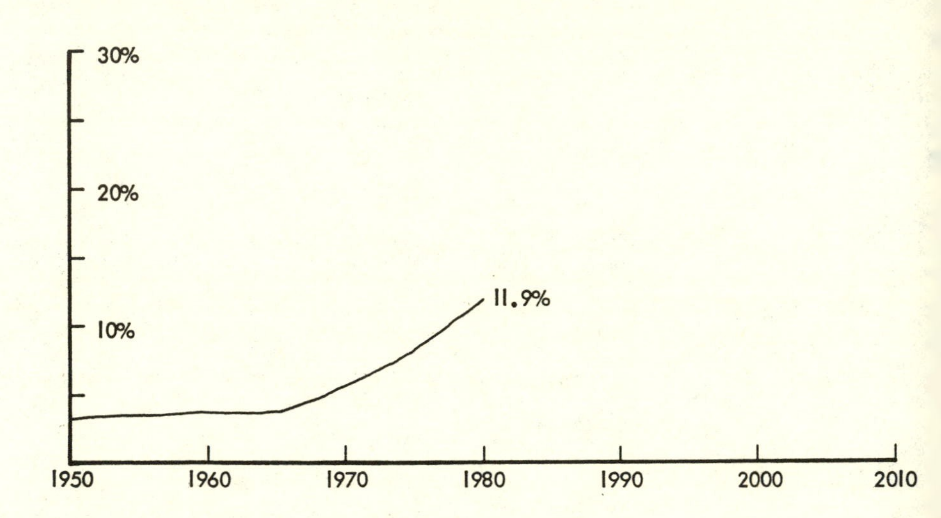

a. Trend Evolution: Please indicate on the above graph how you believe the trend will evolve over the next 30 years.

b. Point Values: Please indicate below your best estimates of the point values (vertical axis values) of the trend in the years 2000 and 2010

2000______________ 2010______________

c. Terminology:

d. Sponsor's Comments:

e. Statistics:

	Round One		Round Three	
Responses:	22		13	
	2000	2010	2000	2010
Low:	13.0%	15.0%	10.0%	8.0%
Mean:	17.3%	20.1%	16.2%	18.2%
High:	25.0%	30.0%	20.0%	25.0%
Std. Dev.:	3.5%	5.4%	2.7%	4.5%

Round One Comments on Item 24 (Foreign Penetration of U.S. Markets)

906. It should continue to increase because world interaction is increasing.

212. Continued maturing of U.S. economy will mean steep rise in import penetration, then leveling off.

211. If the economy of U.S. is going to do well, American manufacturers, now that they have recognized that foreign producers are flooding American markets with superior products, will have to become more competitive (price-quality) to succeed in foreign and domestic markets. I do not feel that third world markets are the answer. Unless they straighten out their internal problems they will not have the income to buy American products.

201. Increasing penetration of media and high technology items. Ease off oil as an increasing percentage.

908. Pessimistic in short term but the growth rate of penetration percentage will slow down toward end of century and may even turn down.

904. American industry will respond better than it has.

207. Principally a decline in oil imports.

903. Increasing interdependence and world trade, along with national specialization -- Japanese cars and electronics, European heavy machinery, Southeast Asia textiles.

Round Three Comments on Item 24 (Foreign Penetration of U.S. Markets)

215. Growing international interdependence will increase impact share.

104. Oil import decline may be offset by increased shift of energy intensive industries to nearby nations. I am guessing that U.S. industry will respond better than it has to foreign competition.

903. World Trade will increase and there will be increased foreign manufacturing of high-labor intensive items and items based on cheaper foreign raw materials, e.g., natural gas.

ITEM 25. VALUE OF U.S. EXPORTS (AS A PERCENTAGE OF U.S. G.N.P.) [26]

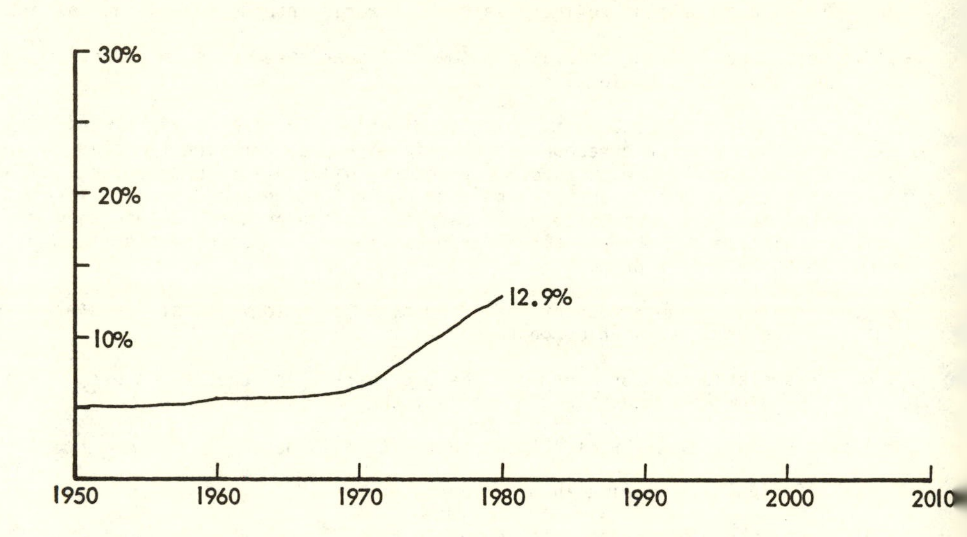

a. Trend Evolution: Please indicate on the above graph how you believe the trend will evolve over the next 30 years.

b. Point Values: Please indicate below your best estimates of the point values (vertical axis values) of the trend in the years 2000 and 2

2000__________ 2010__________

c. Terminology:

d. Sponsor's Comments:

e. Statistics:

	Round One		Round Three	
Responses:	22		13	
	2000	2010	2000	2010
Low:	9.0%	8.0%	10.0%	8.0%
Mean:	16.4%	19.4%	16.2%	18.4%
High:	23.0%	29.0%	20.0%	25.0%
Std. Dev.:	3.4%	5.7%	2.9%	4.6%

Round One Comments on Item 25 (Value of U.S. Exports)

906. I believe there will be marked increase because of increased international linkages.

212. This reflects possible loss of dominance in some high technological fields and less agricultural productivity.

905. I expect a much more interdependent world economy. 30% is typical for European countries today. (2010 value of 30%)

201. Barely equal balance of trade.

908. Going to have to pay for those imports somehow. Also, continued expense of imported energy and raw materials.

903. Continued world interdependence, with U.S. as supplier of high technology

Round Three Comments on Item 25 (Value of U.S. Exports)

211. I do not believe it will reach 30% by 2010. To reach that level would entail large markets in the LDCs or large sales of agricultural exports to Europe and Japan which are now severely restricted. While both of these may take place, I don't think it will be to the extent necessary to expand exports to a 30% ratio of GNP.

903. Increasing world trade and the need to balance imports will increase the value of U.S. exports, especially with high-tech products such as agricultural products, aircraft and computers.

ITEM 26. RATE OF CAPITAL FORMATION (CAPITAL INVESTMENT AS A PERCENTAGE OF U.S. GNP) [27]

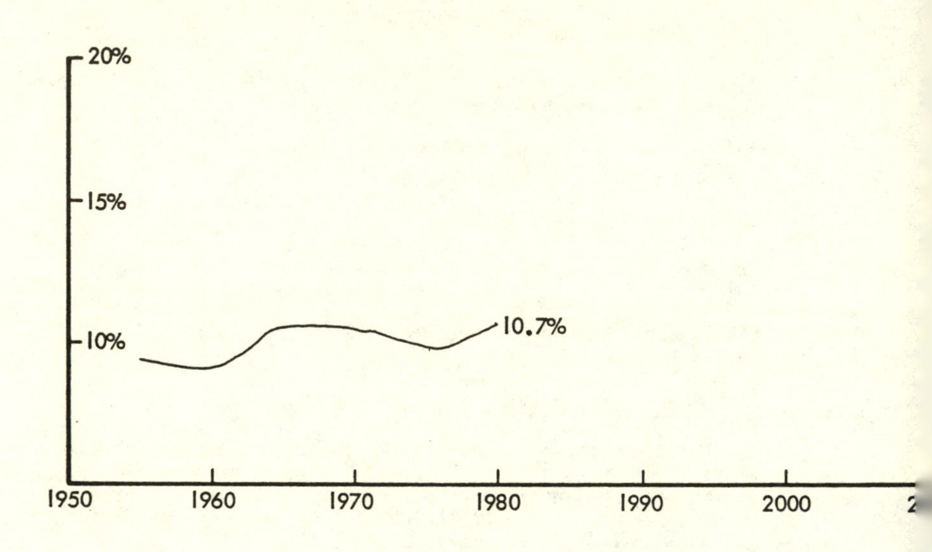

a. Trend Evolution: Please indicate on the above graph how you believe the trend will evolve over the next 30 years.

b. Point Values: Please indicate below your best estimates of the point values (vertical axis values) of the trend in the years 2000 and 2010.

2000____________ 2010____________

c. Terminology:

d. Sponsor's Comments:

e. Statistics:

	Round One		Round Three	
Responses:	22		11	
	2000	2010	2000	2010
Low:	10.0%	8.0%	10.0%	10.0%
Mean:	11.5%	11.3%	12.3%	12.5%
High:	17.0%	15.0%	17.0%	15.0%
Std. Dev.:	1.8%	1.9%	1.8%	1.5%

Round One Comments on Item 26 (Rate of Capital Formation)

205. Will rise a bit initially, then fluctuate without trend.

906. I am not confident this will increase because as we join a world economy, diversification and merger, etc., may replace capital investments for industry.

905. Wealthier country can afford to invest more.

221. Slow down in labor force growth will reduce rate of return on capital.

212. I see a rise in the 80s and dropping back in later decades.

215. It will be cyclical. It is rising now (will rise until about 2000) due to the perception that we are behind in technology and in productivity. By 2000 we will be secure again in our position and be putting less emphasis on capital formation.

223. In real terms, especially if you exclude expenditures for pollution devices, on a private and government level, expenditures for capital -- things that enhance productivity have been going down since 1965. Now we feel pressure from this lack of investment and will push for more capital expenditures. After 1990 we will have updated our "capital plant" and will again reduce the allocation to this area.

903. Capital formation rate will continue to increase at average historic rate.

Round Three Comments on Item 26 (Rate of Capital Formation)

215. The U.S. will go through a cycle of heavy renewal in plant, equipment and infrastructure by the year 2000. This investment bias will have begun giving way to a consumption cycle.

213. I assume a fundamental policy shift to promote increased savings/investment.

903. Increasing domestic emphasis on a policy for capital formation and incentives to save will be required to keep the U.S. competitive.

ITEM 27. CAPITAL ALLOCATION TO ELECTRIC UTILITY INDUSTRY (AS A PERCENTAGE OF TOTAL CAPITAL AVAILABLE) [28]

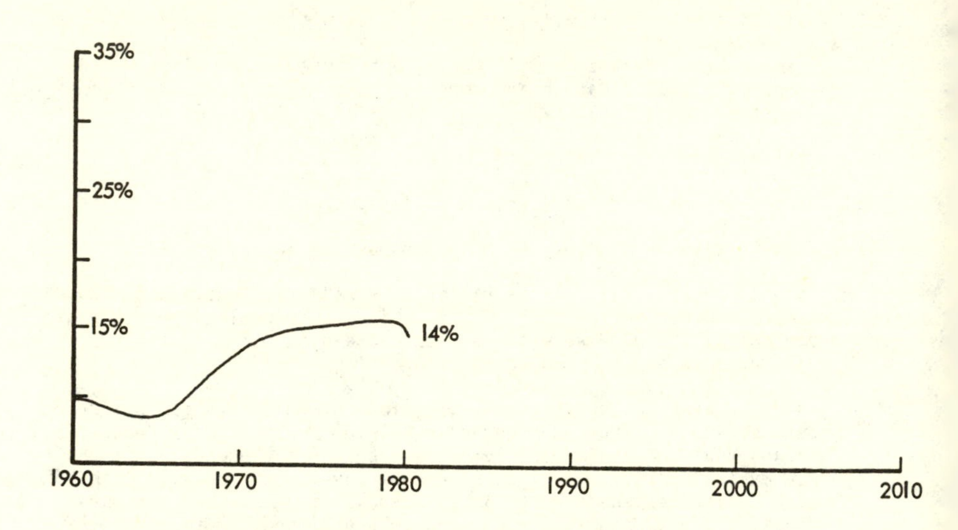

a. Trend Evolution: Please indicate on the above graph how you believe the trend will evolve over the next 30 years.

b. Point Values: Please indicate below your best estimates of the point values (vertical axis values) of the trend in the years 2000 and 201

2000__________ 2010__________

c. Terminology: This ratio is the total amount spent for new plant and equipment by electric utilities divided by the total amount spent for new plant and equipment for all sectors in the U.S. economy, excluding agricultural business, professionals, institutions, and real estate firms.

d. Sponsor's Comments:

e. Statistics:

	Round One		Round Three	
Responses:	23		13	
	2000	2010	2000	2010
Low:	7.0%	5.0%	9.0%	6.0%
Mean:	12.2%	11.9%	12.7%	12.3%
High:	20.0%	22.0%	16.0%	18.0%
Std. Dev.:	3.5%	4.2%	2.4%	3.6%

Round One Comments on Item 27 (Capital Allocation to Electric Utility Industry)

205. Will rise a bit initially, then fluctuate without trend.

905. Two trends: (1) Increased electrification; and (2) Greater efficiency in production.

221. Reluctant regulators, re allowable rate of return and systematically lower load growth, will reduce rate in short-middle range and keep it there.

212. This reflects my bias that centralized electrical generating capacity will be less economical and necessary as the economy moves toward a period of dispersed power generation and less dependency on electricity for many needs (the soft path approach).

215 It will be dropping in the near term because of past over expansion - presently it is 0. By the turn of the century demand will have caught up - the market will begin allocating capital to the electric sector again, but by that time the methods will be less capital intensive (solar).

223 Cyclical - now in a period of oversupply but this will change shortly.

213. Consumer-owned electricity generation should be included in a comprehensive "industry" measure - solar, hydro, coal.

209. We will substitute insulation, etc., for electricity and use more unconventional electricity sources (solar-cells).

908. There will be a falling, with a slight rise in the late 80s as old (more than 30 year old) plants are replaced - then drop to much lower levels.

904. Investment will stay relatively low in the 80s while society decides the appropriate technology mix which the industry should pursue. Then in the 90s and forward, increases in investment will be necessary for new and replacement facilities and response to environmental constraints.

207. Use of excess capacity, then building with many less capital intensive options.

909. The trend is toward more capital intensive technologies (with lower fuel costs). This is mitigated by substantially reduced growth rates. Net effect is a small increase in capital allocation to electric utility industry.

903. After dip through 1980s, pent-up demand and improved interest rates dictate return to investment in electrical generating capacity.

Round Three Comments on Item 27 (Capital Allocation to Electric Utility Industry)

104. Huge fall already this year -- agree with 212, 207 that residual amount is replacements. T & D.

903. Central generating facilities will get larger and more expensive requiring an increasing percentage of capital allocation to both the addition of new plants and the replacement of outmoded inefficient facilities.

ITEM 28. ECONOMIC CENTRALIZATION (PERCENTAGE OF TOTAL U.S. CORPORATE ASSETS CONCENTRATED AMONG THE LARGEST 500 NONFINANCIAL CORPORATIONS) [29]

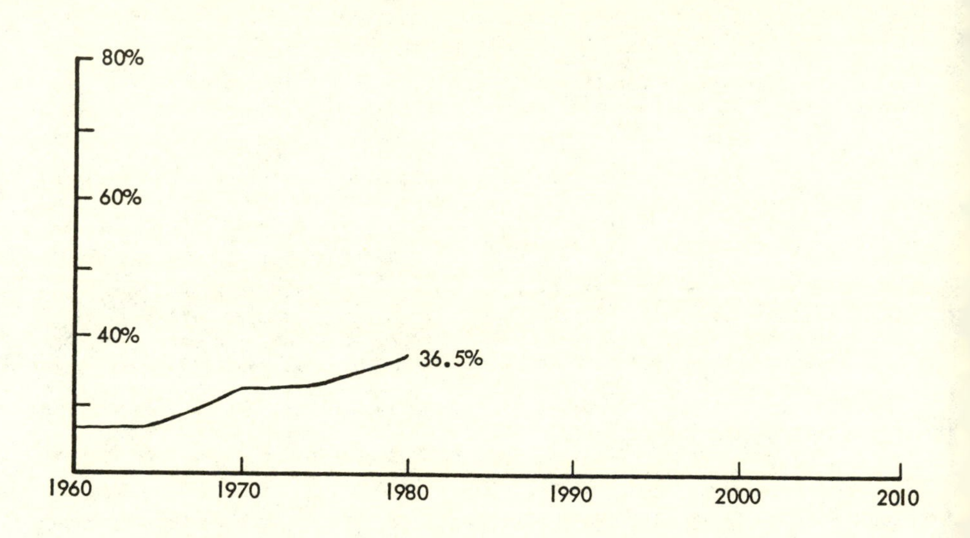

a. Trend Evolution: Please indicate on the above graph how you believe the trend will evolve over the next 30 years.

b. Point Values: Please indicate below your best estimates of the point values (vertical axis values) of the trend in the years 2000 and 2010

2000__________ 2010__________

c. Terminology: Corporations are ranked according to asset value. The assets of financial corporations (i.e., commercial banks) are not included.

d. Sponsor's Comments:

e. Statistics:

	Round One		Round Three	
Responses:	22		13	
	2000	2010	2000	2010
Low:	36.5%	36.5%	31.0%	25.0%
Mean:	42.2%	45.2%	40.7%	41.6%
High:	52.0%	62.0%	52.0%	50.0%
Std. Dev.:	5.3%	8.2%	5.1%	6.5%

Round One Comments on Item 28 (Economic Centralization)

304. Increasing mergers, especially in natural resource or financial services and communications industries.

906. This should increase because interest rates will likely remain quite high. In financial markets larger corporations singly can do better and mergers are desirable.

905. Mature industries tend toward oligopoly.

212. Continued concentration of assets -- "trust busting" -- will play no significant role, alas.

215 Big firms are dinosaurs, but as long as the capital is intact they will remain. Smaller information or technology firms will react to changes in the marketplace (threats and opportunities) much more quickly.

908. Very subject to application of anti-trust laws, of course.

903. The trend to centralization will continue, though at somewhat slower rates than historically. Antitrust policy will continue to be reinterpreted to permit growth, though mergers may from time to time be inhibited.

Round Three Comments on Item 28 (Economic Centralization)

215. Growth in new assets, employment and profits will be in smaller firms tied to new technology.

104. Agree on national resources/financial services companies. Many small outfits are low on asset pole.

903. I expect antitrust laws will be relaxed in order to recognize realities of late 20th and early 21st centuries -- need for large international organizations to handle world trade and compete with Japanese and European giants.

ITEM 29. WORLD TRADE (BILLIONS OF 1980 U.S. DOLLARS) [30]

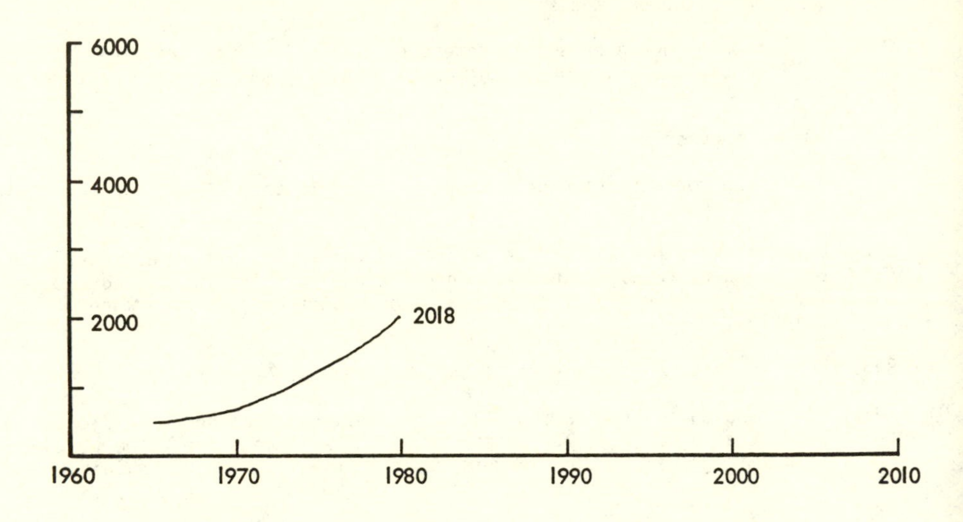

a. Trend Evolution: Please indicate on the above graph how you believe the trend will evolve over the next 30 years.

b. Point Values: Please indicate below your best estimates of the point values (vertical axis values) of the trend in the years 2000 and 2010.

2000____________ 2010____________

c. Terminology: World trade is the value of gross exports for all countries

d. Sponsor's Comments:

e. Statistics:

	Round One		Round Three	
Responses:	22		13	
	2000	2010	2000	2010
Low:	2100	2000	2500	2500
Mean:	4430	5980	4381	6300
High:	6400	11400	6400	11400
Std. Dev.:	1280	2470	1138	2563

Round One Comments on Item 29 (World Trade)

223. It will continue to grow faster than world domestic product but not at the 2/1 growth rate ratios of the last couple of decades.
Growth rates: 1980-2000: 5.0% / year, 2000-2010: 3.5% / year.

206. 6% a year real growth.

304. Possible decline in Middle East trade owing to fast declining crude oil consumption. Increase in protective tariffs.

205. I don't think 1970-80 growth trend can be continued. Say 4% real growth after 1990. (Perhaps 6-7% until 1990 -- a guess.)

906. Again because of increased world interdependence, I believe this must increase.

905. S curve since there is a saturation point.

221. Annual average rates of growth: 1980-1990: 10.0%, 1990-2000: 10.0%, 2000-2010: 7.0%.

212. Continued steep rise, then leveling off based on possible increased economic/technological autonomy of larger developing nations.

201. As world trade grows it will ease off oil as a percentage.

207. Demographics are consistent with U.S. economic prognosis.

903. World trade will continue to grow though slowed national growth will keep trade below rates of the last decade.

Round Three Comments on Item 29 (World Trade)

No additions

ITEM 30. NET VALUE OF U.S. OWNED ASSETS LOCATED ABROAD (BILLIONS OF 1980 DOLLARS) [31]

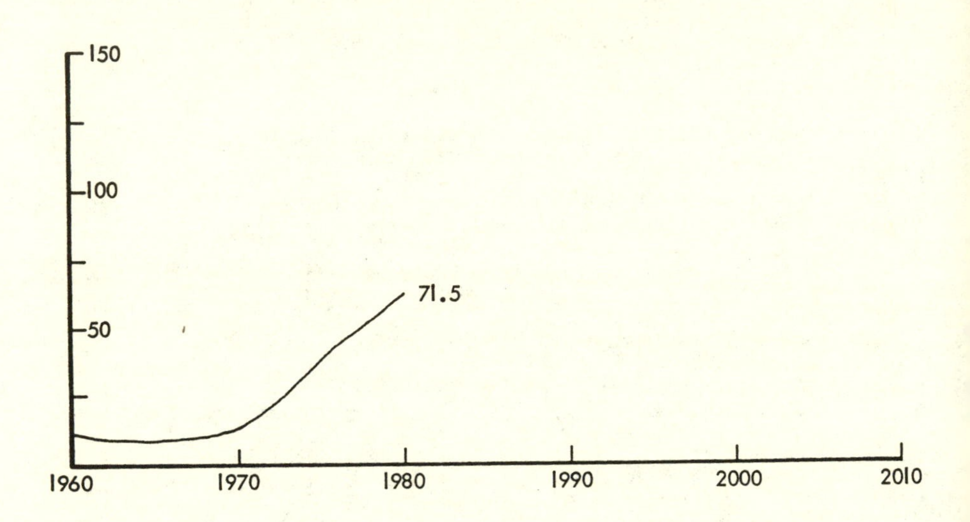

a. Trend Evolution: Please indicate on the above graph how you believe the trend will evolve over the next 30 years.

b. Point Values: Please indicate below your best estimates of the point values (vertical axis values) of the trend in the years 2000 and 2010

2000____________ 2010____________

c. Terminology: Only privately held assets are measured.

d. Sponsor's Comments:

e. Statistics:

	Round One		Round Three	
Responses:	22		13	
	2000	2010	2000	2010
Low:	75.0	75.0	75.0	80.0
Mean:	106.7	128.0	113.1	142.2
High:	155.0	230.0	155.0	230.0
Std. Dev.:	26.5	44.6	24.9	44.2

Round One Comments on Item 30 (Net Value of U.S. Owned Assets Abroad)

205. Same basic pattern as world trade.

906. I expect this increase to be fairly steady and continuous. U.S. firms see decreased labor costs and other advantages to operating abroad.

905. Two trends: (1) Increasing world trade (2) Increased wealth abroad makes U.S. capital less sought after.

212. A continued rise, then a drop in 90s in response to increased nationalism and the desire to retain local control of domestic economies by many developing countries.

209. The steep rise between 1970 and 1980 is primarily due to the devaluation of the dollar.

201. Federal restrictions likely.

908. Will slow down, but should speed up again after "80s doldrums" pass.

903. Growth of U.S. investment abroad will continue. Increasing national limitations abroad will keep such growth below the rates of the 1970s.

Round Three Comments on Item 30 (Net Value of U.S. Owned Assets Located Abroad)

104. Highly judgemental - depends on LDCs political/economic conditions and efficiency of U.S. industrial production.

903. U.S. will continue to encourage foreign investment to permit U.S. to remain competitive in world trade.

ITEM 31. NET VALUE OF FOREIGN OWNED ASSETS IN THE U.S. (BILLIONS OF 1980 DOLLARS) [32]

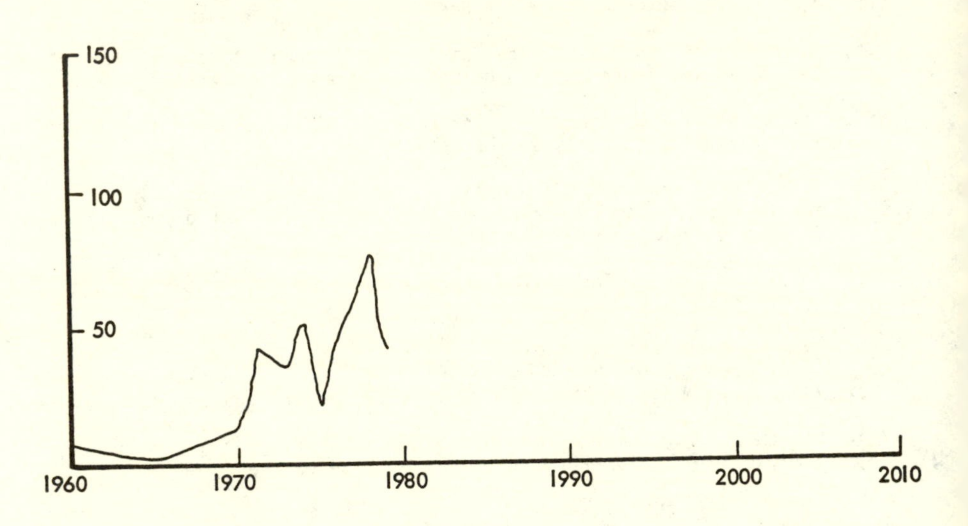

a. Trend Evolution: Please indicate on the above graph how you believe the trend will evolve over the next 30 years.

b. Point Values: Please indicate below your best estimates of the point values (vertical axis values) of the trend in the years 2000 and 2010.

2000__________ 2010__________

c. Terminology:

d. Sponsor's Comments:

e. Statistics:

	Round One		Round Three	
Responses:	22		13	
	2000	2010	2000	2010
Low:	65.0	70.0	75.0	85.0
Mean:	104.1	130.0	102.5	129.9
High:	200.0	300.0	120.0	160.0
Std. Dev.:	36.2	57.8	14.9	28.2

Round One Comments on Item 31 (Net Value of Foreign Owned Assets in the U.S.)

205. Slow growth at first (due to expensive U.S. dollar).

906. I expect this to increase as interdependence increases.

212. This trend is affected by the public perception of amount and implications of foreign investment; the latter will rise in the 90s, then level off.

223. Mid-East will continue to have lots of dollars to invest in the U.S. Also other countries, France or Canada for example, will be less attractive to foreign investors.

201. Will grow with net imports.

903. Foreign investment in the U.S. will continue at essentially historic rates because of size of growth of U.S. market and soundness of U.S. economy.

Round Three Comments on Item 31 (Net Value of Foreign Owned Assets in the U.S.)

903. Capital needs will continue and the U.S. will continue to be open to foreign investment. Because of a large, stable market, we will continue to be attractive to foreign funds.

ITEM 32. CITIZEN SUPPORT OF ENVIRONMENTAL AND CONSERVATION EFFORTS (INDEX NUMBER, 1980=100) [33]

a. Point Values: Please complete the table below to indicate your estimates of the future levels of U.S. citizen interest in environmental and conservation issues, based on a 1980 index level of 100. Please see comments below.

	1965	1970	1975	1980	1990	2000	2010
Level of Interest:	27.3	66.7	84.1	100	____	____	____

b. Terminology: The historical index values were developed by summing annual membership figures for the country's largest environmental and conservation groups, dividing those figures by the non-institutionalized U.S. population 16 years of age or older, and finally dividing the 1965, 1970, and 1975 quotients by the 1980 quotient. The membership figures may include some institutional members (libraries, etc.), and the summing of those figures will multiple-count individuals who belong to more than one of the organizations.

c. Comments: We are interested in estimating the extent to which our society includes individuals whose levels of environmental or conservation interest are high enough that they will take at least nominal steps to further those interests. This is obviously difficult to assess. If you do not believe the process we used to develop the historical index values is appropriate in light of what we wish to assess, disregard that process in making your future estimates (but assume 1980=100).

d. Statistics

	Round One			Round Three		
Responses	78			41		
	1990	2000	2010	1990	2000	2010
Low:	50.0	25.0	25.0	50.0	25.0	25.0
Mean:	108.1	114.6	121.8	105.6	111.6	117.3
High:	200.0	280.0	350.0	150.0	200.0	200.0
Std. Dev.:	20.9	36.6	48.0	18.4	33.3	38.1

Round One Comments on Item 32
(Citizen Support of Environmental and Conservation Efforts)

518. I expect an ultimate decline in interests as people become more concerned with economic (short-run) interests. (2010 value of 80).

411. I think knowledge and interest will increase -- with much activity at local level. I think a number of the organizations may consolidate so index may not be good. (2010 value of 130)

301. The process described leaves me somewhat disatisfied because membership in organizations is sometimes "trendy" -- I have tried to show an index of interest by the general public in the issue. A leveling for a decade or so and then augmenting again as environmental degradation continues. (2010 value of 180)

521. Environmentalists are going to increase. More sophisticated technology will have to be created to meet their demands. (2010 value of 130)

306. Measure seems like a good one. How many leading environmental groups?

322. Better measure would be purchases of energy efficient products. "Conservation" in the parlance of many of the groups you probably counted includes much more than energy.

212. Yes, your process is too limited. I anticipate a surge of interest in these concerns, in tandem with public perception of the desirability of conservation for economic reasons and of environmental preservation. This interest may take many activist forms besides membership in natural groups.

409. Based on trends in California - which tend to lead the nation in these types of issues. There has been a general erosion in the support. As things get better people will lose interest. (2015 value of 25)

408. Rise in 1990 due to reaction to current administration policies, much coming in the form of multiple memberships. Long-term rise based on female activism as more women develop independent policy views. (Note: 1981 figure would be interesting to see.)

303. 1. Dropoff estimated because of usual backlash effects.
2. Mixing "environmental" and "conservation" is questionable.

904. Environmental concerns are not transitory. They have become part of the social fabric and the continual emergence of problems such as toxic substance exposure.

502. The Reagan administration notwithstanding, I believe with high levels of education, ecological awareness will be greater rather than less.

317. Interest will accelerate again at the turn of the century because of the population squeeze on world resources.

909. The "environment movement" has peaked. People will continue to increase their awareness of the environment while taking a more sensible approach to environmental protection. The same can be said about the "conservation movement" trending toward efficient use of energy rather than political rhetoric.

Round One Comments on Item 32
(Citizen Support of Environmental and Conservation Efforts)

104. Less mobile society implies higher environmental awareness.

901. I have based my outlook instead on responses to surveys. I anticipate a greater employment -- income/environmental -- cost squeeze throughout the remainder of the century. In that context, I must forecast declining support.

Round Three Comments on Item 32
(Citizen Support of Environmental and Conservation Efforts)

301. I do not agree with 518, 409, 303 and 909. I still like my figures better. The next 2-3 decades will bring additional pressure to environmental and conservation efforts because:
  1) Higher and higher costs of fuel
  2) All new technologies are "worse" environmentally
  3) Other basic resources (besides energy) will begin to become scarce a
  4) Increased attention to "space ship earth" as a finite environment.

903. I think environmental concern is now institutionalized in U.S. society and will grow through its place in education, communications and the political establishment.

409. Is membership in Audubon or the Sierra Club a measure of "citizen support" Both produce beautiful pictorials which are a tangible "reward" for belonging. I believe that the costs of providing this "reward" will escalate an subscription will drop off. Further, public reaction is starting to shape up along the lines that indicate an awareness of the law of diminishing returns, i.e., haven't we done enough already?

211. I assumed an "interest" in seeing something done about the environment rather than membership. On this basis, my thoughts are closest to comment 909 and 104.

402. (909) Private support for environmental/conservation efforts (as evidence by membership) may diverge from publicly supported positions or public stance, i.e., one's private druthers may erode when faced with pragmatic decisions. Question what this index (membership) really implies. Foresee more interest in regional environmental/conservation concerns and les for larger national/global aspects; thus, local/smaller groups may flouri while national organizations may remain stable/slightly declining.

514. Support for environmental and conservation efforts will decline somewhat with growing citizen recognition that an expanding economy is more important to the welfare of society (especially the disadvantaged) than mos environmentalists and conservationists acknowledge. Hazardous conditions will never be accepted, however, so certain efforts will continue to be strongly supported, e.g., no more Love Canals or Three Mile Islands.

Round Three Comments on Item 32
(Citizen Support of Environmental and Conservation)

330. My original response reflected a slight decline in interest over next decade as economic, world peace, and other issues gain in prominence and there is a slight backlash for the environmental enthusiasm of the 1970s. In the longer scene, I see a rich society willing to consider environmental issues at about today's level or slightly higher.

411. I continue to believe that environmental support will rise as a totally integrated part of our way of life. Local isses, drinking water, solid waste, hazardous waste, open spaces will probably dominate the concerns.

906. I believe the 1980s economic situation will lead to a temporary sacrifice mentality, but the realities of our environmental dependence will return.

104. Tremendous level of local support not counted in national membership, e.g., +300 California anti-nuclear groups in 4 years! and 20 anti-peripheral canal groups in 2 years! Local growth ordinances remain very potent. People care more about local resource squeeze, which accelerates support.

ITEM 33. PRESERVATION OF FEDERAL LANDS (PERCENTAGE OF TOTAL FEDERAL LANDS WITHDRAWN FOR PARKS, WILDLIFE REFUGES AND WILDERNESS AREAS) [34]

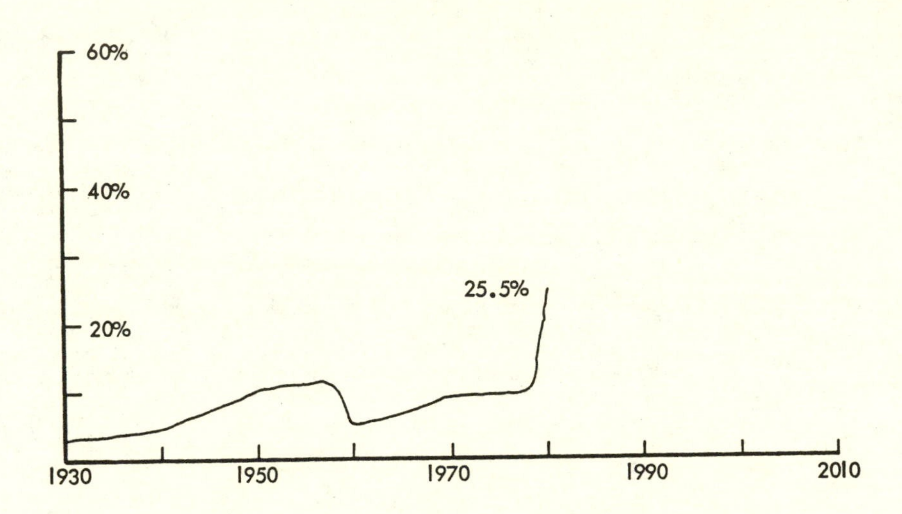

a. Trend Evolution: Please indicate on the above graph how you believe the trend will evolve over the next 30 years.

b. Point Values: Please indicate below your best estimates of the point values (vertical axis values) of the trend in the years 2000 and 201(

2000____________ 2010____________

c. Terminology: To get a measure of public demand to preserve public lan(for recreation and wildlife protection, the cumulative sum of federal land: set aside in the National Park System, National Wildlife Refuge System and other federal lands in the National Wilderness Preservation System is meas-ured as a percentage of total federal lands.

d. Sponsor's Comments: The two sharp changes in the graph indicate the in fluence of Alaskan lands. Prior to 1959, total federal lands (the denomina remained at approximately 410 million acres. Since the addition of Alaska Hawaii, the total has remained near 765 million acres. Between 1978 and 19 approximately 100 million acres of Alaskan lands were added to the three s tems which compise the numerator of the ratio.

e. Statistics:

	Round One		Round Three	
Responses:	17		8	
	2000	2010	2000	2010
Low:	18.0%	12.5%	18.0%	12.5%
Mean:	27.3%	29.2%	27.6%	28.9%
High:	38.0%	40.0%	38.0%	43.0%
Std. Dev.:	5.8%	8.4%	5.9%	8.6%

Round One Comments on Item 33 (Preservation of Federal Lands)

402. Believe there will eventually be (late 1980s) a movement to reduce federal lands in preservation status followed (mid 1990s) by transfer of non-preserved federal lands to private status (yielding higher percentage of preservation lands).

906. With population and economic pressure, I don't believe there will be much support for more preservation.

409. Expect that as time goes on, there will be an equilibrium on leveling off near 12 1/2%. This is based in part on other developed countries including Japan.

908. James Watt is an aberration -- a horrible one. However, future additions will be relatiely minor.

903. I can see a continuing trend to acquire and set aside lands for recreational use at essentially historic rates.

901. I do not think we are going to "get" Watt, and I do think he will "get" land out of protected statuses. There will be a slight rebound in the mid-to-late 1980s, but not back up to the pre-1980 level, which will remain an all-time high.

Round Two Comments on Item 33 (Preservation of Federal Lands)

409. 12 1/2% is still a good prediction of "equilibrium" in land preservation for the contiguous U.S. (410 million acres). Alaska is a special case where the percentage could be considerably higher -- up to 50%. Total U.S. could then be 25%.

411. Clearly I am in the minority here. I think I was thinking too much of general land acquisition projects including state. These are very much on the rise. In general, these are not 100% preservation areas. So my previous note is probably too high on the narrower issues.

903. I think there will be increasing social concern for preservation, as Americal society sees consequences of not preserving public lands, e.g., as in Texas!

ITEM 34. AIR QUALITY IN URBAN AREAS (NUMBER OF AREA DAYS DURING WHICH POLLUTANT STANDARD INDEX EXCEEDED 100) [35]

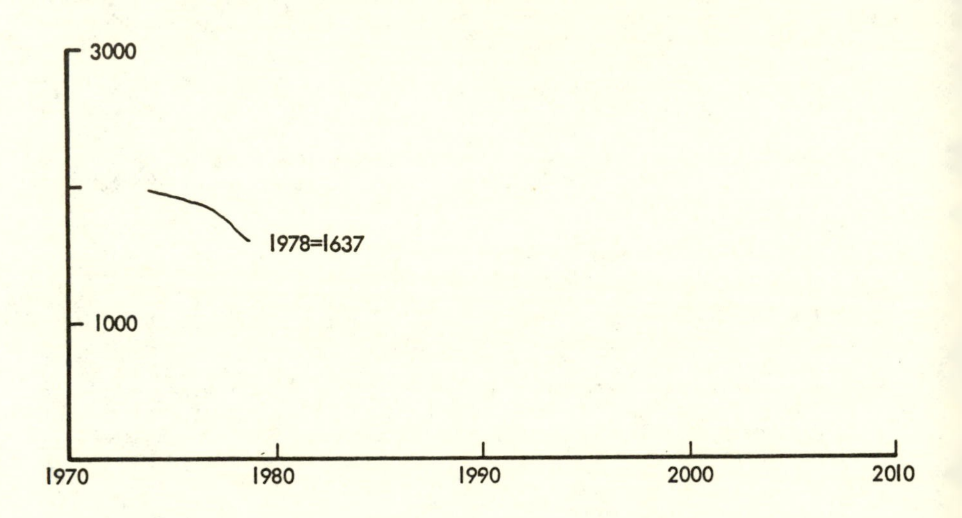

a. Trend Evolution: Please indicate on the above graph how you believe the trend will evolve over the next 30 years.

b. Point Values: Please indicate below your best estimates of the point values (vertical axis values) of the trend in the years 2000 and 2010

2000____________ 2010____________

c. Terminology: The Pollutant Standard Index (PSI) has been developed to get a general measure of pollution in the U.S. The PSI is based on actual measurements in 23 SMSAs (excluding New York). The PSI goes over 100 whenever any one of the 5 national ambient air quality standards is exceeded in any one SMSA. Thus, the graph indicates the number of days of violation of any standard in the 23 SMSAs. The five-year ambient air quality PSI figures have been revised to take into account the change in the ozone ambient air quality standard to .12 ppm.

d. Sponsor's Comments: In completing the trends please assume that the standards will not be changed and that the PSI is an accurate measure to indicate trends in urban air pollution. Please indicate if those assumptions seriously affect your estimate of the trend.

e. Statistics:

	Round One		Round Three	
Responses:	18		8	
	2000	2010	2000	2010
Low:	800	500	1000	900
Mean:	1332	1321	1350	1387
High:	2600	2700	1800	2000
Std. Dev.:	440	540	267	340

Round One Comments on Item 34 (Air Quality in Urban Areas)

402. Opposition between desire for clean air and necessity for increased coal consumption will stabilize air quality. (2010 value of 1350)

409. Attainment of air quality standards is now dependent on the control of mobile source emissions. Unless there is a breakthrough in vehicle emissions, air quality will remain, on the average, at today's levels.

304. If the Clean Air Act remains substantially unchanged, ozone areas will do better.

906. I expect population and industrial pressure and economic necessities will lead to continued growth in air pollution despite recent progress. (2010 value of 2000)

905. Clean air is a luxury that we can afford.

408. Long term improvement comes from replacement of the dirtiest sources as they become uneconomic, which offsets industrial expansion in SMSAs.

908. The major problem with using the PSI is that "standards" pollutants are relatively easy to control and may not measure the most serious pollutants, organics, etc. for which no standards exist. So in human health terms, I am less optimistic than my figures indicate.

904. I do believe the standards will continue to change and become more stringent. We are just seeing the pluses.

909. Sources of NOX are very likely to increase (especially under a low nuclear scenario), sources of hydrocarbon are likely to decrease (improved automobile engines/fuels -- greater fuel efficiency). Net result = small reduction in ozone.

903. Air quality will remain at present levels through the 1980s and gradually improve as technology solves problems of hydrocarbon combustion.

901. I expect the definition will be relaxed. Also real increase is based on assumption of increasing pollutants from energy generation and use.

Round Three Comments on Item 34 (Air Quality in Urban Areas)

409. California has the most restrictive standards and controls, and exceedences of the federal standards (adjusted for meteorological variability) have not declined over the past 3 years. I see little hope that, if restrictions equivalent to California are imposed nationwide, exceedences will decrease.

903. I expect that with increased application of combustion technology, fuel clean-up and mobile source emissions control, the quality of air will probably improve somewhat over the next 30 years.

ITEM 35. FEDERAL EXPENDITURES FOR POLLUTION ABATEMENT AND CONTROL (AS A PERCENTAGE OF TOTAL NON-MILITARY FEDERAL EXPENDITURES) [36]

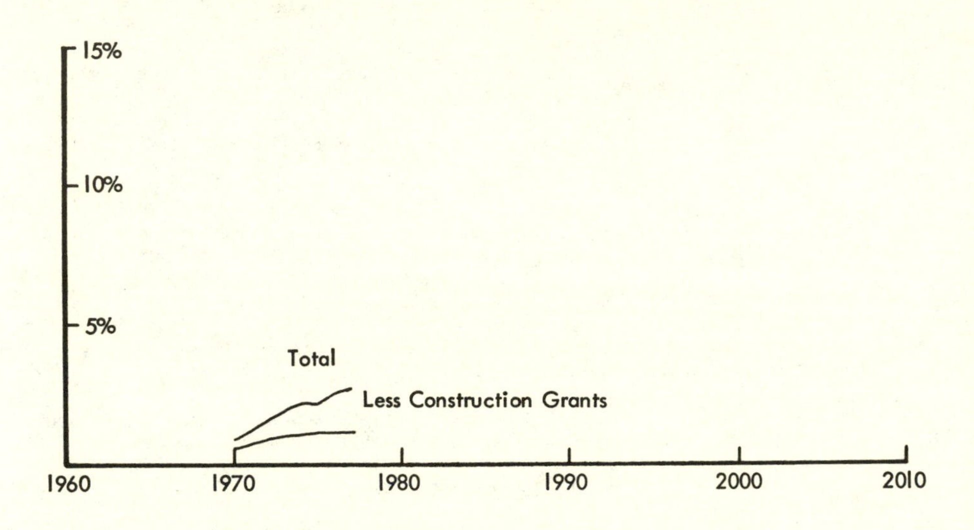

a. Trend Evolution: Please indicate on the above graph how you believe the trends will evolve over the next 30 years.

b. Point Values: Please indicate below your best estimates of the point values (vertical axis values) of the trends in the years 2000 and 2010.

Total Pollution Expenditure 2000________ 2010________

Non Const. Grant Expenditure 2000________ 2010________

c. Terminology: As a measure of federal expenditure for pollution control, t outlays of the Environmental Protection Agency and other Federal Agencies for pollution control and abatement are given as a percentage of total, nonmilitar federal expenditures. For 1970, $.72 billion of a total of $86.6 billion was spent for pollution control, while in 1977 the expenditure was $5.8 billion of total of $220 billion. (Military expenditures and interests on debt have beer subtracted from the total federal expenditures.)

d. Sponsor's Comments: The percentage increase between 1970 and 1977 is larg because of the construction grant program. This program funds local waste wat treatment facilities. Thus, the lower graph indicates a relatively constant p cent of the federal expenditures going to pollution control and abatement once construction grant program is subtracted.

e. Your Comments:

Round One Comments on Item 35
(Federal Expenditures for Pollution Abatement and Control)

402. Move to return initiative to private/state funding sources. Emphasis on "process" and demonstration project funding. (2010 value for total expenditure of 1.5)

906. I expect stabilization because of lessened political support.

905. Construction grants are pork barrel.

408. Small upward trend after mid-80s, phase-down of public works, decentralization of regulatory programs.

909. Greater awareness of environment will cause additional "non-regulatory" real expenditures by feds and privates.

903. Expenditures will not increase at historic rates, but social and political pressure will require continued federal role in pollution control support.

Round Three Comments on Item 35
(Federal Expenditures for Pollution Abatement and Control)

411. Disagree on construction grants -- think the latest program has been mostly O.K.

104. Toxic wastes -- possible federal support for existing source control.

906. I still do not expect significant percentage increases.

903. I expect a continued federal role in pollution abatement and control because many states and municipalities will not be able to cope.

Statistics:

	Round One		Round Three	
Responses:	18		7	

Total

	2000	2010	2000	2010
Low:	1.0%	1.0%	1.0%	1.0%
Mean:	2.3%	2.4%	2.4%	2.5%
High:	4.5%	5.0%	4.5%	4.5%
Std. Dev.:	1.1%	1.3%	1.1%	1.3%

Less Construction Grants

	2000	2010	2000	2010
Low:	0.5%	0.5%	0.5%	0.5%
Mean:	1.8%	1.9%	1.9%	1.7%
High:	4.0%	4.5%	3.0%	3.0%
Std. Dev.:	1.1%	1.3%	0.8%	0.8%

ITEM 36. POLLUTION CONTROL EXPENDITURES BY ELECTRIC UTILITIES (AS A PERCENTAGE OF TOTAL ELECTRIC UTILITY CAPITAL EXPENDITURES) [37]

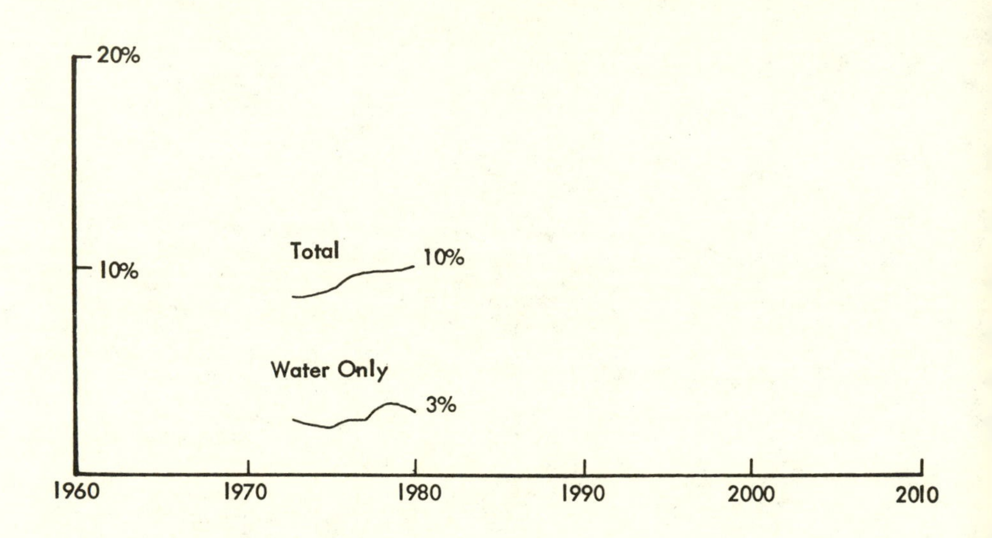

a. Trend Evolution: Please indicate on the above graph how you believe the trends will evolve over the next 30 years.

b. Point Values: Please indicate below your best estimates of the point values (vertical axis values) of the trends in the years 2000 and 2010.

Total Pollution 2000__________ 2010__________

Water Pollution 2000__________ 2010__________

c. Terminology: The percentage expenditure for pollution control is limited to expenditure for plant and equipment. In 1979, for example, total plant and equipment expenditure for electric utilities was $27.7 billion of which $2.7 b lion was spent for control of air, water, and solid waste. $.94 billion was sp for water pollution control.

d. Sponsor's Comments: The U.S. Department of Commerce, Bureau of Economic Analysis, develops the figures plotted above based on an annual survey of electric utilities; the classification of an expenditure as one for pollutio control is somewhat subjective.

e. Your Comments:

Round One Comments on Item 36 (Pollution Control Expenditures by Electric Utilities)

311. Water is the next "big crisis" after energy. This will increase expenditures.

310. By 2010 a larger percent of generation will be nuclear and less clean up will be required.

307. Future expenditures may come from other industries, e.g., those that supply the coal derived gases.

405. Acid rain problem will require sulfur removal from coal, requiring high capital.

330. Don't forget nuclear wastes.

301. Moderate increases by electric utilities as regulatory areas become more stabilized. Perhaps government or other bodies may have to increase funding; for example, acid rain is not a subject for electric utilities.

327. Levels after 2000 as nuclear again begins pushing out coal.

308. Movement toward renewables and advance energy systems beginning in mid to late 80s will reduce pollution control expenditures as will revision of pollution standards to more properly reflect real needs based upon health effects now being properly identified for the first time.

905. Cheaper when not retrofitting.

306. I assume little or no new nuclear in future. Utilities investment is thus dominated by coal, with high pollution control requirements.

322. Increase in pollution control offset by decrease in conventional production investments and increase in non-conventional investments (renewables, cogeneration, etc.).

309. Nuclear world puts less pollution expense on utilities; others may bear the cost. Future society will be more willing to trade slight impurities for power.

409. I don't expect any relaxation of controls of total costs. Expenditures in some areas may level off but total expenditures will continue to rise as new areas of concern require pollution control expenditures. (2010 Total Value of 15%)

408. Slow decline as cheaper equipment becomes available to achieve unchanged standards.

908. Not likely to change a great deal; total will increase if and as coal is used more.

903. Continued pressure to control effluents and increasingly sophisticated technology will require continued increases in pollution control expenditure rates.

319. The environmental ethic is here to stay. Acid rain and CO_2 are not being addressed and must be.

600. Noxious chemicals control will be like earlier sulfur oxide controls.

Round Three Comments on Item 36
(Pollution Control Expenditures by Electric Utilities)

306. It astonishes me that the mean was only slightly above today's. 10% coal will be dominant electricity fuel, and pollution control will equal 1/2 or more of the cost of new coal plants.

300. Definition of "pollution control" may be ambiguous in nuclear case. Relative to other capital expenditures on plants, pollution control should benefit from learning curve to a greater extent.

409. Ref. 311. Water is the ultimate crisis and is closer than generally thought. No southwestern state can now survive a 1 in 100 year drought. Industrial waste water recovery will become mandatory -- once-through cooling will not be affected.

330. Air-electric utilities are facing a $50-$100 billion expenditure within t next two decades to reduce SOx-NOx trace metals emissions. The decision may not come for five years, but it is coming in my opinion. The major cost comes from retrofitting...hence a reduction in % by 2010 as well affected by %.

300. Escalation (cost-inflation) in control technology should lag total escalation factors.

301. Pollution control expenditures will exhibit a slow climb as new effluents are addressed while old nemeses are conquered by more sophisticated environmental control equipment. The earth's ecology is finite...we must continue to pay the price to keep environmental discharges within reasonable and economic grounds. Comment 308 is not appropriate. Renewables all have their own set of environmental and socio-economic problems. Unfortunately, there is "no free lunch" (nuclear waste disposa in the production of energy. Energy production has environmental effects which must be ameliorated.

903. Pressure will continue on utilities to be clean and will require improvec combustion and clean-up technology, all of which require increasing capital investment.

Statistics:

	Round One		Round Three	
Responses:	45		24	

Total

	2000	2010	2000	2010
Low:	9.0%	6.0%	8.0%	5.0%
Mean:	12.7%	13.3%	12.9%	13.8%
High:	23.0%	30.0%	23.0%	30.0%
Std. Dev.:	3.5%	5.1%	3.4%	4.9%

Water Only

	2000	2010	2000	2010
Low:	2.0%	1.0%	2.0%	1.0%
Mean:	4.6%	4.8%	4.4%	5.1%
High:	10.0%	10.0%	10.0%	10.0%
Std. Dev.:	2.9%	2.5%	1.8%	2.3%

ITEM 37. INDUSTRIAL HEALTH RISKS ASSOCIATED WITH ENERGY PRODUCTION (INJURIES AND DEATHS PER GIGAWATT-PLANT YEAR)

This Item has been eliminated from Round 3 because of the very low response rate to it in Round 1.

ITEM 38. RELATIVE SIZES OF U.S. DOMESTIC GOVERNMENTS (FEDERAL, STATE, AND LOCAL GOVERNMENT CURRENT DOMESTIC EXPENDITURES AS A PERCENTAGE OF TOTAL GOVERNMENT CURRENT DOMESTIC EXPENDITURES) [40]

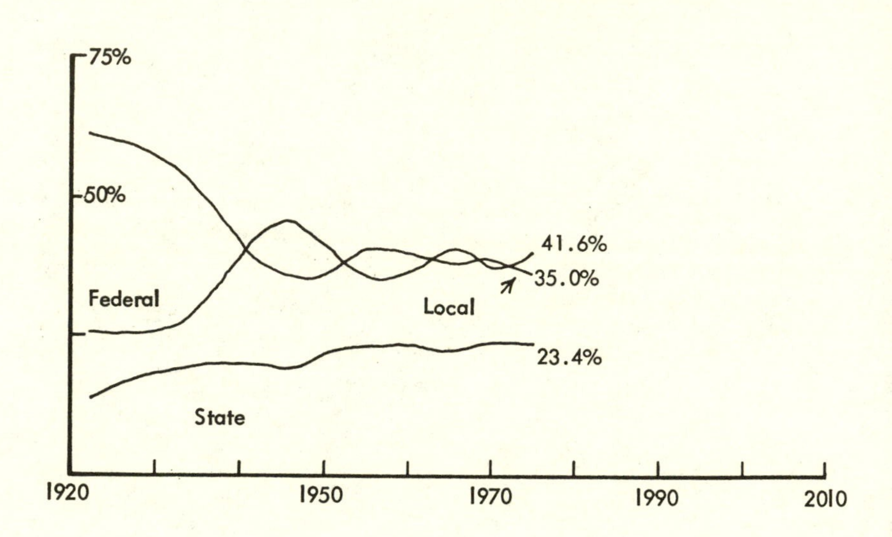

a. Trend Evolution: Please indicate on the above graph how you believe the trends will evolve over the next 30 years.

b. Point Values: Please indicate below your best estimates of the point values (vertical axis values) of the trends in the years 2000 and 2

Local	2000__________	2010__________
State	2000__________	2010__________
Federal	2000__________	2010__________

c. Terminology: Government expenditures exclude expenditures for milit international affairs, and interest on debt. Intergovenmental expenditu are not double-counted.

d. Sponsor's Comments:

e. Your Comments:

Round One Comments on Item 38 (Relative Sizes of U.S. Domestic Governments)

223. The current trend away from federal spending dominance will be reversed by 1990, if not before, with the local governments being the major losers over the long run.

518. The slack left by a presumed relative decline in the federal government will be off-set more by state than by local.

906. I expect continued gravitation toward local and state control.

221. Aging of population will increase federal responsibilities unless there is a major restructuring of relative responsibilities over the next 30 years, which is doubtful.

903. Principal change is reversal of state and federal roles as states increase spending as federal services and staffs decrease relatively.

Round Three Comments on Item 38 (Relative Size of U.S. Domestic Governments)

211. I believe comment 903 will dominate over next 25 years with sharp reversal due after 2010 as social security enters its high jeopardy period.

903. State share will grow and federal share will shrink as government moves closer to undeveloped voter.

Statistics:

	Round One		Round Three	
Responses:	43		16	
Federal	2000	2010	2000	2010
Low:	22.0%	15.0%	32.5%	29.0%
Mean:	40.5%	39.8%	38.9%	37.1%
High:	53.0%	60.0%	44.0%	45.0%
Std. Dev.:	5.6%	9.4%	3.5%	4.9%
State	2000	2010	2000	2010
Low:	20.0%	20.0%	22.0%	21.5%
Mean:	26.0%	27.1%	26.7%	28.7%
High:	32.5%	40.0%	32.5%	40.0%
Std. Dev.:	3.4%	5.1%	3.0%	4.9%
Local	2000	2010	2000	2010
Low:	25.0%	15.0%	30.0%	20.0%
Mean:	34.2%	33.6%	34.5%	34.3%
High:	43.0%	46.0%	42.0%	45.0%
Std. Dev.:	4.3%	7.0%	3.4%	5.6%

ITEM 39. ANNUAL INTER-CITY PASSENGER MILES (BILLIONS) [41]

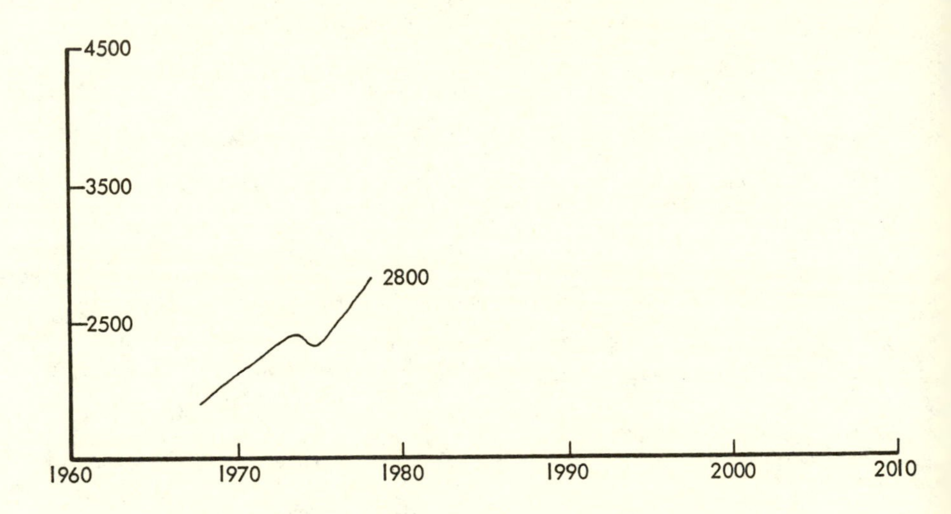

a. Trend Evolution: Please indicate on the above graph how you believe the trend will evolve over the next 30 years.

b. Point Values: Please indicate below your best estimates of the point values (vertical axis values) of the trend in the years 2000 and 2010

2000____________ 2010____________

c. Terminology: Intra-city travel is not included in this measure. All modes of transportation are included.

d. Sponsor's Comments:

e. Statistics:

	Round One		Round Three	
Responses:	57		26	
	2000	2010	2000	2010
Low:	2200	2000	2200	2000
Mean:	3580	3940	3490	3760
High:	4800	6200	4200	5000
Std. Dev.:	750	810	490	720

Round One Comments on Item 39 (Annual Inter-city Passenger Miles)

311. Increasingly mobile population, more business travel, larger population will sustain growth. No real basis except "gut judgment" for this answer

330. Will be reaching asymptote: modal shift to airlines created growth in past decades. Moreover, relative cost of transport will increase, reducing projected traffic. Finally, technology shift (e.g., communication) will reduce need for transport.

525. Should increase with disposable income and small families.

508. Increasing population traveling, combined with an increased cost of traveling.

301. Increasing with population and then decreasing as communication methods improve and fuels get increasingly less available.

327. About same as population increase.

906. I expect continuous increase as a result of population growth and other factors but less rapid growth than in the past.

306. Was 1975-1978 steeper than 1968-1973? Costlier travel and telecommunication will reduce rate of increase.

322. Only applies if CAFE > 60 mpg by 2000.

335. My expectation is that more sophisticated telecommunications technology will substitute for energy intensive business travel for economic reasons.

303. Slope should decrease due to better means of communication, plus higher cost of transportation and lodging.

511. I expect this rate to continue rapidly, despite rising energy prices. But the mode of transportation will undoubtedly shift away from private cars.

337. Increasing costs for liquid fuels and improvements in communications equipment should slow the increase in intercity passenger miles.

903. Intercity travel will continue, though at reduced rates reflecting higher relative energy costs. Communications replaces travel only to some extent, and recreational travel will in any event continue to increase.

102. Long-running tie between transportation and GNP.

901. Increase is essential to the continued functioning of our systems; but rate of increase will decline as a function of increasing costs of movement.

Round Three Comments on Item 39 (Annual Inter-City Passenger Miles)

300. I would expect slower growth, followed by leveling off as communication begins to compete with travel.

903. I see no reason why intercity travel should not continue to grow as society becomes more affluent, leisure increases, and business continues in importance. Telecommunications can replace some, but not a lot, of personal business travel.

ITEM 40. ANNUAL INTER-CITY MASS TRANSIT PASSENGER MILES (AS A PERCENTAGE OF TOTAL INTER-CITY PASSENGER MILES) [42]

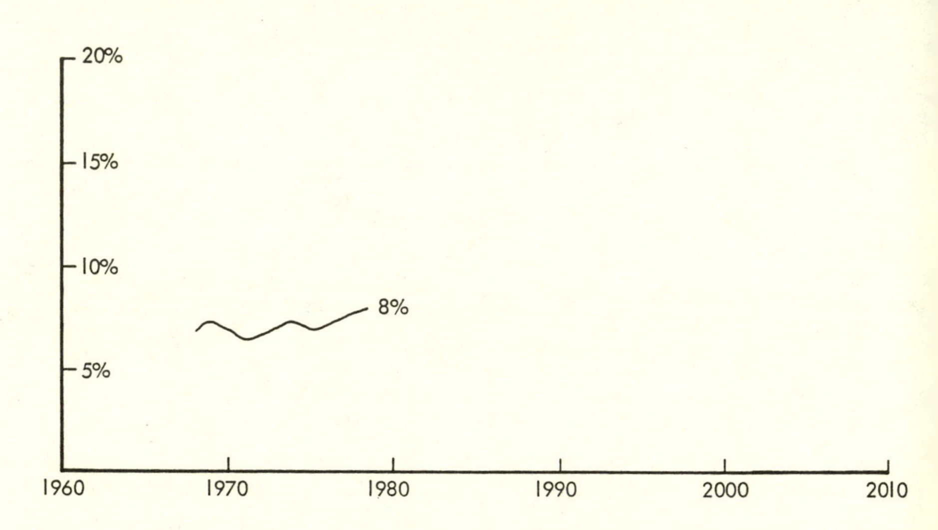

a. Trend Evolution: Please indicate on the above graph how you believe the trend will evolve over the next 30 years.

b. Point Values: Please indicate below your best estimates of the point values (vertical axis values) of the trend in the years 2000 and 2010.

2000__________ 2010__________

c. Terminology: Inter-city bus, class I rail, and domestic air carrier are considered mass transit modes.

d. Sponsor's Comments:

e. Statistics:

	Round One		Round Three	
Responses:	57		26	
	2000	2010	2000	2010
Low:	7.8%	7.6%	7.8%	7.6%
Mean:	11.8%	14.3%	11.8%	14.3%
High:	20.0%	25.0%	20.0%	30.0%
Std. Dev.:	2.8%	4.6%	2.9%	5.2%

Round One Comments on Item 40 (Annual Inter-city Transit Passenger Miles)

311. Americans are "hooked" on personal automobile transportation. They will use mass transportation for long distance or when cost of auto transportation gets too high.

330. Same factors affecting as Item 39. But relative convenience and cost of public transit will improve over auto.

525. People love cars and I don't see them getting out of them. I'd bet there will be a switch to planes, though.

301. Less use of the automobile as gas prices go sky high. More buses/rail/ aircraft also reflect a generally affluent society.

327. Air will increase, rail and bus down.

314. With time, as liquid fuel resources become more scarce, the growth in intercity transit grows.

322. Assumes high efficiency commuter cars compete with mass transit.

335. I have assumed that all other inter-city transit is by personal automobile which should decline.

303. Individual means should slow.

337. Increasing costs for fuel and private automobiles will continue to shift passenger miles to mass transit especially for long distances.

903. Intercity travel via transit will increase as a percentage of total travel as auto costs increase.

901. Increasing percentage will come about more through change in denominator than in numerator.

Round Three Comments on Item 40 (Annual Inter-City Transit Passenger Miles)

300. Should become less popular.

301. The only way that personal automobile transportation can increase is for somebody to invent a suitable and economic substitute to the internal combustion engine for long-trips... I don't see it in time to affect a switch to mass-emergence, just as Europe has done!

903. High energy costs will continue to discourage personal travel and stimulate mass transit. If we ever get new, TGV-type, vehicles, they might be competitive with air for distances of 100-400 miles but this won't happen until 21st century.

ITEM 41. FEDERAL OUTLAYS FOR ENERGY RESEARCH AND DEVELOPMENT (BILLIONS OF 1980 U.S. DOLLARS) [43]

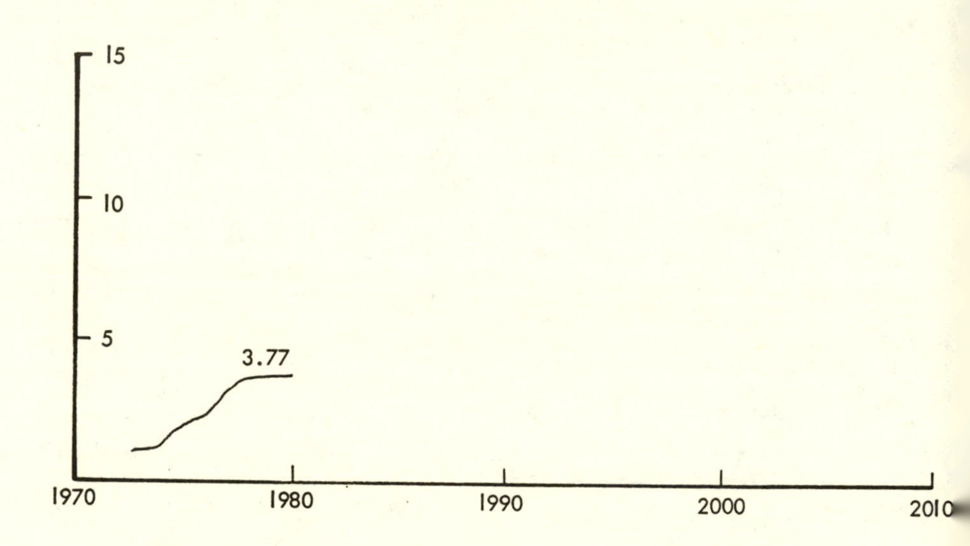

a. Trend Evolution: Please indicate on the above graph how you believe the trend will evolve over the next 30 years.

b. Point Values: Please indicate below your best estimates of the point values (vertical axis values) of the trend in the years 2000 and 20

2000____________ 2010____________

c. Terminology:

d. Sponsor's Comments:

e. Statistics:

	Round One		Round Three	
Responses:	52		29	
	2000	2010	2000	2010
Low:	2.0	1.5	2.5	2.0
Mean:	5.3	6.6	5.2	6.4
High:	15.0	22.0	11.0	18.0
Std. Dev.:	2.4	3.8	1.9	3.2

Round One Comments on Item 41 (Federal Outlays for Energy R & D)

223. Spending will peak during the 1990s on shale, various solar technologies, the breeder and fusion, and decline sharply thereafter as it becomes apparent energy is no longer a concern. Concern and spending will shift to other minerals in short supply, to sealed mining, etc.

311. Amount will probably remain a constant (or slightly decreasing) fraction of GNP.

310. New energy sources are increasingly expensive to develop and private industry can't afford that expense. The breeder and fusion will be expensive to develop.

330. Energy research will be oriented in less of a crisis mode and returned to private sector.

321. Cyclical with changes in Federal administration.

327. This does not include Federal/State utility contributions, which are considered as utilities.

308. Most of the federal effort is expected to be in the area of nuclear fusion with some in advanced fission breeding in the nearer term.

906. I believe there will be a temporary decline, but energy will reemerge as a key issue to be addressed in the early part of the next century.

212. A drop in spending in 80s followed by rise, as private sector is unable or unwilling to fund range of energy R & D society needs.

309. Phone: R&D burden will be borne by industry, not government.

336. Energy and national security will become synonymous with priorities reacting accordingly.

303. More hopeful than anything else. Upswing following the realignment now progressing.

317. Research will have to continue to increase. We have problems we have to solve and they require research dollars.

337. Though Federal research and development expenditures for new sources of energy appear to have leveled out for this Administration, moderate increases in Federal energy research and development will be required. This is especially true if new, high cost, high risk technologies like synthetic fuel production, photovoltaics and fusion power are ever commercialized. The private sector appears to lack the large quantities of venture capital required to commercialize these technologies.

207. Shortrun myopic policy due to oil glut.

903. Need for national preemminence in science and technology will increase federal R & D funding, especially in energy technologies.

319. National security will demand synfuels. Something must be done about nuclear wastes and accident clean-up.

102. Reagan will have his way this decade. Later, everyone will know energy is a problem, or isn't. If it is, industry can move without uncertainty; if it isn't, no one need do anything. So, I don't believe government R&D support will be large.

901. I don't think it will keep pace with inflation through the 1980s.

Round Three Comments on Item 41 (Federal Outlays for Energy R & D)

300. 1980-2000: Non-nuclear R&D will shrink in terms of % of federal budget. Nuclear R&D in constant dollars will remain about even or increase slightly. Beyond 2000, who knows? Other forms of subsidy/incentives likely to appear to induce greater private sector R&D expenditures.

312. Agree with 317 and 102.

330. Key is "federal." I doubt that we will see naivete of the 1970s re: federal as answer to all questions. Energy R&D will be important but it will be private sector demo and R&D carrying the major load except for a few long-run technologies (e.g. fusion) or basic research.

300. Excluding military portion of DOE, expenditures should become decreasing fraction of a) budget b) GNP.

301. The proper government role is to fund high-risk, long time high-potential technologies to achive a pay-off. These include fusion, photovoltaics, etc...(b) to fund technology aimed at solving a safety, health, environmental problem which the private sector cannot approach. This includes acid rain, waste disposal, etc... or (c) to fund technology needed for purposes of national security (for example, nuclear propulsion). This role will continue to be funded... it depends on political approach and most importantly, on ability of federal treasury to absorb; I just can't guess the percentage.

903. I think there will be continued increases in federally-supported energy research and development, though R&D as proportion of GNP will probably not change much. U.S. will need to stay in forefront of energy technology in order to be competitive in increasingly energy-expensive world.

ITEM 42. U.S. ENERGY CONSUMPTION (QUADRILLION BTU'S) [44]

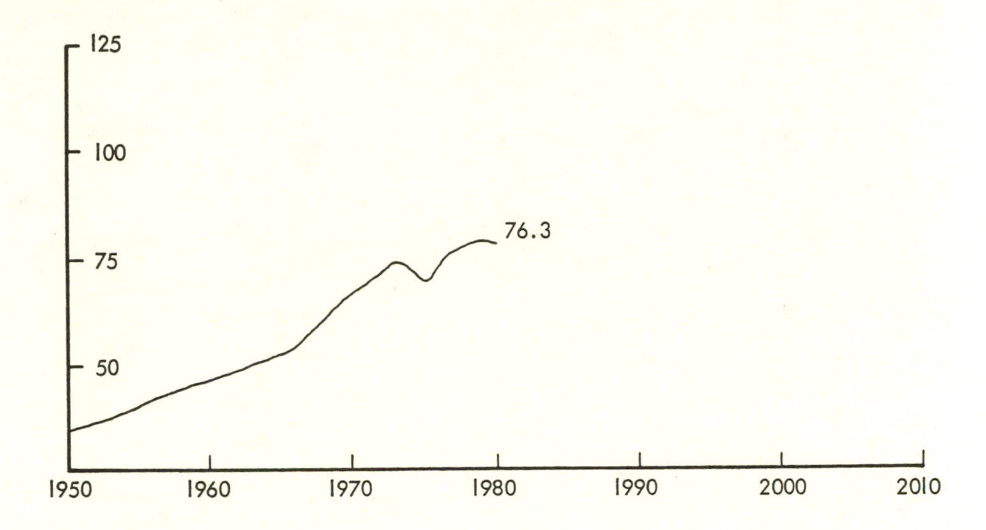

a. Trend Evolution: Please indicate on the above graph how you believe the trend will evolve over the next 30 years.

b. Point Values: Please indicate below your best estimates of the point values (vertical axis values) of the trend in the years 2000 and 2010.

2000____________ 2010____________

c. Terminology: For fossil fuels, consumption prior to first sale is excluded. Nuclear BTUs is reactor heat. Hydropower and biomass BTUs are fossil fuel BTUs for equivalent net electricity generation. Non-utility consumption of alternative fuels is not included above, but should be included in your answers.

d. Sponsor's Comments: DOE's middle world oil price scenario predicts domestic energy consumption of 102.3 quads (2000) and 117.6 quads (2010) [45]. Exxon's December 1980, forecast for the year 2000 is 92.2 quads [46]. DOE's middle world oil prices, in 1980 dollars/barrel, are $65 (2000) and $78 (2010). According to DOE, Exxon's year 2000 price is approximately $51/barrel (1980 dollars).

e. Statistics:

	Round One		Round Three	
Responses:	43		22	
	2000	2010	2000	2010
Low:	68.0	60.0	68.0	60.0
Mean:	92.8	101.5	92.6	99.5
High:	106.0	135.0	106.0	135.0
Std. Dev.:	9.9	15.7	10.7	16.7

Round One Comments on Item 42 (U.S. Energy Consumption)

311. Energy costs will continue to rise. Price elasticity will cause a per capita decrease in personal use that is about equal in total to the extra needed for a growing population. Increase will come from industrial uses to create new jobs for expanding population.

330. Slow economic growth of early 80s will lower prior projection of GNP and hence energy. Moreover, longer term/slower growth in population will have same effect. Finally, prices will drive consumption toward greater efficiency.

327. I assume market pricing is allowed to function for all fuels by 1985.

906. I believe continued conservation will cause projections even lower than expected.

314. Slow, by historical standards, but steady growth.

303. Reasonable to expect growth, but tapered off.

908. The figures I gave are on the low end of my personal prediction -- but elasticity of demand (and speed of change) has surprised us in past decades.

904. All our estimates of the last 5 years have seemed to have been systematically overestimated. In addition, the impact of more dramatic price shift than assumed could be significant. DOE also underestimates the potential conservation.

317. Energy consumption must continue to climb once conservation has tightened the belt. The population is continuing to grow and people will continue to demand a higher standard of living.

903. U.S. energy consumption will increase though at slower than historic rate. After late 80s, I see energy consumption and GNP again moving in lockstep. Conservation efforts will be offset by increased conversion losses.

104. Bad economy ahead for years; therefore, energy consumption could be flat. 70s show the inflection point on the curve.

306. Talent is now geared up to address energy. GNP can use for a long time without driving energy consumption up.

Round Three Comments on Item 42 (U.S. Energy Consumption)

308. Also note that decreasing quality of remaining raw resources and increased difficulty of extraction will increase energy requirements for a given volume of basic goods in spite of conservation.

312. Agree with 314, 303, 317, 903.

304. In periods of abnormal growth (either low or high) we usually overreact in predicting the future; while future growth will be at a lower rate, it will not fall as low as many people now predict.

104. New Swedish govt. study implies U.S. least-cost energy future of 37Q at 1980 activity levels. Efficiency estimates are much higher than ever before. Increased mileage stds/home-conserv/bldg. conservation/appliance efficiency have barely begun to be reflected in energy data. When gas prices increase (ca. 1985), just watch.

330. The conventional wisdom is to forecast < 100 quads for the year 2000. I think that will turn out to be just as wrong as the 160 quad forecast of ten years ago. But I see rate of growth moderating past 2000.

300. Will become more or less proportional go GNP (real), approaching

$$\frac{E<Btu}{G\ (\$1972)} = 45{,}000\ (Btu/\$1972).$$

301. Agree with 330, 314, 303. The U.S. will continue to provide a substantial portion of the world's GNP and use energy... slow, steady growth... perhaps reaching inflection point next century.

903. Expect price effects of oil increases plus increased conservation and efficiency of utilization will reduce energy demand growth.

ITEM 43. U.S. ENERGY CONSUMPTION BY PRIMARY FUEL (HEAT CONTENT OF FUELS IN QUADRILLION BTU) [47]

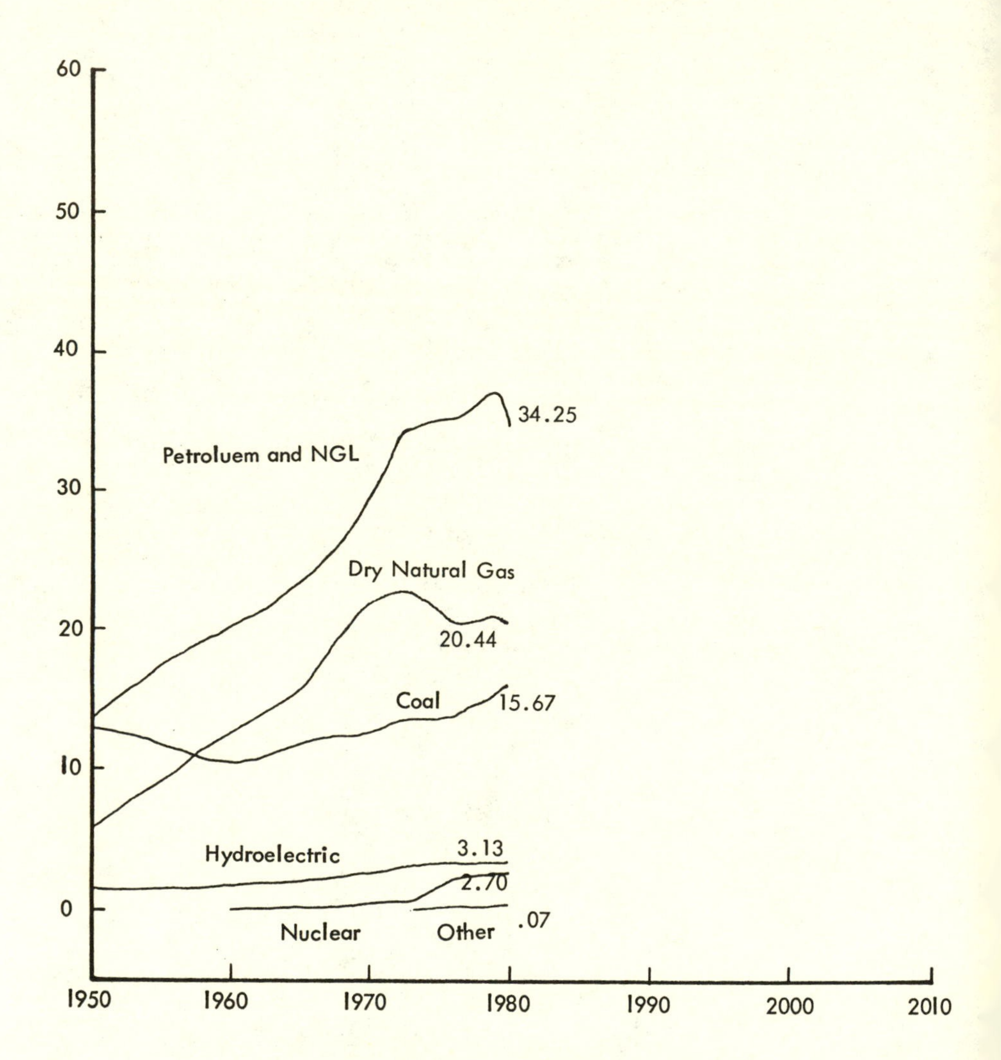

a. Trend Evolution: Please indicate on the above graph how you believe the trends will evolve over the next 30 years.

b. Point Values: Please indicate below your best estimates of the point values (vertical axis values) of the trends in the years 2000 and 2010.

Coal	2000__________	2010__________
Dry Natural Gas	2000__________	2010__________
Petroleum Liquid	2000__________	2010__________
Hydropower	2000__________	2010__________
Nuclear	2000__________	2010__________
Other	2000__________	2010__________

c. Terminology: "Other" is geothermal, solar, biomass, and wind; "other" does not include coal-derived fuels (these are "coal"). Hydropower Btu content is the product of net kWh output from hydropower and the current heat rate factor at fossil fuel steam electric power plants; the nuclear Btu content is the heat currently released in the reactor. In 1979, these figures were 10,435 Btu/kWh and 10,769 Btu/kWh, respectively [48].

d. Sponsor's Comments: DOE's middle world oil price scenario projects the following net U.S. energy consumption values (BTUs).

	2000	2010	
Coal	37	51.3	
Dry Natural Gas	17.2	15.5	
Petroleum liquids	26.5	20.8	
Hydropower	3.5	3.5	
Nuclear	11	16	
Other	7.1	10.5	
	102.3	117.6	[49]

Exxon forecasts approximately 10 quads less energy consumption in the year 2000 and disproportionately less contribution from "hydropower" and "other" [50].

e. Your Comments:

Round One Comments on Item 43 (U.S. Energy Consumption by Primary Fuel)

311. Little additional hydropower and most of that will be low head hydro. Gas and oil will be flat. Coal will grow rapidly as utilities and industry learn to use it in environmentally satisfactory ways but may be limited by "greenhouse effect." Nuclear, including the breeder, and possibly fusion will have to be used to meet the needs of the 21st century. Eventually it will be accepted by the public. Shale oil will be significant beyond 2000.

301. Coal will make an increasing contribution but will reach a level not much greater than 30-35 quads because coal/oil/gas are hydrocarbons whose combustion must be curtailed due to environmental considerations; hence, nuclear and other will have to make increasing contributions.

327. Coal will run into real problems for central station utility use, and by 2000 nuclear will begin to recover. Natural gas will be strong provided market pricing is allowed to work.

308. Doubt that either nuclear or "other" will reach the rate of development envisioned by DOE, given the present nuclear doldrums and the present adverse regulatory, cost, and risk elements facing renewables and advanced energy systems

905. Dry Natural Gas. Abundant sources outside the continental U.S. -- including Ala and Mexico-will supply natural gas. Gas shortage has never materialized. Further, drilling in the U.S. overthrust belts will provide additional supplies.

904. To talk about coal as a homogeneous product may be unhelpful. A separation of conventional coal from advance coal may add to the discussion. The same is true of nuclear. Obviously, I feel DOE overestimates the potential of nuclear and coal and underestimates the potential of solar.

305. Coal goes up rapidly, gas goes up moderately, nuke approximately triple, oil goes down dramatically, solar photovoltaics go up markedly.

909. A growth rate of approximately 1.6% is assumed -- e.g., high conservatism due to high price-high price results in substantial incentives to natural gas, synfuel and other. Nuclear remains limited by non-economic factors (nuclear is a wild card and will grow rapidly if long term inflation is reduced substantiall

903. I expect nuclear to come on slower than DOE hopes because of social concerns.

102. Coal is the swing fuel. It can be large or small depending on what one thinks of other fuels.

319. Nuclear is future's fuel, despite today's evidence. I place 16% of today's plant orders in the "cancellation" category. Waste management mostly affects decisions on new plants. I don't believe utilities can write-off sunk costs in nuclear -- they must complete plants. Time value of money is on the way down so financing will be easier.

306. Imported LNG is unacceptable environmentally. Solar photovoltaic is only solar with long-term promise.

104. Why do people forget that coal transport is a big problem? States will turn on nuclear energy next; voter approval of nucs will become a requirement. FBR does not save enough in fuel costs to merit public consciousness about nuclear.

<u>Round Three Comments on Item 43</u> (U.S. Energy Consumption by Primary Fuel)

303. Expectations of our being able to turn the use of petroleums around are overly optimistic. I don't believe that it can be done.

304. Minor adjustments to allow for some downward adjustment in the amount of coal-generated power.

104. Nuclear cancellations will accelerate and at least 30 existing plants will be decommissioned by that time. Natural gas will be a preferred fuel over coal for may uses; both coal & gas will block out oil.

906. I was clearly too low on WG in first round.

330. My nuclear forecast(13.5%=2000, 17.5%=2010) was probably high--depends on getting construction to 6-7 years by 1990-85. There seems to be a great deal of optimism on natural gas from other respondents, which, to me, seems to ignore cost of transport or is dependent on the pipe dream of non-biogenic gas.

300. I would split the difference with Exxon, but decrease their coal/oil shale contribution.

903. I expect CO2 production and "acid rain" problems will limit coal use, below Round I mean. Probably natural gas will continue to be a premium fuel and exploration will continue.

Statistics:

	Round One		Round Three	
Responses:	36		21	

Coal

	2000	2010	2000	2010
Low:	18.0	17.0	18.0	17.0
Mean:	29.4	35.1	28.9	34.2
High:	53.0	65.0	38.0	40.0
Std. Dev.:	7.4	9.9	5.3	7.7

Dry Natural Gas

	2000	2010	2000	2010
Low:	12.0	10.0	15.0	13.0
Mean:	20.2	20.0	18.7	19.7
High:	32.0	35.0	25.0	26.0
Std. Dev.:	3.8	5.5	5.7	4.1

Petroluem and NGL

	2000	2010	2000	2010
Low:	14.0	10.0	14.0	11.0
Mean:	26.6	22.7	26.6	24.0
High:	43.0	46.0	43.0	46.0
Std. Dev.:	7.2	9.4	7.5	8.9

Hydroelectric

	2000	2010	2000	2010
Low:	3.0	3.0	3.0	3.0
Mean:	4.0	4.3	4.0	4.4
High:	6.0	10.0	6.0	10.0
Std. Dev.:	0.7	1.3	0.8	1.6

Nuclear

	2000	2010	2000	2010
Low:	1.5	0.5	1.5	0.5
Mean:	8.2	11.9	7.9	11.8
High:	16.0	25.0	16.0	24.0
Std. Dev.:	3.8	5.7	4.0	6.0

Other

	2000	2010	2000	2010
Low:	0.0	0.0	1.0	1.2
Mean:	3.8	6.6	5.4	8.4
High:	15.0	20.0	17.5	20.0
Std. Dev.:	3.7	5.6	4.7	5.8

ITEM 44. COMPOSITION OF FUTURE U.S. COAL CONSUMPTION (BY PERCENT)

a. Point Values: Please indicate below your best estimates of the percentages of future coal consumption (from previous question) which will go to each of the end uses listed:

	2000	2010
Boiler fuel (including fluidized bed & combined cycle with coal gasifier):	______	______
Metallurgical coke production:	______	______
Synthetic liquids production:	______	______
Synthetic gas production:	______	______

b. Sponsor's Comments: DOE's middle oil price scenario predicts the following percentages:

	2000	2010
Boiler fuel; unconventional electricity generation:	83.5	69.6
Metallurgical:	5.4	3.9
Synthetic liquids:	10.0	29.5
Synthetic gas:	1.1	2.1

[51]

c. Statistics:

	Round One		Round Three	
Responses:	36		21	
Boiler Fuel	2000	2010	2000	2010
Low:	56.0%	37.0%	56.0%	37.0%
Mean:	82.2%	75.3%	83.4%	76.3%
High:	93.0%	92.0%	98.0%	90.0%
Std. Dev.:	7.7%	12.1%	10.0%	13.1%
Metallurgical Coke	2000	2010	2000	2010
Low:	4.0%	2.5%	4.0%	3.0%
Mean:	5.5%	4.9%	5.5%	4.8%
High:	10.0%	10.0%	10.0%	10.0%
Std. Dev.:	1.4%	1.5%	1.3%	1.7%
Synthetic Liquids	2000	2010	2000	2010
Low:	1.0%	1.0%	1.0%	2.0%
Mean:	6.3%	11.2%	5.8%	9.9%
High:	13.0%	28.0%	10.0%	18.0%
Std. Dev.:	2.9%	6.9%	2.5%	5.2%
Synthetic Gas	2000	2010	2000	2010
Low:	1.0%	1.0%	0.5%	0.5%
Mean:	3.6%	4.8%	3.1%	4.8%
High:	14.0%	21.9%	10.0%	15.0%
Std. Dev.:	3.3%	3.9%	2.7%	4.1%

Round One Comments on Item 44 (Composition of Future U.S. Coal Consumption)

311. Shale oil will be important synthetic liquid -- especially if Eastern shales are developed.

301. Synthetic gas has to be developed first, then liquids will accelerate and surpass gas. Fuel cells have a lot of promise.

327. Assumes market pricing is allowed to work. Substantial synthetic liquids from shale oil will price out coal liquids.

906. I believe that synthetic fuels will hit big in the first part of the century, given current deregulation.

908. I just don't see non-combustion uses of coal growing very much. CO_2 research dramatically changes overall consumption, of course.

337. Significant production of synthetic liquids is expected. Significant quantities of coal will be exported by and after the year 2000. Incremental benefits for MHD and fuel cells (integrated with coal gas) may never be greater than incremental costs.

Round Three Comments on Item 44 (Composition of Future U.S. Coal Consumption)

308. Transportation needs will put great emphasis on production of liquid fuels from coal -- especially by 2010.

330. I may be low on syn-oil <u>if</u> oil shale proves as low cost as sponsor hopes. But even then I doubt it.

300. "Direct" combustion will continue to be lion's share of coal use.

ITEM 45. ENERGY CONSUMPTION BY END-USE SECTOR (EACH SECTOR AS A PERCENTAGE OF TOTAL ENERGY CONSUMPTION) [52]

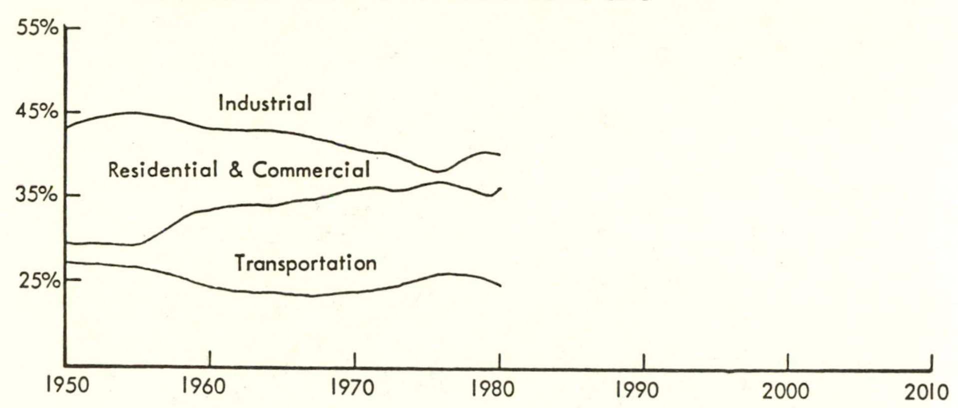

a. This is an Item which was poorly formulated in Round 1. Responses to this Item are heavily influenced by assumptions on Item 42 (Total U.S. Energy Consumption).

Please indicate below your best estimates of the point values (vertical axis values) of the trend in the years 2000 and 2010, under each of two assumptions. For assumption A, assume 85 quadrillion Btu total energy consumption in 2010: for assumption B, assume 115 Btu total energy consumption in 2010.

	A (85 quads)		B (115 quads)	
	2000	2010	2000	2010
Residential/ Commercial	______	______	______	______
Industrial	______	______	______	______
Transportation	______	______	______	______

b. Terminology Electric utility energy consumption is charged to end-use sectors in proportion to each sector's consumption of utility electricity. Please apply this convention to future electricity conversion losses, but allocate synfuels conversion losses to the industrial sector.

c. Sponsor's Comments: DOE's figures, on which the graph is based, were 35.9% (residential/commercial), 39.7% (industrial), and 24.4% (transportation) in 1980. Exxon's figures for 1980 were 33.2%, 42% and 24.8%, respectively. Exxon's year 2000 projections, which allocate synfuel conversion losses to the industrial sector, are:

Residential/Commercial: 30.1%
Industrial: 49.2%
Transportation: 20.7% [53]

d. Comments:

No comments received in either round on item 45.

Statistics:

	Round One		Round Three A (85)		Round Three B (115)	
Responses:	36		24		24	

Industrial

	2000	2010	2000	2010	2000	2010
Low:	35.0%	28.0%	34.5%	35.0%	35.0%	36.0%
Mean:	42.8%	43.9%	40.3%	41.5%	42.9%	43.3%
High:	50.0%	55.0%	50.0%	50.0%	48.0%	50.6%
Std. Dev.:	5.4%	8.3%	4.8%	4.3%	4.7%	4.4%

Residential & Commercial

	2000	2010	2000	2010	2000	2010
Low:	30.0%	25.0%	26.1%	23.8%	30.0%	30.0%
Mean:	34.7%	35.7%	33.7%	34.4%	33.5%	33.5%
High:	45.0%	55.0%	38.0%	40.0%	40.0%	40.0%
Std. Dev.:	4.5%	7.3%	3.8%	5.1%	3.2%	3.2%

Transportation

	2000	2010	2000	2010	2000	2010
Low:	15.0%	10.0%	20.0%	18.0%	20.0%	18.0%
Mean:	22.8%	21.1%	23.3%	22.9%	24.6%	24.6%
High:	27.6%	28.0%	34.0%	30.0%	35.0%	32.2%
Std. Dev.:	2.7%	3.9%	3.9%	3.8%	4.9%	4.9%

ITEM 46. NET U.S. ELECTRICITY GENERATION (BILLION KWH) [54]

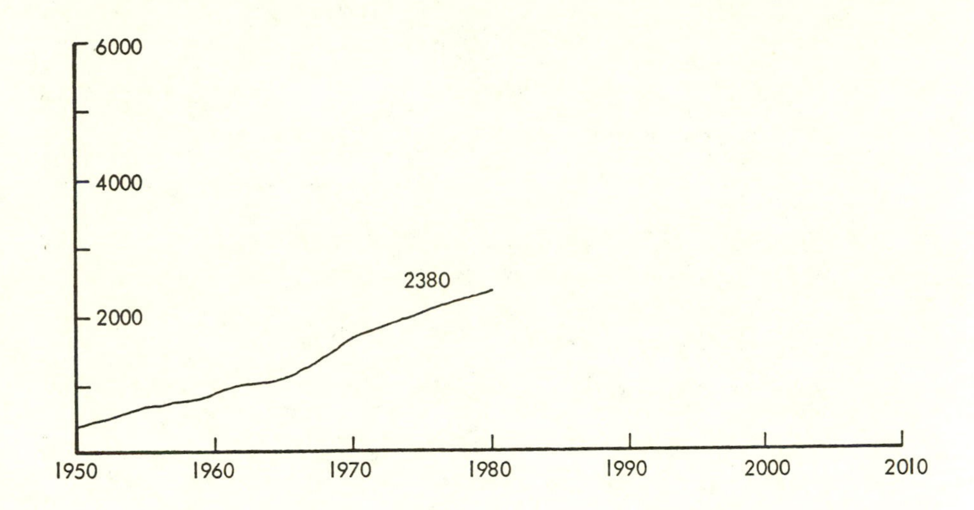

a. Trend Evolution: Please indicate on the above graph how you believe the trend will evolve over the next 30 years.

b. Point Values: Please indicate below your best estimates of the point values (vertical axis values) of the trend in the years 2000, 2010, and 2020.

2000__________ 2010__________ 2020__________

c. Terminology: Net electricity generation excludes generating plant consumption. Industrially generated power is included, as are line losses.

d. Sponsor's Comments: DOE's middle world oil price scenario predicts net U.S. electricity generation of 3,957 billion kWh in 2000 and 4,836 billion kWh in 2010. [55]

e. Statistics

	Round One			Round Three		
Responses		36			22	
	2000	2010	2020	2000	2010	2020
Low:	2300	2200	2200	2300	2200	2200
Mean:	3680	4470	4870	3760	4450	5120
High:	5000	6700	6500	4700	6100	7450
Std. Dev.:	620	1020	1480	610	990	1350

Round One Comments on Item 46 (Net U.S. Electricity Generation)

311. Continued growth. Conservation due to price increases will be more than offset by increased population, new uses of electricity (for reliability and environmental reasons) and larger industrial base.

330. Electrification likely to be higher than DOE's forecast because: fuels to be used are less likely to be oil and gas and more likely to be coal and nuclear. Coal use growth in industrial sector will be difficult leading to utility electricity replacement and shift of technology (e.g. infra-red, micro wave, induction heating) toward "tuned" energy. Pollutio control will reduce individual firm combustion.

304. Electric energy will form an increasing proportion of U.S. energy supply due to substitution of electricity for other energy forms.

327. With reasonable market pricing, electricity will displace many other fuels nuclear will be increasing part of generation.

308. Most scenarios fail to consider the future socio/environmental/economic attractiveness of substitution of electricity for end-use applications. This will begin to be recognized in the late 1980s.

306. I anticipate a 1% per year electricity growth.

336. Electrification is the alternative to petroleum and natural gas uses in mass transport.

317. The demand for electrical energy will continue to climb. It is cheap and efficient and will continue to be used for more and more purposes.

903. Growth of economy and increasing electrification of society, e.g. electri cars, homes, dictates continued growth in electrical generating capacity.

104. Electricity prices will have to escalate more quickly than primary fuel prices if most utilities are to survive. No industry in America has been so weak so long. Large capital carrying costs and 33% conversion efficiency means electricity prices rise, consumption stays nearly flat.

Round Three Comments on Item 46 (Net U.S. Electricity Generation)

303. I certainly agree with the "substitution philosophy" in most of the comments.

300. #104 has a good point for a "short-term," "one time only" adjustment but not for the long term. Electricity prices relative to other energy prices will tend to decrease over the long run.

104. Electricity conservation potential is for underestimates; cost-effective electricity use today would be 60% of the current level, at utility marginal costs, 40%.* This far outweighs increased activity/wealth/ substitution potential.
* With today's conservation technology

300. Electricity's share of total energy (on primary equivalent basis) increases to about 50% by 2020. (Being about 30% currently.)

301. I assumed about 2.6% growth rate to 2000; 2.25% to 2010; and 1% to 2020 but for this growth to occur (as modest as it is) utility companies will need help from the capital needed for expansion...it is in the hands of the PUCs.

903. I agree that energy demand growth will continue. Primary fuel costs plus premium for conversion will tend to limit use of electricity in preference to other fuels. Small likelihood of extensive electric car utilization before 2000.

ITEM 47. NET ELECTRIC UTILITY POWER SUPPLY BY ENERGY SOURCE (EACH SOURCE'S OUTPUT AS A PERCENTAGE OF TOTAL) [56]

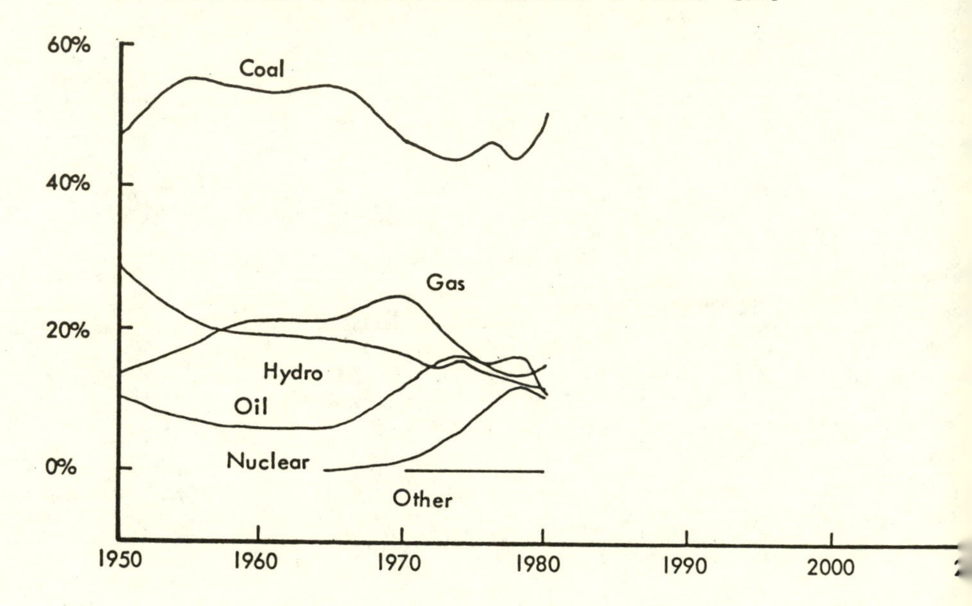

a. This is another item which was poorly formulated in Round 1. Responses to this item are heavily influenced by assumptions on Item 46 (Total U.S. Electricity Generation).

Please indicate below your best estimates of the point values (vertical a values) of the trend in the years 2000 and 2020 under each of two assumpti For assumption A, assume 2010 electricity generation of 3400 billion KWh. For assumption B, assume 5400 billion KWh in 2010.

	A (3400 billion KWh)		B (5400 billion KWh)	
	2000	2020	2000	2020
Coal:	______	______	______	______
Oil:	______	______	______	______
Gas:	______	______	______	______
Nuclear:	______	______	______	______
Other: (includes alternative fuels and non-utility generation sold to utility grids)	______	______	______	______

b. Terminology: "Oil" and "gas" include coal-derived liquids and gases. Please include in "other" electricity purchased from non-utility generators, regardless of input fuel; solar, wind, OTEC, biomass, geothermal, etc. are also "other.

c. Sponsor's Comments: DOE's middle world oil price scenario forecasts the following 2000, 2010, and 2020 percentages:

	1980	2000	2010	2020	
Coal:	50.8	60	54.5	53.1	
Oil:	10.8	1.5	1.2	1.0	
Gas:	15.1	0	0	0	
Nuclear:	11.0	25.9	29.7	29.6	
Hydro:	12.1	8.4	7.3	6.1	
Other:	0	4.4	6.7	10.7	[57]

d. Your Comments:

Round One Comments on Item 47 (Net Electric Utility Power Supply by Energy Source

311. My predictions for nuclear are based on the assumptions of:
 (1) Public acceptance of nuclear power
 (2) Satisfactory resolution of the politics of nuclear waste disposal.

906. I expect other sources to develop over time and cost to remain more of a staple than DOE projects. Other, including solar, should begin to have a major effect by early next century.

306. I didn't include cogeneration -- was I supposed to?

317. Unused hydro resources will be exploited and then its contribution will level off. Other sources will gradually start to pick up the load.

903. I see nuclear uninhibited by social concerns. Oil and gas will be hard to finally squeeze out, but coal coming on strong as an available and secure fuel.

600. Oil and gas-fired generation will be a larger share of the pie than expect These are easy fuels to use if capital costs are to be avoided, especially since fuel costs flow through to rate-payers. I think F.U.A. will itself "backed out."

612. Coal can be as big as society will allow, but I think its percent is as hi as society wants, now.

618. Alternatives, except geothermal, are not cheap if you consider management and capital expenses.

Round Three Comments on Item 47 (Net Electric Utility Power Supply by Energy Sou

303. (In reference to comment 317) True, there isn't much hydro left.

303. (In reference to comment 618) If you consider capital, geothermal isn't cheap either.

306. There will probably be new nuclear plants installed after 1990-92. My 5% nuclear in 2020 is the last group of reactors installed during 1985-1992. But if demand for electric does increase rapidly in your assumption B (which is 65% over my projection), nuclear to contribute 5-10% in 2020 through new installations after 1990.

300. Nuclear will grow more slowly than above, while oil and gas phase-out will lag. Coal derived electricity is changed to coal by my accounting.

Statistics:

	Round One		Round Three A(3400)		Round Three B(5400)	
Responses:	36		23		23	

Coal

	2000	2020	2000	2020	2000	2020
Low:	47.0%	43.0%	45.0%	43.0%	46.0%	43.0%
Mean:	56.8%	52.5%	56.7%	58.0%	56.5%	54.4%
High:	67.0%	65.0%	75.0%	81.0%	71.0%	75.0%
Std. Dev.:	5.8%	8.2%	9.2%	10.2%	7.7%	8.1%

Oil

	2000	2020	2000	2020	2000	2020
Low:	0.5%	0.0%	2.0%	0.0%	1.0%	0.0%
Mean:	5.3%	1.8%	7.2%	4.1%	5.1%	3.4%
High:	10.0%	5.0%	19.0%	16.0%	19.0%	16.0%
Std. Dev.:	2.7%	1.7%	5.3%	4.6%	4.6%	5.2%

Gas

	2000	2020	2000	2020	2000	2020
Low:	2.0%	1.0%	1.0%	0.0%	2.0%	0.0%
Mean:	6.7%	2.3%	6.4%	4.1%	5.0%	3.2%
High:	15.0%	7.0%	15.0%	16.0%	15.0%	16.0%
Std. Dev.:	3.9%	2.2%	4.4%	4.7%	3.6%	4.0%

Nuclear

	2000	2020	2000	2020	2000	2020
Low:	18.0%	24.0%	6.0%	5.0%	6.0%	8.0%
Mean:	21.1%	26.6%	19.9%	21.8%	23.7%	28.7%
High:	31.0%	45.0%	40.0%	45.0%	45.0%	47.0%
Std. Dev.:	5.7%	6.6%	8.5%	11.0%	10.4%	10.4%

Hydro

	2000	2020	2000	2020	2000	2020
Low:	4.0%	3.0%	Not Specified in Round Three			
Mean:	10.0%	8.8%				
High:	18.0%	20.0%				
Std. Dev.:	3.5%	4.6%				

Other

	2000	2020	2000	2020	2000	2020
Low:	0.0	0.0%	2.0%	3.0%	3.0%	3.0%
Mean:	2.2	7.6%	9.8%	13.1%	8.9%	11.9%
High:	10.0%	33.0%	15.0%	22.0%	15.0%	25.0%
Std. Dev.:	3.8%	6.4%	4.1%	4.8%	4.3%	5.5%

ITEM 48. NET ELECTRIC UTILITY POWER SUPPLY FROM UNCONVENTIONAL SOURCES (EACH SOURCE'S NET OUTPUT AS A PERCENTAGE OF TOTAL UNCONVENTIONAL ELECTRICITY SUPPLY)

a. This is the final item which was poorly formulated for Round 1.

Of the net electric utility power supply characterized as "other" in the preceding question, please give your best estimates of the percentages (total=100%) which will be attributable to the following sources in the years 2000 and 2020. Do this under each of two assumptions:
Assumption A- "other" comprises 4% of electricity supply in 2020
Assumption B- "other" comprises 15% of electricity supply in 2020.

	A ("other"=4%)		B ("other"=15%)	
SOURCE	2000	2020	2000	2020
Non-utility gen.	______	______	______	______
Solar thermal	______	______	______	______
Solar photovoltaic	______	______	______	______
Biomass	______	______	______	______
Wind	______	______	______	______
Geothermal	______	______	______	______
"Other" (Please specify)	______	______	______	______

b. Terminology:

c. Sponsor's Comments:

d. Comments:

Round One Comments on Item 48 (Net Electric Utility Power -- Unconventional Sources)

311. Solar will grow rapidly in late 1980s and 90s but not for utilities. Non-utility generation will develop where and when it makes sense. Garbage burners could become very significant as burial sites become scarce. Geothermal -- most sites that are readily developed will be developed by 2000. OTEC is a "dog"! Low efficiency means large (expensive equipment) in a hostile environment.

330. Why worry over these minor sources? Only major issue is whether utility or industrial customer will cogenerate if that turns out to be an important source.

327. Most cogeneration (from coal, oil, and gas) will be achieved by 2000.

326. Much of the non-utility generation will come from the other sources on the list.

906. I expect solar to be very dominant by the turn of the century because all of the other sources mentioned are either marginal in availibility or in likelihood of adoptation.

905. Cogeneration is the only significant source.

904. The EPRI discussions of these issues and the notion of the 5-phases of R&D are helpful here.

317. Geothermal and biomass will be the only technologies that will make a significant contribution through the turn of the century.

337. Long term use of biomass for electricity should fall because of competing demand for land for food production. Biomass from wood and waste should continue being used for production of alcohols.

903. Geothermal and OTEC are probably the sources that, while site-limited, have the greatest growth potential.

102. Unless utility-supplied power becomes unreliable, it is inefficient for industry to buy boilers larger than needed for process heat purposes. Also, industry can't compete with utilities for economical production of electricity in condensing steam plants. In Europe, most co-generation is in utility-owned plants.

600. Solar thermal looks like a loser.

Round Three Comments on Item 48 (Net Electric Utility Power-Unconventional Sources)

303. There sure are a lot of people who think co-generation and non-utility generation are the same things!

906. I can simply see no way that solar can have a major effect.

903. Most of these sources, e.g. geothermal, wind, are site-specific. I see little likelihood of things like tidal and OTEC playing a major role, but they'll be experimented with.

Statistics:

	Round Three A (4%)		Round Three B (15%)	
Responses:	18		18	

Non-Utility Generation

	2000	2020	2000	2020
Low:	0.0%	0.0	0.0%	0.0%
Mean:	44.0%	37.3%	49.8%	47.6%
High:	85.0%	85.0%	90.0%	85.0%
Std. Dev.	26.3%	27.3%	31.4%	31.4%

Solar Thermal

	2000	2020	2000	2020
Low:	0.0%	0.0%	0.0%	0.0%
Mean:	8.1%	8.8%	8.5%	8.8%
High:	45.0%	45.0%	45.0%	45.0%
Std. Dev.:	12.6%	12.5%	13.0%	12.6%

Solar Photovoltaic

	2000	2020	2000	2020
Low:	0.0%	0.0%	0.0%	0.0%
Mean:	6.7%	10.5%	8.4%	10.8%
High:	45.0%	45.0%	45.0%	45.0%
Std. Dev.:	12.3%	13.0%	13.0%	13.0%

Biomass

	2000	2020	2000	2020
Low:	0.0%	0.0%	0.0	0.0%
Mean:	11.4%	12.1%	9.0%	8.3%
High:	50.0%	50.0%	30.0%	30.0%
Std. Dev.:	14.5%	15.9%	10.8%	9.8%

Wind

	2000	2020	2000	2020
Low:	0.0%	0.0%	0.0%	0.0%
Mean:	13.6%	15.7%	11.8%	11.8%
High:	40.0%	44.0%	40.0%	44.0%
Std. Dev.:	11.6%	13.7%	12.0%	12.6%

Geothermal

	2000	2020	2000	2020
Low:	0.0%	0.0%	0.0%	0.0%
Mean:	15.5%	19.8%	12.3%	13.1%
High:	35.0%	50.0%	30.0%	34.0%
Std. Dev.:	11.7%	16.8%	9.8%	10.3%

Other

2000	2020	2000	2020
16 responses=0% 2 responses=10%	16 responses=0% 2 responses=5%	17 responses=0% 1 response=5%	17 responses=0% 1 response =5%

ITEM 49. INDUSTRY ELECTRICITY GENERATION (AS A PERCENTAGE OF TOTAL GENERATION) [58]

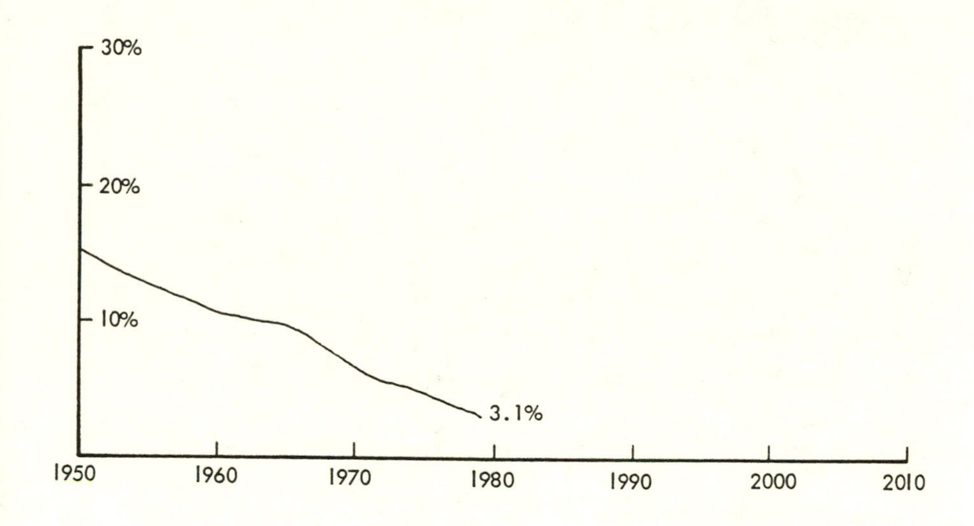

a. Trend Evolution: Please indicate on the above graph how you believe the trend will evolve over the next 30 years.

b. Point Values: Please indicate below your best estimates of the point values (vertical axis values) of the trend in the years 2000 and 2010.

2000 ______________ 2010 ______________

c. Terminology:

d. Sponsor's Comments: Industrial electricity generation was 50% of all generation in 1900. EPRI Journal (July/August 1981, p.10) puts current industrial generation at 5% of total, and says 15% is possible by 2000 [59]. Industrial cogeneration increases, if they materialize,would likely drive the trend upwards.

e. Statistics:

	Round One		Round Three	
Responses:	50		29	
	2000	2010	2000	2010
Low:	2.0%	1.0%	2.0%	2.1%
Mean:	7.3%	9.9%	7.1%	10.1%
High:	20.0%	25.0%	20.0%	25.0%
Std. Dev.:	4.3%	5.9%	4.3%	6.2%

Round One Comments on Item 49 (Industry Electricity Generation)

223. Industrial cogeneration is the driving force of the increase according to EPRI. One must be very optimistic about cogeneration prospects to get even close to the EPRI 15% figure.

311. Cogeneration should be developed where it makes sense and economics will prevail in the long run. The artificial stimulus given to cogeneration by PURPA is a "flash in the pan" that will soon die out as maintenance problems and low reliability make even "marginal rates" uneconomical.

321. Will remain constant unless artificially stimulated by the Federal government or others.

319. Failure of electric power industry could sharply increase this.

212. The proportion of industrial to total generation rises sharply in the 90s due both to greater cogeneration and drop in total production of electricity.

317. Cogeneration will be a big thing until well after the century.

903. Tendency to generate industrial electricity will probably be restrained by capital costs and pollution control requirements. These problems will probably be left to utilities to cope with.

600. Off-the-shelf equipment for industry electricity generation does not exist, really. I can't see them going to all the trouble, unless blackouts and curtailments increase a lot.

Round Three Comments on Item 49 (Industry Electricity Generation)

303. Why should industry raise capital if they can get someone else to do it.

300. I expect paper-pulp-lumber industries to go nearly 100% cogeneration. Expect little additional move in this direction elsewhere.

104. Dow, Princeton and other studies estimate 210 GWe economic today.

300. See item 48. #49 may increase to the point where industry's fraction can stay level.

903. I suspect cogeneration will play an increasingly important role, but never a major one.

ITEM 50. NATIONAL AVERAGE UTILITY ELECTRICITY END-USER PRICES (1980 CENTS/KWH) [60]

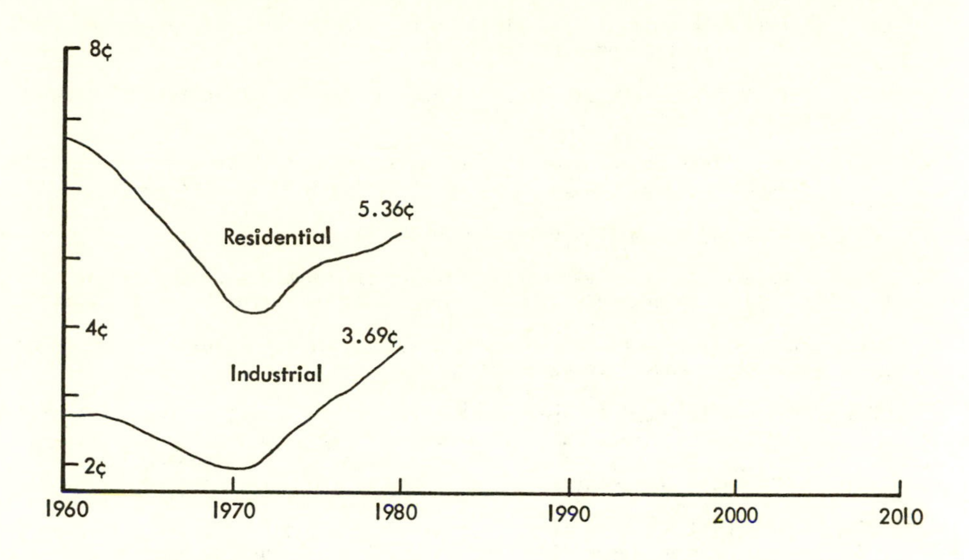

a. Trend Evolution: Please indicate on the above graph how you believe the trends will evolve over the next 30 years.

b. Point Values: Please indicate below your best estimates of the point values (vertical axis values) of the trends in the years 2000 and 2010.

Residential	2000__________	2010__________
Industrial	2000__________	2010__________

c. Terminology:

d. Sponsor's Comments: DOE's middle world oil price scenario predicts electricity prices (1980 cents) of:

	2000	2010
Residential	6.33	6.3
Industrial	4.74	4.71

[61]

e. Your Comments:

Round One Comments on Item 50 (National Average Utility Electricity End-user Prices)

311. PUCs will not allow a $.016 differential to continue to exist. Regulations will drive capital costs through the roof, and power costs will go right along with them.

330. I think DOE's forecast does not provide for enough internalization of externalities.

301. DOE's prices do not support the investment needed to supplement nation's electricity use. Prices must climb in terms of about 2-3% per year.

205. DOE may be optimistic between now and 2000.

906. I expect increases to continue faster than DOE expects, but also for rates to converge due to increased public pressure for equalization.

212. Price will rise higher than DOE predicts due to increases in costs of production and distribution.

309. Prices will rise in parallel this centruy because fuel resources are limited or delivery is constrained. FBR will halt price rise next century. Industry will continue to get a price discount because it contributes jobs, and products to the benefit of society at large.

905. Prices will not rise significantly because electricity will be supplied by coal and nuclear. Coal prices won't go up much except for moderate transportation cost increases. Nuclear power won't increase in cost. There is a surplus of uranium. Finally, there is a slowdown in electricity growth due to current surpluses in generating capacity.

303. Industry will continue to subsidize residential. Rise in rates will probably be faster than inflation in the beginning of the period.

317. Depends on emphasis on conservation and on substitution of electricity (generated by coal, nuclear and other) for oil and gas.

207. Just roll over of old capital will have an increasing effect.

903. Tendency will be to eliminate industrial block pricing and equilibrate electricity costs in long run.

612. Electricity price escalations lag primary fuel prices by about a decade because of long-term contracts, fuel substitution, etc. This, plus terri financial shapes of most utilities today, means real price escalation wil increase sharply during the next 15 years. Time of use pricing may help, but it will help industry first.

Round Three Comments on Item 50
(National Average Utility Electricity End-User Prices)

306. Slow growth will save cost of power from big increase after 1990. Although capital costs of power plants will not really level off, the fixed charges will decline as % of costs as rate of plant building slows.

215. Real price will rise till mid 1990s, then decline as current construction program is finished and further expansion based on lower cost alternative.

104. Low growth-no growth will keep rates down.
Overcapacity will increase rates.
Overdue conversion to coal and pollution control will increase rates.

300. "Bus-Bar" costs will be capital cost dominated by 2000. Rate differentials will narrow.

301. Cost of capital and increased fuel costs will drive prices--I entirely disagree with comment 905. Comment 309 seems to come from the same person who predicted that nuclear-based electricity would be so cheap that only a metering charge was necessary.

903. Reflects increases in fuel and capital costs.

Statistics:

	Round One		Round Three	
Responses:	50		28	
Residential				
	2000	2010	2000	2010
Low:	3.5	5.0	5.3	5.3
Mean:	6.7	7.5	6.9	7.6
High:	10.0	12.0	10.0	10.0
Std. Dev.:	1.3	1.5	0.9	1.2
Industrial				
	2000	2010	2000	2010
Low:	3.0	3.7	3.7	3.7
Mean:	5.3	5.9	5.5	6.2
High:	7.5	9.0	7.5	9.0
Std. Dev.:	1.0	1.3	1.0	1.3

ITEM 51. AVERAGE U.S. REFINER ACQUISITION COST OF CRUDE OIL (1980 DOLLARS/BARREL; ALL SOURCES) [62]

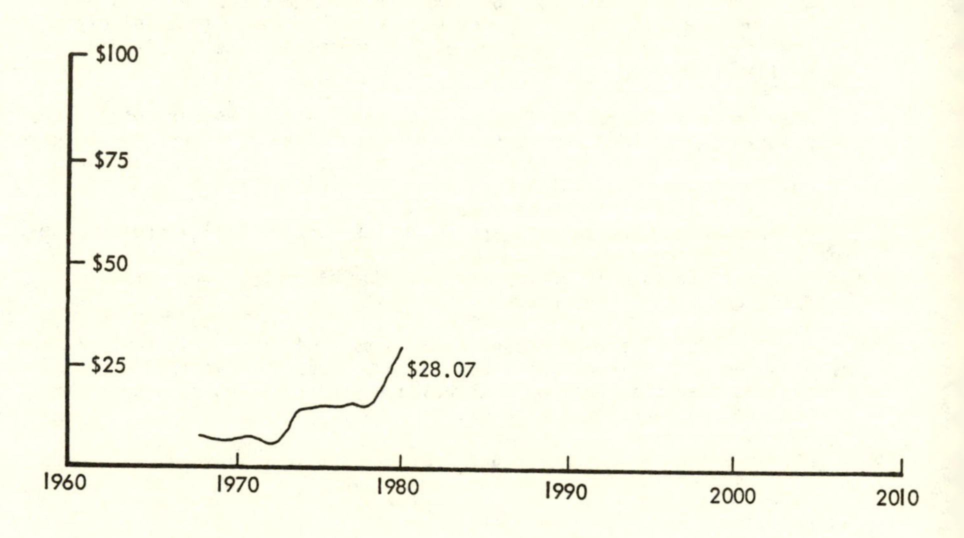

a. Trend Evolution: Please indicate on the above graph how you believe the trend will evolve over the next 30 years.

b. Point Values: Please indicate below your best estimates of the point values (vertical axis values) of the trend in the years 2000 and 2010.

2000____________ 2010____________

c. Terminology:

d. Sponsor's Comments: For comparison purposes: 1980 average refiner acquisition cost for imported crude was $33.89/BBL.

e. Statistics:

	Round One		Round Three	
Responses:	50		29	
	2000	2010	2000	2010
Low:	$25.00	$25.00	$25.00	$30.00
Mean:	$51.76	$61.05	$51.10	$59.90
High:	$90.00	$130.00	$90.00	$105.00
Std. Dev.:	$14.30	$21.60	$15.10	$19.60

Round One Comments on Item 51 (Average U.S. Refiner Acquisition Cost of Crude Oil)

223. By the year 2000, we will be learning to live without OPEC in the U.S. and in enough other countries to severely weaken if not destroy the cartel.

311. As Third World countries develop, the need for petroleum will escalate rapidly. Every time there is a crisis in oil supply, there will be dramatic increases that will not be reversed (eg.1973 Israel-Egypt war and 1979 Iran Crisis). Oil costs will increase at least enough to follow inflation and somewhat more as depletion begins to affect supply.

321. Long term conservation will create an oversupply causing price increases at the same rate as inflation.

327. OPEC will fall to competing energy sources for all uses except transportation, where efficient use will reduce need.

906. I expect continued increases with some stabilization at the turn of the century as other sources develop.

212. Price will peak in 90s then fall due to conservation measures and availability of alternative energy sources.

309. Prices will escalate greatly because foreign sources of oil are unstable; U.S. oil prices will be world prices and most of the world does not have cheaper substitutes available to the U.S.

207. $P(\leq \$20) \leq .02$, $P(\geq 200) \geq .02$

909. 1% real escalation based on improved crude price of 1980.

903. Tendency will be for crude prices on world basis to rise at least at inflation rate.

102. LDCs set the price ceiling. World will not let high oil prices crush LDCs and many are on the edge now. OPEC will not set its production too low, because OPEC unity requires that each country enjoy high cash flow. $50-$60/barrel in 2010 (today's dollars) seems reasonable.

612. Next Middle East war may not affect oil flows because Arab world as a whole has come to understand its continued power is tied to reliability of supply.

Round Three Comments on Item 51 (Average U.S. Refiner Acquisition Cost of Crude Oil)

308. Oil prices are likely to stabilize for next decade or two (assuming no Middle East war) but eventual declines in reserves will bring renewal of relative price increases.

300. "Sporadic" rises, which are difficult to predict. Real price should double (if not more) by 2000.

301. My estimate of 2% real increase still seems right. Every estimate of fuel prices (oil & gas) that I have seen over the last 7-10 years has underestimated the actual values achieved.

903. I expect world costs will be held reasonably level in constant dollars and will rise at inflation rate or thereabout. This implies comparatively low inflation rate, or alternatively, restraining effect of possible alternative sources.

ITEM 52. ENERGY EFFICIENCY IN MANUFACTURING (BTU'S PER DOLLAR VALUE ADDED) [65]

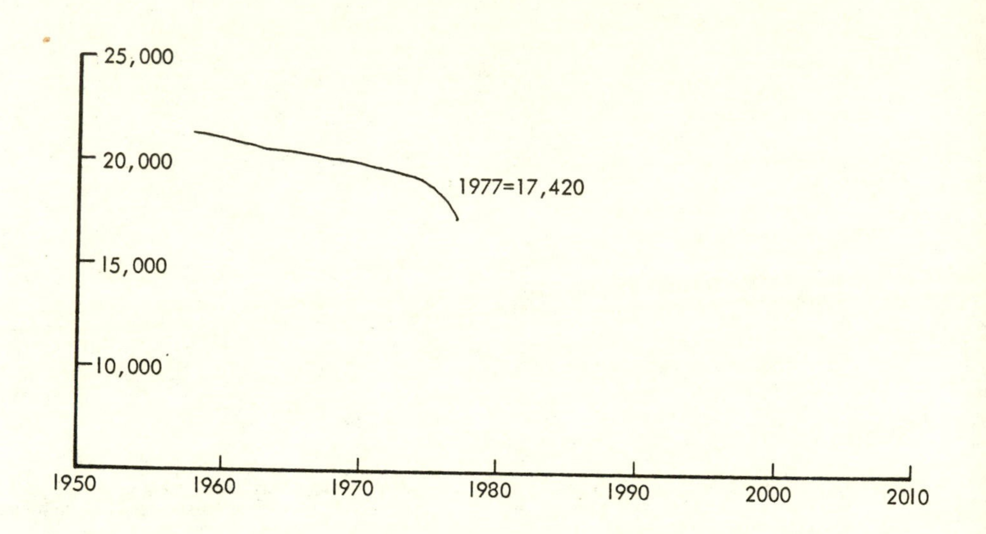

a. Trend Evolution: Please indicate on the above graph how you believe the trend will evolve over the next 30 years.

b. Point Values: Please indicate below your best estimates of the point values (vertical axis values) of the trend in the years 2000 and 2010.

2000______________ 2010______________

c. Terminology:

d. Sponsor's Comments: BTU total accounts for all energy used by all manufacturing industries. The dollar total (1980 dollars) sums the total market value added to goods by the process of manufacture.

e. Statistics:

	Round One		Round Three	
Responses:	49		24	
	2000	2010	2000	2010
Low:	9000	8000	10000	8000
Mean:	13380	12420	13300	11900
High:	17000	18000	15000	15000
Std. Dev.:	1880	2430	1400	2100

Round One Comments on Item 52 (Energy Efficiency in Manufacturing)

311. The final dip in the curve is due to actions taken in response to the 73 crisis; there are the "easy" things like insulation, and low cost modifications. The second dip will occur when fuel prices rise to the point that new high efficiency (but more costly) equipment can be economically justified. Countering this will be a trend toward automation which will increase energy consumption.

330. Easy energy fixes taken. Moreover, energy (particularly electricity) is still a bargain compared to labor or capital.

321. The GNP- Energy use relationship was firmly constant prior to 1973. After 1973, and the impact of the oil embargo, this relationship no longer held. At this time it is difficult to predict future GNP-energy use trends. However, increased efficiency is necessitated by higher energy costs.

327. This is not a good question. Except for a few industries, energy is a minor item. Answer is based on energy-using industries, petrochemical, electro-metallurgical, paper, chemical, glass.

906. Increased computerized technology should increase efficiency.

212. Marked improvements in efficiency will continue, leveling off after 2000.

223. Phone: Energy will no longer be a problem. Labor will. There will be a shift in the energy-labor trade-off due to baby boom passing and a lack of skilled entry level workers.

903. Efficiency of energy utilization will continue to rise as automation and energy conservation continue to pay off.

102. Continued U.S. movement away from processing of raw materials to fabrication of finished products will keep trend headed down, regardless of efficiency improvements. Most "high tech" manufacturing does not consume a lot of energy.

Round Three Comments on Item 52 (Energy Efficiency in Manufacturing)

300. I do not expect dramatic changes in this index. Energy per se may go up in manufacturing because of automation and more conversion losses in process industries. Energy per capita & energy per GNP, however, will decline.

104. Agree with 102. As long as energy is an input to production, it is not a bargain and it is a problem.

301. Hard to predict--increased efficiency will be driven by higher energy costs--the cost of capital (if continued high) will also limit the improvement that can be introduced.

903. My faith in continuing efficiency of utilization and conservation continues.

ITEM 53. PUBLICLY-OWNED UTILITY ELECTRICITY GENERATION (AS A PERCENTAGE OF TOTAL UTILITY ELECTRICITY GENERATION) [66]

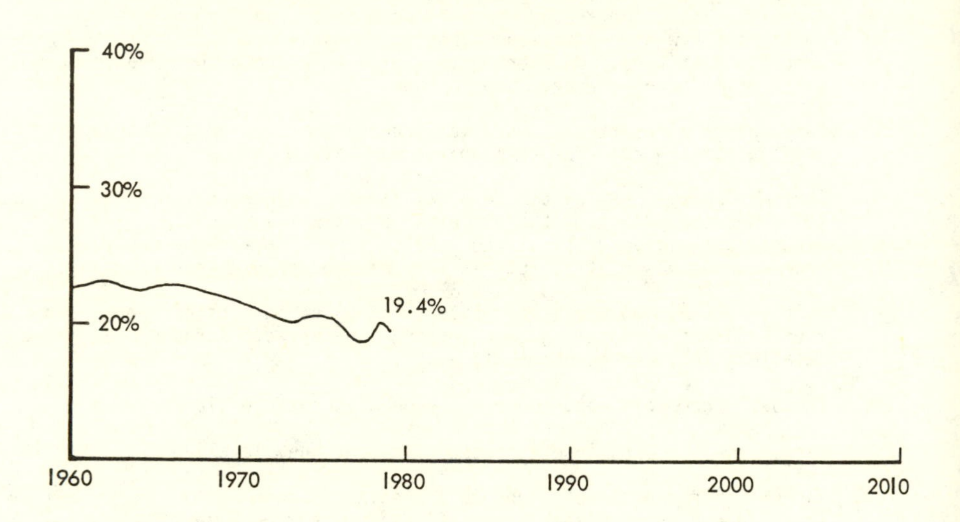

a. Trend Evolution: Please indicate on the above graph how you believe the trend will evolve over the next 30 years.

b. Point Values: Please indicate below your best estimates of the point values (vertical axis values) of the trend in the years 2000 and 2010

2000____________ 2010____________

c. Terminology: Publicly-owned utility generation does not include co-operative generated electricity (which is included in the total).

d. Sponsor's Comments: "Nationalization" of some or all of the industry would obviously drive the trend upward.

e. Statistics:

	Round One		Round Three	
Responses:	50		29	
	2000	2010	2000	2010
Low:	11.0%	8.0%	11.0%	8.0%
Mean:	21.7%	23.6%	21.5%	23.2%
High:	40.0%	40.0%	40.0%	55.0%
Std. Dev.:	4.9%	9.0%	5.6%	10.1%

Round One Comments on Item 53 (Publicly-Owned Utility Electricity Generation)

223. Relative financing capabilities will be the primary determining factor.

311. This trend will be determined by government policy. The trend toward big public utilities will continue (PASNY is the forerunner of many state power authorities). Public utility commissions responding to public pressure are driving investor utilities to the wall. They cannot raise money to build the required new plants. Many will be in bankruptcy by 1990. Others will sell their plants to public authorities (like Con Edison sold IP#3 and Astoria #6) and become distribution companies. Only a publicly-owned utility will have access to money required to build new plants.

330. Trends to sell off generation capacity, use financial strength of non-investor owned sector will slow up in next decade. Beyond that, I assume regulation will either get its act together or deregulation will take place.

327. State, municipal, co-op, and federal ownership is increasing today at a rate few within the industry or the administration understand.

906. The capital involved suggests to me that few new public utilities will be created.

212. Slowly increasing proportion will be publicly owned, mostly by municipalities.

904. I believe it will increase slightly because of capital pressures and increasing regulation -- not nationalization.

317. Public utilities will continue to be a part of the generating industry at about the same level.

903. I see the proportion staying at or near historical trend -- nationalization very unlikely, and state intervention especially so.

102. When the capital-intensive plants come on line, utilities will start to do much better economically. PUCs everywhere are allowing impressive RORs, so you can't say regulation will "get" the investor-owned utilities. In all I don't think government wants to generate power, or will have to.

Round Three Comments on Item 53 (Publicly-Owned Utility Electricity Generation)

300. Expect to see greater "concentration" in industry with time, but whether it will go public or private, cannot say.

303. There has been some reaction to rates wherein municipalities have bought out investor-owned segments. This may continue for a few years but will drop off significantly as long as costs of money increase. No nationalization seen.

300. Likely to increase in very near future due to poor fiscal climate.

903. I continue to think nationalization unlikely, and that proportion will not change much over forecast period.

ITEM 54. LEAD TIMES FOR CENTRAL ELECTRIC POWER PLANTS (YEARS)

a. Means are given in parentheses below for each round. There were 57 responses to round one and 28 responses to round three.

b. Point Values: Please indicate below your best estimates of the lead times (in years) for each of the listed central electric power plants, if the decision to build were made in each of the following years:

	1990	(RD1,RD3)	2000	(RD1,RD3)
Coal/conventional boiler:	______	(8.9,9.0)	______	(8.3,8.7)
Coal/fluidized bed:	______	(8.8,8.9)	______	(8.2,8.2)
Combined cycle/coal gasifier	______	(8.8,9.1)	______	(8.3,8.4)
Oil or Gas/boiler:	______	(8.1,8.1)	______	(7.6,7.7)
Oil or Gas/turbine:	______	(3.9,3.7)	______	(3.7,4.4)
Nuclear/fission LWR:	______	(11.2,10.8)	______	(9.9,9.4)
Nuclear/fusion:	______	(16.9,15.2)	______	(13.8,13.6
Geothermal:	______	(8.3,8.2)	______	(7.7,7.1)
Biomass:	______	(6.7,6.5)	______	(6.2,6.1)
Solar/thermal:	______	(7.3,7.8)	______	(7.1,6.2)
Solar/terrestrial photovoltaic:	______	(7.0,5.9)	______	(5.8,5.4)
Solar/power satellite:	______	(16.5,15.2)	______	(14.7,14.1
Wind:	______	(4.3,4.0)	______	(3.8,3.6)

c. Terminology: Lead time is the time between the initial decision to build and full scale operation of the plant. LWR=light water reactor.

d. Sponsor's Comments: DOE's Annual Report to Congress estimates the following lead times:

Coal/conventional boiler:	12	Geothermal:	12
Coal/fluidized bed:	10	Biomass:	8
Combined cycle/coal gasifier:	10	Solar/thermal:	--
Oil or Gas/boiler:	12	Solar/terrestrial photovoltaic:	8
Oil or Gas/turbine:	4	Solar/power satellite:	--
Nuclear/fission LWR:	13	Wind:	4
Nuclear/fusion:	--		

e. Your Comments:

Round One Comments on Item 54 (Lead Times for Central Electric Power Plants)

330. The cost of human-imposed delay will be perceived as unacceptable by 1990. By 2000 we should have balanced speed with due deliberation.

327. I assume above that by 1990 we've become a rational nation once again in the regulatory area. If lawyers continue to dominate the decision process, add 10 years to each estimate.

326. All plants will be considerably smaller than today.

308. Rationality in regulation will begin to prevail for survival reasons.

905. I expect lead times for nuclear/LWR to fall in line with engineering time tables. Delay currently is not caused by engineering aspects, but by regulatory delay and a decrease in demand and need for rapid construction. Lead times should decrease as there is greater concensus in the regulatory process and as projections fall in line with actual demand. (Nuclear LWR values of 9 in 1990 and less in 2000)

409. Nuclear/LWR as lead times related to federal regulation are reduced, other delays including those related to a larger state and local government role, will increase (1990 and 2010 values for Nuclear/LWR of 15 and 10 years).

904. I'm assuming decision to build means technology is proven, EIS completed, etc.

337. On solar/terrestrial photovoltaic: Modular Construction--some units will be on line before entire plant is completed.

903. Streamlining of licensing procedures and delivery schedules will speed things up to a degree. I don't see fusion central stations in this century.

319. Delay will decrease in future as utilities become more aware of the need to influence appointments to regulatory and public opinion. This was never necessary before and utilities have been slow coming up to steam, so to speak.

612. During the war, you could build a plant in 3 years. It is physically possible to build any plant in 5 years. Regulation will come more in line as consumer costs rise.

Round Three Comments On Item 54 (Lead Times for Central Electric Power Plants)

321. I felt that some lead times should not be quantified either because:
(A) a large commercial facility has never been built
(B) discouragement by federal law.

409. Geothermal/Biomass/Wind/Solar/Thermal are actual recorded lead times. By 1990 and 2000 they could come down in certain locations, but go up in others.

411. I believe the lead time will decrease due to regulatory streamlining.

903. Safety considerations will continue to slow down construction, especially on "new" technologies like nuclear, solar satellite.

ITEM 55. CAPTIAL COSTS OF COAL AND NUCLEAR ELECTRIC CAPACITY

a. Statistics

	Round One			Round Three		
Responses	46			20		
Coal						
	1990	2000	2010	1990	2000	2010
Low:	$600	$600	$600	$700	$800	$800
Mean:	$1041	$1148	$1185	$1050	$1090	$1170
High:	$1500	$2000	$2500	$1600	$1800	$2200
Std. Dev.:	$220	$350	$405	$210	$210	$320
Nuclear,LWR						
Low:	$750	$750	$700	$900	$800	$700
Mean:	$1380	$1510	$1575	$1435	$1550	$1660
High:	$2200	$2800	$3000	$2000	$2800	$3000
Std. Dev.:	$310	$450	$565	$325	$460	$560
Nuclear,FBR						
Low:	$1000	$950	$900	$1800	$1350	$1000
Mean:	$2275	$2325	$2255	$2400	$2500	$2400
High:	$3000	$3200	$3500	$3000	$5000	$4000
Std. Dev.:	$640	$770	$780	$480	$840	$800

b. Point Values: Please indicate below your best estimates of the capital costs per kilowatt capacity, exclusive of interest charges, of large (750-1000 MWe) coal and nuclear power plants begun in each of the specified years. Please assume national average construction costs and bituminous, medium surlfur coal. Please specify dollars used, if not using 1980 dollar

	1990	2000	2010
Coal	______	______	______
Nuclear, LWR	______	______	______
Nuclear, FBR	______	______	______

c. Terminology: LWR=light water reactor; FBR=fast breeder reactor; coal=steam boiler system with lowest life-cycle cost.

d. Sponsor's Comments: DOE forecasts "mature" plant costs of $948, $1286, and $2013 (1980 dollars) per KW for coal, LWR, and FBR, respectively [68]. The National Academy of Sciences placed coal and nuclear plant costs at approximately $795 and $730 per KW in 1978 [69]. The Solar Energy Research Institute forecast year 2000 per KW costs of $969 and $1277 (1980 dollars) for coal and nuclear plants, respectively [70].

Round One Comments on Item 55 (Capital Costs of Coal and Nuclear Electric Capacity)

223. Technological improvement will more than offset other factors which will tend to push prices upward. Most potential for technology exists in breeders and least in coal.

311. Nuclear costs $1500/KW today; Coal with scrubbers is $1200 to $1300 today. Nobody knows what a fast breeder will cost.

310. There is no reason why nuclear plant costs in 1980 dollars can't be reduced markedly. The cost experience curve should hold if we quit changing the design during the design phase and the construction phase and get rid of a lot of unnecessary complexity.

330. Interest during construction is basic reduction. Plus, there are economics to be achieved in LWR. In FBR, we would learn.

321. We have no specific values. However, we strongly believe that the capital cost of a coal plant and its fuel gas desulfurization equipment will fall in the range of 80-90% of the capital cost of a nuclear LWR station. We have no feel for a FBR.

304. WPPss 4/5 = $5200/KW, Shoreham = $3200/KW, Diablo Canyon= $1400/KW, Bailly = $2700/KW, Pilgrim 2 = $3100/KW (All 1981 estimates in today's dollars for 1982-1990 operation. Cal Energy Commission est. in 1978 for Sundesert= $1600/KW for 84 operation (1978 $s) No one is building or ordering new plants now; ergo, costs are virtually unknown or irrelevant for 1990-2000-2010. +80 cancellations since 73.

906. I expect costs will be in proportion to DOE, but as always higher than initially projected.

309. U.S. and world will be driven to nuclear power over next 30 years. This means LWR costs will go down because of experience, and dwindling public opposition. FBRs will still be low on learning curve (hence, expensive).

317. Nuclear LWR will continue to be more and more expensive and difficult to site, build and bring on line in a reliable manner.

306. Very few people think of these costs in real, no interest, cost terms. And no wonder: 7% real (cost & interest) escalation doubles the real cost of a plant in 10 years. Lots of construction costs exceed general inflation by more than 3 or 4% annually.

Round Three Comments on Item 55 (Capital Costs of Coal and Nuclear Electric Capacity)

306. The group's responses reflect either a naive, touching optimism re: "learning" in nuclear design and construction, or an alarming ignorance of the actual relative current capital costs of nuclear and coal (about 1.5 or 1.6 to 1, 1980 completions, incl. scrubbers).

ITEM 56. SOLAR PHOTOVOLTAIC ARRAY COSTS (1980 DOLLARS/PEAK WATT OUTPUT) [71]

a. Point Values: Please indicate below your best estimates of the costs per peak watt of solar photovoltaic arrays (exclusive of supporting electronics and hardware) in each of the specified years; if you are using other than constant 1980 dollars, be sure to specify dollars used.

	1990	2000	2010
Cost/per peak watt:			

b. Sponsor's Comments: Current costs are about $9/peak watt [72]. DOE forecasts $.71/peak watt in 1986, in 1980 dollars [73]. The National Academy of Sciences says most experts believe $1 to $2/peak watt (1978 dollars) will be feasible, apparently by the late 1980s [74]. The Solar Energy Research Institute forecasts a year 2000 cost of $.17 to $.67/peak watt, apparently in 1979 dollars [75].

c. Statistics

	Round One			Round Three		
Responses	37			22		
	1990	2000	2010	1990	2000	2010
Low:	$0.50	$0.30	$0.20	$0.50	$0.40	$0.10
Mean:	$2.70	$1.68	$1.28	$2.32	$1.40	$1.00
High:	$8.00	$4.00	$4.00	$5.00	$4.00	$4.00
Std. Dev.:	$1.88	$1.33	$1.17	$1.55	$0.90	$1.00

Round One Comments on Item 56 (Solar Photovoltaic Array Costs)

311. Reliable- low maintenance solar will be expensive

310. The cost will only come down with mass production. The cost of government subsidy is too high for the quantities that would be involved. Therefore, increased production will not take place.

330. Academic question -- will not be competitive. No way can it be reduced below $1 a peak watt---more likely $3.

304. A major technological breakthrough is needed before the highly optimistic cost reductions alluded to can be realized.

321. We do not have specific values. However, we do not feel as optimistic as your sponsors. We do not believe the costs will drop this low without artificial stimulation.

327. Array costs will go up after a low about 1990 because normal price rises will exceed technology improvements as situation matures. Look at automobiles for example.

906. I believe all of the costs estimated for solar are much too optimistic and the costs will decrease less and less rapidly than anticipated.

317. A breakthrough in the manufacture of these cells will take place around the turn of the century.

909. Cell costs will constitute a "minor" portion of the cost for either central station or distributed PVs - Mature technology goal (1990-2000) for PV systems is about $2400 per kw peak ($2.40 per peak watt).

903. This is a technology where dramatic improvements will occur in next 30 years, as in solid state chip technology.

600. Miniaturization helped semi-conductor computers, but it does not apply to solar cost reductions.

104. I have no reason to doubt government projections, they have been on the money, to date.

Round Three Comments on Item 56 (Solar Photovoltaic Array Costs)

104. JPL says $2.80/w at new facilities (under construction) in 1982.

ITEM 57. SUPERCONDUCTING GENERATOR (YEAR OF FIRST COMMERCIAL APPLICATION)

a. Statistics:

	Round One	Round Three
Responses	36	17
Answered "Never"	16.6%	11.7%
Low:	1990	1995
Mean:	2005	2006
High:	2020	2020
Std. Dev:	9	8

b. Point Values: By what year do you believe there is a 90% probability that the first electricity generator with a superconducting rotor and a rating at or above 1,200 MVA (3,600 rpm) will enter commercial operation? If you do not believe this will happen before 2020, please answer "never."

Year of 90% probability:________________

c. Sponsor's Comments: Westinghouse and EPRI are now developing a 300 MVA demonstration superconducting generator which is scheduled for factory tests in 1984. At least six smaller superconducting generator efforts are currently under way here or abroad [76].

d. Your Comments:

Round One Comments on Item 57 (Superconducting Generator)

311. Commercial application by 1995 assumes a full blown development program is initiated and sustained, starting today.

330. If not by 1995, then probably never.

304. The superconducting generator will achieve substantial application in sizes smaller than 1200 KVA because of its lower losses and other cost benefits.

327. Add 10 years if the regulatory process for coal and nuclear plants isn't taken out of the hands of lawyers.

308. The question is not of technical viability but rather of economic viability The superconducting generator may well be a product not really needed nor wanted.

906. I believe engineering difficulties will lead to delays.

909. From a technical point of view, a 1200 MVA superconducting generator could be in operation by 1990, with 1995 being a "practical" date for planning. Superconducting generation of approximately 1000 MVA (or less) have a reasonable (50%+) probability by 2000.

Round Three Comments on Item 57 (Superconducting Generator)

903. I expect there are situations where the superconducting generator will be economically feasible and will be tried on a commercial basis.

ITEM 58. INTRACITY ELECTRIC-POWERED PERSONAL TRANSPORTATION

a. Statistics

	Round One	Round Three
Responses	43	23
Answered "Never"	29.7%	34.0%
Low:	1993	2000
Mean:	2007	2009
High:	2020	2020
Std. Dev:	8	7

b. Point Values: By what year do you believe there is a 90% probability that 10% of all personal intracity passenger miles will be traveled in personal vehicles powered at least 70% by electricity? If you do not believe this will occur by the year 2020, please answer "never."

Year of Probability: ____________

c. Sponsor's Comments: Currently, the principal manufacturer of battery-powered cars, vans, and pickup trucks offers vehicles which travel 70-100 miles between charges and retail for $12,000-15,000 [76].

d. Comments:

Round One Comments on Item 58
(Intracity Electric-Powered Personal Transportation)

311. Even then, it will need a breakthrough in batteries to achieve this goal.

330. If electricity is to be major fuel for intracity travel, it will probably be through mass transit-fixed bed. If personal electric transit is to be important, will take institutional change or technology change -- e.g., battery station, guided highways, etc.

308. It is most likely to be a hybrid vehicle employing recharging at rest.

909. Recent studies sponsored by this company indicate that internal combustion vehicles will retain a cost edge over electric vehicles (battery) at least through 2000. Two possible exceptions are "intracity" short start-stop trips and LIM (Linear Electric Induction Motor) for freeway use.

903. The practical electric car will require an inexpensive lightweight high-density capacity battery, just beginning to emerge from laboratory. I see such cars as being experimental through the next decade and reaching the market only after 1990.

600. I figure batteries have to improve three-fold, and this is not likely.

618. People don't understand the substantial health and environmental hazards of batteries.

306. They are quiet and not obviously polluting. People will gladly pay dearly for this option.

102. I figure 10% of sales (i.e. 1.3 million) will be electric in 2000. People will buy electric like they buy diesel -- regardless of bottom-line disadvantage. Of course, I see real electricity price growth to be almost zero after 2000.

Round Two Comments on Item 58
(Intracity Electric Powered Personal Transportation)

311. (In reference to comment 306) Corona discharge on motor brushes can create a serious ozone problem according to some "experts."

411. Are we going to get electricity from coal, nuclear fusion? Problems with R&D finances makes this very unrealistic.

903. The electric car, as originally introduced, will be a "second" car, useful for shopping and commuting. As internal combustion cars get smaller and lighter, the contrast between hydrocarbon-fueled and electrics on the highway will decrease. One thing that will make electrics come on faster, in addition to efficient batteries, will be comparatively low-priced and secure nuclear electricity.

References

[1] U.S. Department of Commerce, Bureau of the Census, Statistical Abstract of the United States: 1979, Table 10, page 4, (citing the Census of the Population: 1970, vol. I, and 1980 Census of Population and Housing).

[2] Statistical Abstract of the United States: 1979, (SAUS 1979), Table 14, page 17, (based on the U.S. Census of Population: 1940, 1950, 1960, and 1970 and Current Population Reports, series P-25, No. 810).

[3] S.A.U.S. 1979, Table 16, page 17 (based on U.S. Census of Population: 1960-1970, Vol. I and Current Population Reports, series P-20, No. 336).

[4] S.A.U.S. 1979, Table 644, page 392 (taken from U.S. Bureau of Labor Statistics, Employment and Earnings, monthly).

[5] U.S. Department of Commerce, Bureau of the Census, Social Indicators III, Table 7/11, page 354 (based on U.S. Department of Labor, 1979 Employment and Training Report of the President).

[6] S.A.U.S. 1979, Table 644, page 392 and Table 653, page 397 (citing U.S. Bureau of Labor Statistics, Employment and Earnings, monthly).

[7] J. Singelman, From Agriculture to Services: The Transformation of Industrial Employment, Beverly Hills: Sage, 1978, Appendix A.1.

[8] Economic Report of the President 1981, page 262 (citing the Department of Commerce, Bureau of Census, Current Population Reports, series P-60).

[9] Social Indicators III, Table 9/24, page 491 (citing Bureau of Census, Current Population Reports, series P-60, no. 1201).

[10] S.A.U.S. 1980, Table 769, page 463 (citing U.S. Bureau of Census, Current Population Reports, series P-23, no. 28; and P-60, nos. 81 and 125).

[11] National Public Radio, All Things Considered, September 1981.

[12] S.A.U.S. 1979, Table 29, page 29 (citing Bureau of Census, U.S. Census of Population 60 and 70, Vol. I, and Current Population Reports, series P-25, no. 800 and 1980 Census of Population, Age, Sex, Race, and Spanish Origin of the Population).

[13] S.A.U.S. 1979, Table 27, page 28 and S.A.U.S. 1974, Table 41, page 34 (citing Bureau of Census, Census of Population '30, '40, '50, '60, and '70 and Current Population Reports, series P-25, no. 800).

[14] S.A.U.S. 1979, Table 125, page 88 (citing U.S. Immigration and Naturalization Service, Annual Report).

[15] Michael S. Teitelbaum, "Right versus Right: Immigration and Refuge Policy in the United States," Foreign Affairs 59: 21-59 (Fall, 1980).

[16] S.A.U.S. 1979, Table 58, page 44 (citing Bureau of Census, Current Population Reports, series P-20, no. 327 and earlier issues).

[17] Ibid.

[18] S.A.U.S. 1979, Table 231, page 145 (citing Bureau of Census, U.S. Census of Population 1960 and Current Population Reports, series P-20, nos. 207 and 295).

[19] Social Indicators III, Table 11/13, page 559 (citing John P. Robinson, "Changes in America's Use of Time, 1965-1975," Report of Communication Research Center, Cleveland State University, 1976).

[20] Economic Report of the President 1981, page 293 (citing Department of Labor, Bureau of Labor Statistics, Monthly Labor Review).

[21] United Nations, Department of International Economic and Social Affairs, Statistical Yearbook, annual.

[22] S.A.U.S. 1980, Table 724, page 437 (based on Department of Commerce, Bureau of Economic Analysis, The National Income and Product Accounts and The Survey of Current Business).

[23] S.A.U.S. 1979, Table 1185, page 687 (citing Bureau of Census, Census of Agriculture 1964 an 1969, Vol II, 1974 Vol. I).

[24] S.A.U.S. 1980, Table 594, page 366 (citing U.S. Office of Management and Budget, The Budget of the U.S., annual).

[25] S.A.U.S. 1980, Table 1522, page 860 (based on Bureau of Economic Analysis, Survey of Current Business, June-September, 1980).

[26] Ibid.

[27] Economic Report of the President 1981, page 260 (citing Department of Commerce, Bureau of Economic Analysis, The National Income-Product Accounts and The Survey of Current Business).

[28] S.A.U.S. 1980, Table 940, page 562 (citing Bureau of Economic Analysis, Survey of Current Business, monthly).

[29] TIME, INC., Fortune, directory of the largest 500 Industrial Corporations, annually.

[30] Economic Report of the President 1981, page 350 (citing International Monetary Fund and the Organization for Economic Cooperation and Development, Revenue Statistics, annual).

[31] Bureau of Economic Analysis, Survey of Current Business, Table 5-1, and Business Statistics 1977.

[32] Ibid.

[33] National Audubon Society, Sierra Club, and National Wildlife Federation: Personal Communication, September 1981.

[34] U.S. Department of the Interior, Bureau of Land Management, Public Land Statistics.

[35] Council on Environmental Quality, 1980 Report, page 152.

[36] Office of Management and Budget, Special Analysis, Budget of the U.S. Government, annual.

[37] Bureau of Economic Analysis, Survey of Current Business, June 1981, page 20.

[40] S.A.U.S. 1979, Tables 473, 476, 483, and 494 (citing U.S. Bureau of the Census, Governmental Finances - GF no. 5 and unpublished data, and U.S. Bureau of the Census, Census of Governments: 1972).

[41] National Council on Transportation, National Transportation Statistics Annual Report, September 1980, page 54.

[42] Ibid.

[43] S.A.U.S. 1980, Table 1064, page 625 (citing U.S. National Science Foundation, Federal Funds for Research and Development, annual).

[44] U.S. Energy Information Administration, 1980 Annual Report to Congress, Volume 2 Data, DOE/ETA - 0173 (80)/2 (Washington, D.C.: U.S. Government Printing Office, 1981), Table 3.

[45] Ibid. Volume 3: Forecasts, page 124.

[46] Exxon Company, U.S.A., Energy Outlook 1980-2000 (Houston, Texas: Exxon Public Affairs Department, December 1980), page 4.

[47] 1980 Annual Report to Congress, Volume 2, Table 3.

[48] Ibid., page 232.

[49] 1980 Annual Report to Congress, Volume 3, Table 4.3.

[50] Energy Outlook 1980-2000, page 5.

[51] 1980 Annual Report to Congress, Volume 3, Table 4.19.

[52] 1980 Annual Report to Congress, Volume 2, Table 4.

[53] Energy Outlook 1980-2000, page 4.

[54] 1980 Annual Report to Congress, Volume 2, Tables 64 and 67.

[55] 1980 Annual Report to Congress, Volume 3, Table 4.15.

[56] 1980 Annual Report to Congress, Volume 2, Table 65.

[57] 1980 Annual Report to Congress, Volume 3, Table 4.15.

[58] 1980 Annual Report to Congress, Volume 2, Tale 67.

[59] Mary Wayne and Robert Mauro, "Plugging Co-generation into the Grid," EPRI Journal, July/August 1981, pages 6-14.

[60] 1980 Annual Report to Congress, Volume 2, Table 74, and page 227.

[61] 1980 Annual Report to Congress, Volume 3, Table 4.4

[62] 1980 Annual Report to Congress, Volume 2, Table 40 and page 227.

[63] 1980 Annual Report to Congress, Volume 3, Table 4.1.

[64] Ibid., Table 4.6.

[65] S.A.U.S. 1980, Table 1434, page 805 and Table 1453, page 820 (citing U.S. Bureau of Census, Census of Manufactures and Annual Survey of Manufactures).

[66] Edison Electric Institute, Statistical Yearbook of the Electric Utility Industry/1979, J. David Bailey, editor (Washington, D.C.: EEI, 1980), page 16.

[67] 1980 Annual Report to Congress, Volume 2, page 287.

[68] Ibid.

[69] National Academy of Sciences, Energy in Transition 1985-2000, Report of the Committee on Nuclear and Alternative Energy Systems (San Francisco, CA: W.H. Freeman and Co., 1979), pages 161-264.

[70] Solar Energy Research Institute, Building a Sustainable Energy Future, Committee Print, U.S. House of Representatives' Committee on Energy and Natural Resources (1981), page 927.

[71] 1980 Annual Report to Congress, Volume 3, page 304.

[72] Ibid.

[73] Office of Technology Assessment, Application of Solar Technology to Today's Energy Needs (Washington, D.C.: U.S. Government Printing Office, 1978), page 394.

[74] Energy in Transition, page 368.

[75] Building a Sustainable Energy Future, page 946.

[76] J.S. Edmonds and W.R. McCown, "Coming: Large Superconducting Generators," Electrical World (December 1980), pages 69-72.

[77] Personal Communication, Tom Henigan, Treasurer of Jet Industries, Austin, Texas, September 1981.

APPENDIX C

Example Statistical Descriptions of Six Potential Futures

The development of the assessment scenarios (Chapter II) began with the selection and superposition of themes by which one might characterize different electric power futures. Next, values for exogenous variables, such as population and RGNP growth rates, were selected to be compatible with each of the six sets of themes. Then, hypothetical electrical power policies were developed that would be consistent with the sets of themes and exogenous variable values.

To force more rigorous analysis for this last effort, detailed quantitative longitudinal descriptions of particularly important exogenous variables and electric power supply mixes were developed. These descriptions were quantitative exemplars, not predictions, of values that might come to pass. This appendix presents the descriptions developed for each scenario.

PATH A	198Ø	199Ø	2ØØØ	201Ø	198Ø	198Ø	198Ø	198Ø	199Ø	199Ø	199Ø	199Ø
TOT Q	76.ØØ	75.67	89.13	1ØØ		76.ØØ				75.67		
ELEC%	.3314	.375	.4Ø6	.46		.3314				.375		
NONUT%	.Ø286	.Ø333	.Ø387	.Ø45		.Ø286				.Ø333		
Q ELEC	25.19	28.38	36.19	46		25.19				28.38		
	CAPACITY FACTORS				GEN %	BKWH	GW	CAP %	GEN %	BKWH	GW	CAP %
COAL	.56	.58	.62	.64	.4872	1162.	236.9	.3993	.51	1372.	27Ø.1	.4Ø98
NUKE	.56	.59	.65	.7	.1Ø53	251.2	51.2Ø	.Ø863	.15	4Ø3.6	78.Ø9	.1185
O&G	.323	.28	.23	.175	.2482	592.Ø	2Ø9.2	.3527	.175	47Ø.9	192.Ø	.2913
HYDRO	.49	.49	.48	.48	.117	279.1	65.Ø1	.1Ø96	.11	296.Ø	68.96	.1Ø46
OTHER	.35	.38	.53	.62	.ØØ25	5.963	1.945	.ØØ33	.ØØ74	19.91	5.982	.ØØ91
STORE	Ø	Ø	Ø	Ø	Ø	−4	12	.Ø2Ø2	Ø	−8	22.7	.Ø344
IMPORT	Ø	Ø	Ø	Ø	.Ø113	26.95	Ø	Ø	.Ø15	4Ø.36	Ø	Ø
UT TOT	.4582	.4641	.499Ø	.5251	.9715	2313.	576.3	.9715	.9674	2593.	637.8	.9677
NON-UT	.46	.48	.52	.54	.Ø286	68.21	16.93	.Ø285	.Ø333	89.51	21.29	.Ø323
TOTAL	.4582	.4647	.4998	.5258	1.ØØØ	2381.	593.2	1	1.ØØ1	2683.	659.1	1
HT RTE	3.1	3.1	3.1	3.1		3.1				3.1		
RGNP	2633	3247	415Ø	5Ø53		2633				3247		
US POP	226.5	246.5	261.7	275.1		226.5				246.5		
BTU/$	28864	233Ø5	21477	1979Ø		28864				233Ø5		
KWH/$	.9Ø43	.8262	.8244	.86Ø7		.9Ø43				.8262		
$/POP	11625	13172	15858	18368		11625				13172		
OIL $	33.89	36	44	55		33.89				36		
ELEC $	.Ø37	.Ø51	.Ø56	.Ø6		.Ø37				.Ø51		

	2ØØØ	2ØØØ	2ØØØ	2ØØØ	2Ø1Ø	2Ø1Ø	2Ø1Ø	2Ø1Ø	AV RTE PER YR
TOT Q		89.13				1ØØ			.9191
ELEC%		.4Ø6				.46			
NONUT%		.Ø387				.Ø45			
Q ELEC		36.19				46			2.Ø28
	GEN %	BKWH	GW	CAP %	GEN %	BKWH	GW	CAP %	BKWH
COAL	.54	1853.	341.2	.4367	.561	2447.	436.5	.4633	2.513
NUKE	.185	635.Ø	111.5	.1427	.198	863.7	14Ø.8	.1495	4.2Ø3
O&G	.1	343.2	17Ø.3	.218Ø	.Ø59	257.4	167.9	.1782	−2.74
HYDRO	.1Ø5	36Ø.4	85.71	.1Ø97	.Ø92Ø	4Ø1.2	95.41	.1Ø13	1.217
OTHER	.Ø172	59.Ø3	12.72	.Ø163	.Ø3	13Ø.9	24.Ø9	.Ø256	1Ø.84
STORE	Ø	−11	3Ø.7	.Ø393	Ø	−13	36	.Ø382	4.ØØ7
IMPORT	.Ø15	51.48	Ø	Ø	.Ø15	65.43	Ø	Ø	3.ØØ1
UT TOT	.9622	3288.	752.2	.9627	.955Ø	4153.	9ØØ.7	.956Ø	1.97Ø
NON-UT	.Ø387	132.8	29.15	.Ø373	.Ø45	196.3	41.5Ø	.Ø44Ø	3.586
TOTAL	1.ØØ1	3421.	781.4	1	1.	4349.	942.2	1	2.Ø28
HT RTE		3.1				3.1			
RGNP		415Ø				5Ø53			2.197
US POP		261.7				275.1			.65Ø1
BTU/$		21477				1979Ø			−1.25
KWH/$		.8244				.86Ø7			−.165
$/POP		15858				18368			1.537
OIL $		44				55			1.627
ELEC $		.Ø56				.Ø6			1.624

PATH B	198Ø	199Ø	2ØØØ	201Ø	198Ø	198Ø	198Ø	198Ø	199Ø	199Ø	199Ø	199Ø
TOT Q	76.ØØ	77.Ø9	1Ø4.9	13Ø		76.ØØ				77.Ø9		
ELEC%	.3314	.3835	.475	.55		.3314				.3835		
NONUT%	.Ø286	.Ø224	.Ø162	.Ø1Ø1		.Ø286				.Ø224		
Q ELEC	25.19	29.56	49.83	71.5		25.19				29.56		
	CAPACITY FACTORS				GEN %	BKWH	GW	CAP %	GEN %	BKWH	GW	CAP %
COAL	.56	.59	.62	.64	.4872	1162.	236.9	.3993	.495	1388.	268.5	.3991
NUKE	.56	.6	.68	.72	.1Ø53	251.2	51.2Ø	.Ø863	.1837	514.9	97.97	.1457
O&G	.323	.29	.26	.18	.2482	592.Ø	2Ø9.2	.3527	.1771	496.4	195.4	.29Ø5
HYDRO	.49	.49	.49	.49	.117	279.1	65.Ø1	.1Ø96	.1Ø13	284.Ø	66.15	.Ø984
OTHER	.35	.38	.53	.62	.ØØ25	5.963	1.945	.ØØ33	.ØØ83	23.27	6.989	.Ø1Ø4
STORE	Ø	Ø	Ø	Ø	Ø	-4	12	.Ø2Ø2	Ø	-8	22.7	.Ø337
IMPORT	Ø	Ø	Ø	Ø	.Ø113	26.95	Ø	Ø	.Ø125	35.Ø4	Ø	Ø
UT TOT	.4582	.4742	.5377	.5867	.9715	2313.	576.3	.9715	.978	2732.	657.7	.9778
NON-UT	.46	.48	.52	.54	.Ø286	68.21	16.93	.Ø285	.Ø224	62.79	14.93	.Ø222
TOTAL	.4582	.4744	.5374	.5862	1.ØØØ	2381.	593.2	1	1.ØØØ	2795.	672.6	1
HT RTE	3.1	3.1	3.1	3.1		3.1				3.1		
RGNP	2633	33Ø9	4611	5851		2633				33Ø9		
US POP	226.5	249	269.6	29Ø.2		226.5				249		
BTU/$	28864	23297	2275Ø	22218		28864				23297		
KWH/$	.9Ø43	.8447	1.Ø22	1.155		.9Ø43				.8447		
$/POP	11625	13289	171Ø3	2Ø162		11625				13289		
OIL $	33.89	38	48	6Ø		33.89				38		
ELEC $	.Ø37	.Ø52	.Ø51	.Ø5		.Ø37				.Ø52		

	2ØØØ	2ØØØ	2ØØØ	2ØØØ	2Ø1Ø	2Ø1Ø	2Ø1Ø	2Ø1Ø	AV RTE PER YR
TOT Q		1Ø4.9				13Ø			1.8Ø6
ELEC%		.475				.55			
NONUT%		.Ø162				.Ø1Ø1			
Q ELEC		49.83				71.5			3.539
	GEN %	BKWH	GW	CAP %	GEN %	BKWH	GW	CAP %	BKWH
COAL	.4655	2199.	4Ø4.8	.4Ø45	.45	3Ø49.	543.9	.4132	3.268
NUKE	.3149	1487.	249.7	.2495	.4Ø5	2744.	435.1	.33Ø5	8.297
O&G	.Ø9	425.1	186.6	.1865	.Ø35	237.2	15Ø.4	.1143	-3.ØØ
HYDRO	.Ø857	4Ø4.8	94.3Ø	.Ø942	.Ø65	44Ø.4	1Ø2.6	.Ø779	1.533
OTHER	.Ø142	67.Ø7	14.45	.Ø144	.Ø2	135.5	24.95	.Ø19Ø	1Ø.97
STORE	Ø	-12.1	34	.Ø34Ø	Ø	-16	45	.Ø342	4.729
IMPORT	.Ø138	65.18	Ø	Ø	.Ø15	1Ø1.6	Ø	Ø	4.524
UT TOT	.984	4634.	983.8	.9832	.99	6691.	13Ø2.	.989Ø	3.6Ø4
NON-UT	.Ø162	76.51	16.8Ø	.Ø168	.Ø1Ø1	68.44	14.47	.Ø11Ø	.Ø1Ø8
TOTAL	1.ØØØ	4711.	1ØØ1.	1	1.ØØØ	676Ø.	1316.	1	3.539
HT RTE		3.1				3.1			
RGNP		4611				5851			2.697
US POP		269.6				29Ø.2			.8295
BTU/$		2275Ø				22218			-.868
KWH/$		1.Ø22				1.155			.8198
$/POP		171Ø3				2Ø162			1.852
OIL $		48				6Ø			1.922
ELEC $		.Ø51				.Ø5			1.ØØ9

PATH C	1980	1990	2000	2010	1980	1980	1980	1980	1990	1990	1990	1990
TOT Q	76.00	77.47	104.0	130		76.00				77.47		
ELEC%	.3314	.389	.458	.55		.3314				.389		
NONUT%	.0286	.035	.05	.045		.0286				.035		
Q ELEC	25.19	30.14	47.64	71.5		25.19				30.14		
	CAPACITY FACTORS				GEN %	BKWH	GW	CAP %	GEN %	BKWH	GW	CAP %
COAL	.56	.58	.61	.64	.4872	1162.	236.9	.3993	.5	1429.	281.2	.4045
NUKE	.56	.59	.64	.67	.1053	251.2	51.20	.0863	.156	445.8	86.26	.1241
O&G	.323	.29	.24	.2	.2482	592.0	209.2	.3527	.1855	530.1	208.7	.3001
HYDRO	.49	.49	.49	.49	.117	279.1	65.01	.1096	.1013	289.5	67.44	.0970
OTHER	.35	.5	.56	.65	.0025	5.963	1.945	.0033	.008	22.86	5.220	.0075
STORE	0	0	0	0	0	-4	12	.0202	0	-8.6	22.7	.0326
IMPORT	0	0	0	0	.0113	26.95	0	0	.0142	40.58	0	0
UT TOT	.4582	.4673	.4944	.5489	.9715	2313.	576.3	.9715	.965	2749.	671.5	.9658
NON-UT	.46	.48	.48	.47	.0286	68.21	16.93	.0285	.035	100.0	23.79	.0342
TOTAL	.4582	.4678	.4936	.5448	1.000	2381.	593.2	1	1	2849.	695.3	1
HT RTE	3.1	3.1	3.1	3.1		3.1				3.1		
RGNP	2633	3325	4780	6398		2633				3325		
US POP	226.5	249	269.6	290.2		226.5				249		
BTU/$	28864	23299	21759	20319		28864				23299		
KWH/$	.9043	.8569	.9422	1.057		.9043				.8569		
$/POP	11625	13353	17730	22047		11625				13353		
OIL $	33.89	39	50	65		33.89				39		
ELEC $	.037	.05	.065	.07		.037				.05		

	2000	2000	2000	2000	2010	2010	2010	2010	AV RTE PER YR
TOT Q		104.0				130			1.806
ELEC%		.458				.55			
NONUT%		.05				.045			
Q ELEC		47.64				71.5			3.539
	GEN %	BKWH	GW	CAP %	GEN %	BKWH	GW	CAP %	BKWH
COAL	.525	2371.	443.7	.4260	.535	3625.	646.5	.4564	3.865
NUKE	.1715	774.5	138.1	.1326	.22	1490.	253.9	.1793	6.116
O&G	.1173	529.7	252.0	.2419	.06	406.5	232.0	.1638	-1.25
HYDRO	.0857	387.0	90.16	.0866	.065	440.4	102.6	.0724	1.532
OTHER	.0325	146.8	29.92	.0287	.055	372.6	65.44	.0462	14.78
STORE	0	-12.1	34	.0326	0	-15	42	.0296	4.504
IMPORT	.018	81.28	0	0	.02	135.5	0	0	5.530
UT TOT	.95	4278.	987.8	.9484	.955	6455.	1342.	.9477	3.480
NON-UT	.05	225.8	53.70	.0516	.045	304.9	74.05	.0523	5.117
TOTAL	1	4504.	1042.	1	1	6760.	1417.	1	3.539
HT RTE		3.1				3.1			
RGNP		4780				6398			3.004
US POP		269.6				290.2			.8295
BTU/$		21759				20319			-1.16
KWH/$		.9422				1.057			.5199
$/POP		17730				22047			2.156
OIL $		50				65			2.195
ELEC $		.065				.07			2.148

PATH D	1980	1990	2000	2010	1980	1980	1980	1980	1990	1990	1990	1990
TOT Q	76.00	76.33	99.57	130		76.00				76.33		
ELEC%	.3314	.375	.402	.42		.3314				.375		
NONUT%	.0286	.054	.085	.12		.0286				.054		
Q ELEC	25.19	28.62	40.03	54.6		25.19				28.62		
	CAPACITY FACTORS				GEN %	BKWH	GW	CAP %	GEN %	BKWH	GW	CAP %
COAL	.56	.59	.63	.65	.4872	1162.	236.9	.3993	.49	1330.	257.3	.3884
NUKE	.56	.59	.65	.68	.1053	251.2	51.20	.0863	.14	380.0	73.52	.1110
O&G	.323	.275	.26	.25	.2482	592.0	209.2	.3527	.18	488.6	202.8	.3061
HYDRO	.49	.49	.49	.49	.117	279.1	65.01	.1096	.1	271.4	63.23	.0954
OTHER	.35	.38	.53	.62	.0025	5.963	1.945	.0033	.01	27.14	8.154	.0123
STORE	0	0	0	0	0	-4	12	.0202	0	-8	22.7	.0343
IMPORT	0	0	0	0	.0113	26.95	0	0	.026	70.57	0	0
UT TOT	.4582	.4655	.5157	.5574	.9715	2313.	576.3	.9715	.946	2560.	627.7	.9474
NON-UT	.46	.48	.53	.57	.0286	68.21	16.93	.0285	.054	146.6	34.86	.0526
TOTAL	.4582	.4662	.5169	.5589	1.000	2381.	593.2	1	1	2706.	662.6	1
HT RTE	3.1	3.1	3.1	3.1		3.1				3.1		
RGNP	2633	3277	4335	5367		2633				3277		
US POP	226.5	247.6	264.8	280		226.5				247.6		
BTU/$	28864	23293	22969	24222		28864				23293		
KWH/$	.9043	.8258	.8730	.9618		.9043				.8258		
$/POP	11625	13235	16371	19168		11625				13235		
OIL $	33.89	36	39.5	45		33.89				36		
ELEC $	.037	.0435	.0511	.06		.037				.0435		

	2000	2000	2000	2000	2010	2010	2010	2010	AV RTE PER YR
TOT Q		99.57				130			1.806
ELEC%		.402				.42			
NONUT%		.085				.12			
Q ELEC		40.03				54.6			2.613
	GEN %	BKWH	GW	CAP %	GEN %	BKWH	GW	CAP %	BKWH
COAL	.507	1923.	348.5	.4170	.53	2742.	481.5	.4567	2.903
NUKE	.123	466.6	81.95	.0981	.092	476.0	79.90	.0758	2.154
O&G	.12	455.3	199.9	.2392	.083	429.4	196.1	.1860	-1.06
HYDRO	.102	387.0	90.15	.1079	.085	439.7	102.4	.0972	1.527
OTHER	.023	87.26	18.79	.0225	.04	206.9	38.10	.0361	12.55
STORE	0	-9.6	27	.0323	0	-11.4	32	.0303	3.553
IMPORT	.04	151.8	0	0	.05	258.7	0	0	7.830
UT TOT	.915	3462.	766.3	.9169	.88	4541.	930.1	.8821	2.274
NON-UT	.085	322.5	69.46	.0831	.12	620.8	124.3	.1179	7.639
TOTAL	1	3784.	835.8	1	1	5162.	1054.	1.	2.613
HT RTE		3.1				3.1			
RGNP		4335				5367			2.402
US POP		264.8				280			.7093
BTU/$		22969				24222			-.583
KWH/$		.8730				.9618			.2056
$/POP		16371				19168			1.681
OIL $		39.5				45			.9496
ELEC $		.0511				.06			1.624

PATH E	1980	1990	2000	2010
TOT Q	76.ØØ	72.9	8Ø.92	8Ø
ELEC%	.3314	.375	.423	.55
NONUT%	.Ø286	.Ø4	.Ø9	.18
Q ELEC	25.19	27.34	34.23	44
	CAPACITY FACTORS			
COAL	.56	.58	.63	.65
NUKE	.56	.59	.6	.62
O&G	.323	.29	.23	.14
HYDRO	.49	.49	.49	.5
OTHER	.35	.5	.56	.64
STORE	Ø	Ø	Ø	Ø
IMPORT	Ø	Ø	Ø	Ø
UT TOT	.4582	.4686	.4925	.5119
NON-UT	.46	.48	.53	.58
TOTAL	.4582	.469Ø	.4956	.523Ø
HT RTE	3.1	3.1	3.1	3.1
RGNP	2633	3198	4259	6393
US POP	226.5	242.3	248.4	25Ø
BTU/$	28864	22795	19ØØØ	12514
KWH/$	.9Ø43	.8Ø82	.7598	.65Ø7
$/POP	11625	13199	17146	25572
OIL $	33.89	39	5Ø	65
ELEC $	.Ø37	.Ø53	.Ø6	.Ø7

	198Ø	198Ø	198Ø	198Ø
TOT Q		76.ØØ		
ELEC%		.3314		
NONUT%		.Ø286		
Q ELEC		25.19		
	GEN %	BKWH	GW	CAP %
COAL	.4872	1162.	236.9	.3993
NUKE	.1Ø53	251.2	51.2Ø	.Ø863
O&G	.2482	592.Ø	2Ø9.2	.3527
HYDRO	.117	279.1	65.Ø1	.1Ø96
OTHER	.ØØ25	5.963	1.945	.ØØ33
STORE	Ø	-4	12	.Ø2Ø2
IMPORT	.Ø113	26.95	Ø	Ø
UT TOT	.9715	2313.	576.3	.9715
NON-UT	.Ø286	68.21	16.93	.Ø285
TOTAL	1.ØØØ	2381.	593.2	1
HT RTE		3.1		
RGNP		2633		
US POP		226.5		
BTU/$		28864		
KWH/$		.9Ø43		
$/POP		11625		
OIL $		33.89		
ELEC $		.Ø37		

	199Ø	199Ø	199Ø	199Ø
TOT Q		72.9		
ELEC%		.375		
NONUT%		.Ø4		
Q ELEC		27.34		
	GEN %	BKWH	GW	CAP %
COAL	.4817	1249.	245.8	.39Ø7
NUKE	.155	4Ø1.8	77.75	.1236
O&G	.1785	462.8	182.2	.2896
HYDRO	.118	3Ø5.9	71.27	.1133
OTHER	.ØØ8	2Ø.74	4.735	.ØØ75
STORE	Ø	-8	22.7	.Ø361
IMPORT	.Ø188	48.74	Ø	Ø
UT TOT	.96	2481.	6Ø4.4	.96Ø8
NON-UT	.Ø4	1Ø3.7	24.66	.Ø392
TOTAL	1	2585.	629.1	1
HT RTE		3.1		
RGNP		3198		
US POP		242.3		
BTU/$		22795		
KWH/$		.8Ø82		
$/POP		13199		
OIL $		39		
ELEC $		.Ø53		

	2ØØØ	2ØØØ	2ØØØ	2ØØØ
TOT Q		8Ø.92		
ELEC%		.423		
NONUT%		.Ø9		
Q ELEC		34.23		
	GEN %	BKWH	GW	CAP %
COAL	.4858	1577.	285.8	.3835
NUKE	.1185	384.8	73.21	.Ø982
O&G	.1Ø91	354.3	175.8	.2359
HYDRO	.119	386.4	9Ø.Ø2	.12Ø8
OTHER	.Ø4Ø5	131.5	26.81	.Ø36Ø
STORE	Ø	-11	3Ø.7	.Ø412
IMPORT	.Ø371	12Ø.5	Ø	Ø
UT TOT	.91	2944.	682.4	.9155
NON-UT	.Ø9	292.2	62.95	.Ø845
TOTAL	1	3236.	745.3	1
HT RTE		3.1		
RGNP		4259		
US POP		248.4		
BTU/$		19ØØØ		
KWH/$		.7598		
$/POP		17146		
OIL $		5Ø		
ELEC $		.Ø6		

	2Ø1Ø	2Ø1Ø	2Ø1Ø	2Ø1Ø	AV RTE PER YR
TOT Q		8Ø			.1712
ELEC%		.55			
NONUT%		.18.			
Q ELEC		44			1.877
	GEN %	BKWH	GW	CAP %	BKWH
COAL	.417	174Ø.	3Ø5.6	.3358	1.355
NUKE	.Ø95	396.5	73.ØØ	.Ø8Ø2	1.534
O&G	.Ø5	2Ø8.7	17Ø.2	.1869	-3.42
HYDRO	.11	459.1	1Ø4.8	.1152	1.673
OTHER	.Ø95	396.5	7Ø.72	.Ø777	15.Ø2
STORE	Ø	-13.6	38	.Ø417	4.164
IMPORT	.Ø53	221.2	Ø	Ø	7.269
UT TOT	.82	34Ø9.	762.3	.8376	1.3Ø1
NON-UT	.18	751.2	147.9	.1624	8.325
TOTAL	1	416Ø.	91Ø.2	1.	1.877
HT RTE		3.1			
RGNP		6393			3.ØØ1
US POP		25Ø			.3296
BTU/$		12514			-2.75
KWH/$		.65Ø7			-1.Ø9
$/POP		25572			2.663
OIL $		65			2.195
ELEC $		.Ø7			2.148

PATH F	1980	1990	2000	2010	1980	1980	1980	1980	1990	1990	1990	1990
TOT Q	76.00	73.85	80.89	80		76.00				73.85		
ELEC%	.3314	.375	.384	.38		.3314				.375		
NONUT%	.0286	.042	.053	.06		.0286				.042		
Q ELEC	25.19	27.69	31.06	30.4		25.19				27.69		
	CAPACITY FACTORS				GEN %	BKWH	GW	CAP %	GEN %	BKWH	GW	CAP %
COAL	.56	.58	.61	.62	.4872	1162.	236.9	.3993	.4885	1283.	252.5	.3906
NUKE	.56	.59	.62	.65	.1053	251.2	51.20	.0863	.14	367.7	71.14	.1100
O&G	.323	.27	.21	.15	.2482	592.0	209.2	.3527	.1758	461.7	195.2	.3019
HYDRO	.49	.49	.48	.48	.117	279.1	65.01	.1096	.118	309.9	72.20	.1117
OTHER	.35	.38	.53	.62	.0025	5.963	1.945	.0033	.0083	21.80	6.548	.0101
STORE	0	0	0	0	0	-4	12	.0202	0	-8	22.7	.0351
IMPORT	0	0	0	0	.0113	26.95	0	0	.0275	72.22	0	0
UT TOT	.4582	.4615	.4690	.4663	.9715	2313.	576.3	.9715	.9581	2508.	620.3	.9594
NON-UT	.46	.48	.52	.54	.0286	68.21	16.93	.0285	.042	110.3	26.23	.0406
TOTAL	.4582	.4623	.4715	.4701	1.000	2381.	593.2	1	1.000	2618.	646.5	1
HT RTE	3.1	3.1	3.1	3.1		3.1				3.1		
RGNP	2633	3169	3717	4111		2633				3169		
US POP	226.5	242.3	247.8	250		226.5				242.3		
BTU/$	28864	23304	21762	19460		28864				23304		
KWH/$	.9043	.8262	.7901	.6991		.9043				.8262		
$/POP	11625	13079	15000	16444		11625				13079		
OIL $	33.89	36	44	55		33.89				36		
ELEC $	.037	.055	.07	.065		.037				.055		

	2000	2000	2000	2000	2010	2010	2010	2010	AV RTE PER YR
TOT Q		80.89				80			.1712
ELEC%		.384				.38			
NONUT%		.053				.06			
Q ELEC		31.06				30.4			.6292
	GEN %	BKWH	GW	CAP %	GEN %	BKWH	GW	CAP %	BKWH
COAL	.505	1488.	278.4	.3938	.535	1544.	284.2	.4104	.9508
NUKE	.1525	449.3	82.73	.1170	.145	418.3	73.47	.1061	1.715
O&G	.12	353.6	192.2	.2718	.075	216.4	164.7	.2378	-3.30
HYDRO	.119	350.6	83.38	.1179	.13	375.1	89.20	.1288	.9904
OTHER	.0142	41.84	9.011	.0127	.025	72.13	13.28	.0192	8.665
STORE	0	-9.6	27	.0382	0	-11	31	.0448	3.429
IMPORT	.0363	106.9	0	0	.03	86.55	0	0	3.966
UT TOT	.947	2780.	672.7	.9515	.94	2701.	655.8	.9471	.5182
NON-UT	.053	156.2	34.28	.0485	.06	173.1	36.59	.0529	3.153
TOTAL	1	2937.	707.0	1	1	2874.	692.4	1	.6292
HT RTE		3.1				3.1			
RGNP		3717				4111			1.496
US POP		247.8				250			.3296
BTU/$		21762				19460			-1.31
KWH/$		.7901				.6991			-.854
$/POP		15000				16444			1.163
OIL $		44				55			1.627
ELEC $		.07				.065			1.896

APPENDIX D

Data Sheets for Exhibits in Chapter V

Exhibits V.1 and V.2

YEAR	AVNOM CENTS KWH-R 500/M	AVNOM CENTS KWH-I 2E5/M	1972' GNP PRICE DEF'R	AV'80 CENTS KWH-R 500/M	AV'80 CENTS KWH-I 2E5/M	GNP PRICE DEF'R '70=1	AVNOM RES $ INDEX '70=1
1935	2.77	1.54	27.73	17.84	9.92	30.32	131.90
1936	2.47	1.51	27.91	15.81	9.66	30.52	117.62
1937	2.30	1.48	29.24	14.05	9.04	31.97	109.52
1938	2.21	1.43	28.56	13.82	8.94	31.23	105.24
1939	2.17	1.43	28.38	13.66	9.00	31.03	103.33
1940	2.11	1.41	29.03	12.98	8.68	31.74	100.48
1941	2.08	1.41	31.24	11.89	8.06	34.16	99.05
1942	2.07	1.41	34.29	10.78	7.35	37.50	98.57
1943	2.05	1.43	36.08	10.15	7.08	39.45	97.62
1944	2.05	1.44	36.93	9.92	6.97	40.38	97.62
1945	2.04	1.43	37.82	9.64	6.75	41.36	97.14
1946	2.03	1.44	43.84	8.27	5.87	47.94	96.67
1947	2.01	1.45	49.60	7.24	5.22	54.24	95.71
1948	2.02	1.50	53.03	6.80	5.05	57.99	96.19
1949	2.04	1.55	52.49	6.94	5.28	57.40	97.14
1950	2.02	1.51	53.56	6.74	5.04	58.57	96.19
1951	2.00	1.51	57.09	6.26	4.72	62.43	95.24
1952	2.02	1.52	57.92	6.23	4.69	63.34	96.19
1953	2.04	1.58	58.82	6.20	4.80	64.32	97.14
1954	2.05	1.58	59.55	6.15	4.74	65.12	97.62
1955	2.06	1.58	60.84	6.05	4.64	66.53	98.10
1956	2.07	1.60	62.79	5.89	4.55	68.66	98.57
1957	2.08	1.62	64.93	5.72	4.46	71.00	99.05
1958	2.09	1.64	66.04	5.65	4.44	72.21	99.52
1959	2.10	1.64	67.60	5.55	4.33	73.92	100.00
1960	2.12	1.65	68.70	5.51	4.29	75.12	100.95
1961	2.13	1.67	69.33	5.49	4.30	75.81	101.43
1962	2.13	1.68	70.61	5.39	4.25	77.21	101.43
1963	2.13	1.72	71.67	5.31	4.29	78.37	101.43
1964	2.12	1.71	72.77	5.20	4.20	79.57	100.95
1965	2.08	1.71	74.36	5.00	4.11	81.31	99.05
1966	2.07	1.70	76.76	4.82	3.96	83.94	98.57
1967	2.07	1.71	79.06	4.68	3.86	86.45	98.57
1968	2.07	1.71	82.54	4.48	3.70	90.26	98.57
1969	2.06	1.72	86.79	4.24	3.54	94.90	98.10
1970	2.10	1.75	91.45	4.10	3.42	100.00	100.00
1971	2.23	1.89	96.01	4.14	3.51	104.99	106.00
1972	2.40	2.07	100.00	4.28	3.70	109.35	114.19
1973	2.51	2.20	105.75	4.24	3.72	115.64	119.62
1974	2.82	2.60	115.82	4.35	4.01	126.65	134.29
1975	3.59	3.44	125.79	5.09	4.89	137.55	170.76
1976	3.85	3.70	132.34	5.20	4.99	144.71	183.43
1977	4.17	4.11	140.05	5.32	5.25	153.14	198.67
1978	4.44	4.49	150.42	5.27	5.33	164.48	211.33
1979	4.61	4.70	163.42	5.04	5.14	178.70	219.52
1980	5.50	5.46	178.64	5.50	5.45	195.34	261.90
1981	6.52	6.38	195.51	5.96	5.83	213.79	310.57

Sources: Bureau of the Census, *Historical Statistics of the United States*, U.S. Department of Commerce (Washington, D.C.: U.S. Government Printing Office, 1975), S-115, 117; Energy Information Administration, *1981 Annual Report to Congress*, vol. 2, U.S. Department of Energy (Washington, D.C.: U.S. Government Printing Office, 1982), Table 72.

GNP: Joint Economic Committee, *1980 Supplement to Economic Indications*, U.S. Congress (Washington, D.C.: U.S. Congress (Washington, D.C.: U.S. Government Printing Office, 1980), Table 3; Bureau of Economic Analysis, *Survey of Current Business*, U.S. Department of Commerce, October 1982.

Exhibit V.3

YEAR	NET UT. BKWH/YR GENER'N	MMKW UTIL. CAP'Y	KWH/KW UTIL. CAP'Y
1930	91	32.4	2808.6
1931	87	33.7	2581.6
1932	79	34.4	2296.5
1933	82	34.6	2369.9
1934	87	34.1	2551.3
1935	95	34.4	2761.6
1936	109	35.1	3105.4
1937	119	35.6	3342.7
1938	114	37.5	3040.0
1939	128	38.9	3290.5
1940	142	39.9	3558.9
1941	165	42.4	3891.5
1942	186	45.1	4124.2
1943	218	48.0	4541.7
1944	228	49.2	4634.1
1945	222	50.1	4431.1
1946	223	50.3	4433.4
1947	256	52.3	4894.8
1948	283	56.6	5000.0
1949	291	63.1	4611.7
1950	329	68.9	4775.0
1951	371	75.8	4894.5
1952	399	82.2	4854.0
1953	443	91.5	4841.5
1954	472	102.6	4600.4
1955	547	114.5	4777.3
1956	601	120.7	4979.3
1957	632	129.1	4895.4
1958	645	142.6	4523.1
1959	710	156.8	4528.1
1960	753	168.0	4482.1
1961	792	180.7	4383.0
1962	852	191.1	4458.4
1963	917	210.5	4356.3
1964	984	222.3	4426.5
1965	1055	236.1	4468.4
1966	1144	247.8	4616.6
1967	1214	269.3	4508.0
1968	1329	291.1	4565.4
1969	1442	313.3	4602.6
1970	1532	341.6	4484.8
1971	1613	368.9	4372.5
1972	1750	398.6	4390.4
1973	1861	442.4	4206.6
1974	1867	477.6	3909.1
1975	1918	508.3	3773.4
1976	2038	531.2	3836.6
1977	2124	560.2	3791.5
1978	2206	579.2	3808.7
1979	2247	598.3	3755.6
1980	2286	613.5	3726.2
1981	2293	634.5	3613.9

Sources: Bureau of the Census, *Historical Statistics of the United States*, U.S. Department of Commerce (Washington, D.C.: U.S. Government Printing Office, 1975), S-36; Energy Information Administration, *1981 Annual Report to Congress*, vol. 2, U.S. Department of Energy (Washington, D.C.: U.S. Government Printing Office, 1982), table 64.

Exhibit V.4

YEAR	NETGEN UT+IMP BKWH	NETGEN INDR'L BKWH	TOTAL ELECT. AVAIL. BKWH	INDR'L % OF TOTAL BKWH
1935	95	24	119	19.91
1936	109	27	136	19.63
1937	119	28	146	18.82
1938	114	28	142	19.83
1939	128	34	161	20.88
1940	142	38	180	21.17
1941	165	44	208	20.89
1942	186	47	233	20.23
1943	218	50	268	18.60
1944	228	51	280	18.37
1945	222	49	271	17.98
1946	223	46	270	17.22
1947	256	52	307	16.81
1948	283	54	337	16.07
1949	291	54	345	15.64
1950	329	60	389	15.32
1951	373	63	436	14.39
1952	402	64	466	13.70
1953	445	72	517	13.84
1954	474	73	547	13.34
1955	552	82	634	12.93
1956	605	84	689	12.21
1957	635	85	720	11.79
1958	648	80	728	10.95
1959	713	86	799	10.74
1960	758	89	847	10.49
1961	794	87	881	9.90
1962	853	92	945	9.71
1963	917	95	1012	9.35
1964	986	100	1086	9.19
1965	1055	102	1157	8.84
1966	1146	105	1251	8.40
1967	1214	103	1317	7.82
1968	1329	107	1436	7.42
1969	1443	111	1554	7.12
1970	1534	108	1642	6.59
1971	1616	103	1719	6.00
1972	1757	105	1862	5.61
1973	1875	105	1980	5.31
1974	1880	101	1981	5.11
1975	1924	85	2009	4.25
1976	2047	87	2134	4.08
1977	2141	87	2228	3.92
1978	2227	79	2306	3.41
1979	2268	73	2341	3.11
1980	2307	68	2375	2.86
1981	2315	62	2377	2.61

Sources: Bureau of the Census, *Historical Statistics of the United States*, U.S. Department of Commerce (Washington, D.C.: U.S. Government Printing Office, 1975), S-36, 40; Energy Information Administration, *1980 Annual Report to Congress*, vol. 3, U.S. Department of Energy (Washington, D.C.: U.S. Government Printing Office, 1981), table 67; Energy Information Administration, *1981 Annual Report to Congress*, vol. 2, U.S. Department of Energy (Washington, D.C.: U.S. Government Printing Office, 1981), table 64.

Exhibit V.5

YEAR	FOSSIL FUEL BTU/KWH (UTIL.)
1930	19800
1931	18800
1932	18450
1933	18150
1934	17950
1935	17850
1936	17800
1937	17850
1938	17450
1939	16700
1940	16400
1941	16550
1942	16100
1943	16000
1944	15850
1945	15800
1946	15700
1947	15600
1948	15738
1949	15033
1950	14030
1951	13641
1952	13361
1953	12889
1954	12180
1955	11699
1956	11456
1957	11365
1958	11090
1959	10879
1960	10701
1961	10552
1962	10493
1963	10438
1964	10407
1965	10384
1966	10399
1967	10396
1968	10371
1969	10457
1970	10508
1971	10536
1972	10479
1973	10429
1974	10481
1975	10383
1976	10369
1977	10449
1978	10495
1979	10470
1980	10489
1981	10506

Sources: Bureau of the Census, Historical Statistics of the United States, U.S. Department of Commerce (Washington, D.C.: U.S. Government Printing Office, 1975), S-107; Edison Electric Institute, *Statistical Yearbook of the Electric Utility Industry* (New York: Edison Electric Institute, various years).

Exhibits V.6 and V.7

YEAR	AV NOM CENTS/ MMBTU INPUT	1972' GNP PRICE DEF'R	AV '80 CENTS/ MMBTU INPUT	H-WHTN CONST. INDEX 67BASE	H-WHTN CONST. INDEX 72BASE
1954	24.4	59.55	73.20	72	53
1955	24.3	60.84	71.35	74	55
1956	25.4	62.79	72.26	81	60
1957	27.1	64.93	74.56	86	64
1958	26.8	66.04	72.49	88	65
1959	26.1	67.60	68.97	89	66
1960	26.2	68.70	68.13	89	66
1961	26.7	69.33	68.80	88	65
1962	26.4	70.61	66.79	89	66
1963	25.7	71.67	64.06	89	66
1964	25.3	72.77	62.11	90	67
1965	25.2	74.36	60.54	94	70
1966	25.4	76.76	59.11	96	71
1967	25.8	79.06	58.30	100	74
1968	26.3	82.54	56.92	104	77
1969	27.2	86.79	55.99	110	81
1970	31.3	91.45	61.14	119	88
1971	37.5	96.01	69.77		95
1972	41.1	100.00	73.42		100
1973	48.4	105.75	81.76		107
1974	89.0	115.82	137.27		127
1975	108.3	125.79	153.80		149
1976	115.3	132.34	155.64		158
1977	131.8	140.05	168.12		169
1978	145.3	150.42	172.56		179
1979	155.4	163.42	169.87		197
1980	197.0	178.64	197.00		215
1981	230.2	195.51	210.34		

Sources: Price: Edison Electric Institute, *Statistical Yearbook of the Electric Utility Industry* (New York: Edison Electric Institute, various years).

GNP: Bureau of Economic Analysis, *Survey of Current Business*, U.S. Department of Commerce (October 1982), p. 43.

Handy-Whitman Index: U.S. Department of Commerce (Washington, D.C.: U.S. Government Printing Office, 1975), N-131; Bureau of the Census, *Statistical Abstracts of the United States*, U.S. Department of Commerce (Washington, D.C.: U.S. Government Printing Office, 1979 and 1981), price, wage scale, and cost indexes for construction.

Exhibit V.8

YEAR	MOODY'S 24 UTIL NOM ROE	INFLAT'N BY 1972 GNP DEFR	MOODY'S 24 UTIL REAL ROE
1950	8.8	2.1	6.7
1951	7.9	6.6	1.3
1952	8.4	1.4	7.0
1953	8.8	1.6	7.2
1954	9.1	1.2	7.9
1955	9.7	2.2	7.5
1956	9.7	3.2	6.5
1957	9.4	3.4	6.0
1958	9.8	1.7	8.1
1959	9.9	2.4	7.5
1960	10.2	1.6	8.6
1961	10.3	.9	9.4
1962	10.7	1.8	8.9
1963	10.8	1.5	9.3
1964	11.1	1.5	9.6
1965	11.7	2.2	9.5
1966	12.1	3.2	8.9
1967	12.2	3.0	9.2
1968	11.5	4.4	7.1
1969	11.4	5.1	6.3
1970	10.8	5.4	5.4
1971	10.8	5.0	5.8
1972	11.0	4.2	6.8
1973	10.5	5.8	4.7
1974	10.4	8.8	1.6
1975	10.3	9.3	1.0
1976	10.6	5.2	5.4
1977	11.0	5.8	5.2
1978	10.7	7.4	3.3
1979	11.0	8.6	2.4
1980	10.7	9.3	1.4
1981	12.4	9.4	3.0

Sources: Nominal ROE: Moody's Investors Service, *Moody's Public Utility Manual* (New York: Moody's Investors Service, Vol. II, 1982), a13 and a14.

GNP Inflation: Bureau of Economic Analysis, *Survey of Current Business*, U.S. Department of Commerce (October 1982), p. 43.

Index

Acid rain, 139, 160. *See also* Coal-related health and environmental concerns
Adams, Edward Dean, 233, 234
Air pollution, 159-61. *See also* Pollution
Allowance for Funds Used During Construction (AFUDC), 132, 141. *See also* Construction
Alternate sources of electricity supply, 133-34; biomass, 153; fusion, 64, 111; geothermal, 111; hydroelectric, 64, 149; low head hydropower, 111; ocean thermal, 6, 111; "soft energy" alternatives, 197, 202; solar photovoltaics, 19, 67, 68, 70, 86, 111, 153; solar satellites, 61, 75-76. *See also* Conservation of power
Arab oil embargo, effects of, 5, 176, 246; on price, 247; on RD&D programs, 111, 214-15; on regulation, 102, 147
Assessment Methodology, 16-17, 22-23, 263-85
Atomic Energy Commission (AEC), 129, 211, 243, 246
Atomic Industrial Forum, 129
"Average Future" Scenario: description of, 7, 24-25; economy, 200; policy issues, 19, 44-45, 53, 112, 205; population distribution, 196, 202, 204; power plant cancellation, 139, 140; power plant utilization, 137, 185; price regulation, 141, 144; types of appropriate RD&D, 224-25

Biomass, 153. *See also* Alternate sources of electricity supply
Bonneville Power Administration (BPA), 85, 241
Breeder technology, 56, 112. *See also* Nuclear power
Brown, C.E.C., 234
Brush, Charles Francis, 230

California Energy Commission, 134, 142
California Gas and Electric Corporation, 237
Cancellation costs of power plants, 45, 47, 48. *See also* Costs
Capitalized costs of power plants, 44, 71. *See also* Costs
Cataract Construction Company, 233, 236
Catastrophic losses of third parties, 45, 47. *See also* Utility insurance
Central cities, 200-201. *See also* Increased electrification; Population distribution

Chemical waste, 163. *See also* Pollution

Citizen advisory boards, 109. *See also* Regulators

Clean Air Act of 1970, 159, 162, 246

Coal, 7; based electricity, 19, 45, 58, 97; emissions control, 49, 65; fuel supply vulnerability, 180; generating dependence on, 69, 96, 193, 196; mining, 153; plants, public opposition to, 68, 96; wastes, 45, 46, 85, 163. *See also* Coal-related health and environmental concerns; Pollution

Coal-related health and environmental concerns, 148; acid rain, 139, 160; pollution, 66; public awareness of, 65, 96; RD&D programs, 56, 59, 63. *See also* Coal, wastes

Cogenerators. *See* Electricity generators

Common carrier, 47, 100. *See also* Grid management

Commonwealth Edison Company, 239

Competition: among power suppliers, 103; in allocation of resources, 148

Conservation of power, 72, 133; efficiency, 58, 62; programs, 60, 78, 111, 247

Construction, 72, 73; Allowance for Funds Used During Construction (AFUDC), 132, 141; cancellation, 44, 129, 139-40; Construction Work in Progress (CWIP), 46, 47, 49, 108, 132-33, 136, 141; nuclear plants, 59, 60, 129; oil and gas plants, 60

Construction Work in Progress (CWIP). *See* Construction

Control and communication vulnerabilities, 187. *See also* Fuel supply vulnerabilities

Cooperatives, 47. *See also* Electricity generators

Costs: of cancellation, 45, 47, 48; of capitalization, 44, 71; of construction (*see also* Construction), 56, 127, 132; control, 73; of electric power, 44, 46, 48, 92, 121-22, 136; of energy, 82; of fuel, 130; historic trends, 160-68; recovery of capital investments, 68, 125-26

Customer classes, 109

Davy, Sir Humphrey, 230

Demand for electricity. *See* Electricity

Department of Energy (DOE), 111, 215, 220, 246

Deregulation, 47, 70, 103, 104. *See also* Regulation

Disruptions, power system: natural events, 167-68; political (*see also* Arab Oil Embargo), 176-78; technical failures, 175-76; terrorist-related, 169-75. *See also* Fuel supply vulnerabilities

Diversification of utilities, 133; history of, 134; into new business areas (*see also* Public Utility Regulatory Policy Act), 12, 45, 47, 142-43; regulation of (*see also* Regulation), 47

East Central Area Reliability Coordinating Agreement (ECAR), 180

Economic growth, electric power requirements, 23, 71, 82, 90

"Economic Malaise" Scenario: description of, 8, 29-30; economy, 144; energy consumption, 33; population distribution, 198, 200, 204; policy issues, 19, 48-49, 109, 185; power plant cancellation, 140; power plant utilization, 137; price regulation, 138, 139, 141-42; types of appropriate RD&D, 226-27

Economic Regulatory Administration, 246

Edison, Thomas A., 231-32, 234

Efficiency, equity, and risk, balance of, 13, 106-9

Electricity: access to, 46, 242; demand decreases, 247; demand for (various Scenarios), 55, 57, 59, 60, 61, 64, 66, 71, 73; demand growth, 10, 91, 193, 242; early development of, 230-36; end use, 5n, 72, 112, 228; price deregulation, 35, 45, 143;

price regulation, 134; "right" to electricity, 49, 242. *See also* Economic growth
Electricity generators: cogenerators, 45, 46, 56, 63, 95, 99, 134, 144; cooperatives, 47; non-utility, 9, 47, 64, 67, 69; small power producers, 95, 104, 144; super-conducting electricity, 86; third-party power producers, 56, 63, 65, 99
Electricity intensity, 33
Electric Power Research Institute (EPRI): creation of, 215, 245; programs, 111; RD&D support, 114, 220
Electric Reliability Council of Texas (ERCOT), 180
Electric Research Council (ERC), 211
Electric transportation, 6, 86
End use, 5n, 72, 112, 228. *See also* Electricity
Energy consumption, 8, 9, 33, 60, 67, 71, 88-89
Energy Reorganization Act of 1974, 246
Energy Research and Development Administration (ERDA), 111, 215, 246
Energy Supply and Environmental Coordination Act of 1974, 246
Environmental pollution, 78, 129. *See also* Pollution
Environmental protection, 57, 125, 148; citizen support of, 107; RD&D programs, 82, 111; regulation, 102, 138, 214. *See also* Environmental Protection Agency
Environmental Protection Agency (EPA), 102, 245
Evershed, Thomas, 234
Excess capacity, 68, 71, 124

Faraday, Michael, 230
Federal Energy Regulatory Commission (FERC), 102, 246
Federal lands, 153-54
Federal Power Commission (FPC), 214, 215, 245, 246
Federal Water Pollution Control Act of 1972, 161
Fuel costs, 130. *See also* Costs
Fuel supply vulnerabilities, 179-80; coal, 180; control and communication, 187; dynamic, 188-89; generation, 184-85; nuclear, 181-84; oil and gas, 181; transmission and distribution, 185-87. *See also* Disruptions, power system
Fusion power, 64, 111. *See also* Alternate sources of electricity supply

Galvani, Luigi, 230
General Accounting Office (GAO), 220-22
General Electric Company, 231, 235, 238, 242
Generating vulnerabilities, 184-85. *See also* Fuel supply vulnerabilities
Geothermal energy, 111. *See also* Alternate sources of electricity supply
Government policies (general), 113; to encourage RD&D, 227-29; issues, 48
Government regulation of utilities: increase in, 45, 72; states' roles, 61, 105. *See also* Regulation; Regulators
Grid management, 103, 244. *See also* Common carrier
Gross National Product (GNP), 8, 30, 31-32, 55, 60, 67, 71; energy intensity, 8, 32, 88-90, 92

Handy-Whitman Index of Public Utility Construction Costs, 127
Hartford Electric Light Company of Connecticut (HELCO), 238
Health and environmental concerns, 147-48, 160; policies, 85; public awareness of, 45, 56, 96; RD&D programs, 66; regulation, 72; waste disposal, 61, 63; worker health, 108, 154-57. *See also* Coal-related health and environmental concerns
Henry, Joseph, 230
High voltage transmission capabilities, 112
Hydrocarbon-fueled plants, 57, 65,

124, 137. *See also* Alternate sources of electricity supply
Hydroelectric power, 64, 149. *See also* Alternate sources of electricity supply

Increased electrification, impacts of: on central cities, 201-2; on industrial distribution, 199-200; on population distribution, 196-98; on regional divisiveness, 205-6; on service sector, 203-4
Industrial consumers, 58, 89. *See also* Stakeholder categories
Industrial distribution, 198-99. *See also* Increased electrification
Industrial productivity, 111
Insull, Samuel, 238, 241
Integration of power systems, 68, 130, 244
International impacts: of carbon dioxide buildup, 160-61; on regulation, 46, 102

Land resources, 148, 153-54
Lifeline rates, 49. *See also* Electricity
Lilienthal, David, 243, 246
Load management, 68, 93, 111. *See also* Grid management
Low head hydropower, 111. *See also* alternate sources of electricity supply

McCone, John A., 246
"Mega-plant" Scenario: description of, 7, 26-27; economy, 31, 55, 60-61, 83, 144; policy issues, 35, 44, 45-46, 47, 112; population distribution, 196, 200, 202, 204; power plant cancellation, 139; power plant utilization, 98, 137, 138, 140, 152, 185; price regulation, 138, 141; types of appropriate RD&D, 225-26; utility diversification, 143; utility regulation, 103, 105, 109, 139
Methodology of the Assessment, 16-17, 22-23, 263-85
Mid-American Interpool Network (MAIN), 180
Mid-Atlantic Area Council (MAAC), 179
Mid-Continent Area Reliability Coordinating Agreement (MARCA), 180

National Electric Reliability Council (NERC), 211, 244
National Energy Conservation Policy Act of 1978, 99
National importance of electricity, 90
National Power Survey of the Federal Power Commission (NPS-FPC), 215
Niagara Falls power facility, 233-36
Non-utility generators. *See* Electricity generators
Northeast Power Coordinating Council (NPCC), 179
Nuclear fuel cycle, 65, 108. *See also* Nuclear power
Nuclear power, 65, 66, 97, 111, 242-44; accidents, 45, 178; breeder technology, 56, 112; fuel cycle, 65, 108; fuel supply, 69, 72, 181-84; health and safety, 57; issues, 44, 65, 170; Nuclear Regulatory Commission, 129, 246; policies, 84; public opposition to, 60, 68, 96; social concern, 56, 163, 170, 178; technology, 56, 112; wastes, 85, 162, 243. *See also* Construction, nuclear plants
Nuclear Regulatory Commission (NRC), 129, 246
"Nuclear Resurgence" Scenario: description of, 7, 25-26; economy, 55-57, 144; policy issues, 45, 47, 53; population distribution, 200, 202, 204, 206; power plant cancellation, 139, 140; power plant utilization, 137, 183, 185; price regulation, 138, 139; regulatory policies, 138; types of appropriate RD&D, 225; utility diversification, 143
Nuclear Wastes Policy Act of 1982, 162

Occupational health, 154-57. *See also* Health and environmental concerns
Ocean thermal power, 6, 111. *See also* Alternate sources of electricity supply
Oil and gas: price, 46, 125; suppliers, 62, 181

Pacific Gas and Electric Company, 237
Paine, Sidney B., 239
Pixii, Hippolyte, 230
Pollution: air, 159-61; chemical waste, 163; control issues, 49, 129; environmental, 78, 129; residuals management (waste), 158-59; trace elements, 161; water, 161-62
Population distribution, 193-98. *See also under* "Average Future" Scenario; "Economic Malaise" Scenario; "Mega-plant" Scenario; "Nuclear Resurgence" Scenario; "Post-industrial" Scenario; "Small Coal Plants" Scenario
Population size, 8, 9
"Post-industrial" economy, 92
"Post-industrial" Scenario: description of, 8, 28-29; economy, 31, 32, 67-68; energy consumption, 33, 34; fuel supply vulnerability, 187; policy issues, 35, 44, 47-48, 104; population distribution, 83, 197, 202, 204, 206; power plant cancellation, 140; power plant utilization, 98, 137, 181, 185; price regulation, 138, 141; types of appropriate RD&D, 226; utility diversification, 100, 101, 143; utility regulation, 105, 138, 139, 144
Power equipment manufacturers, 58. *See also* Construction
Power generation mixes, 49. *See also* Alternate sources of electricity supply
Power Plant and Industrial Fuel Use Act of 1978, 246
Power plant utilization, 136-42. *See also under* "Average Future" Scenario; "Economic Malaise" Scenario; "Mega-plant" Scenario; "Nuclear Resurgence" Scenario; "Post-industrial" Scenario; "Small Coal Plants" Scenario
Power shortages, 56, 58, 60. *See also* Fuel supply vulnerabilities
Power system disruptions. *See* Disruptions, power system
Price deregulation of electricity, 35, 45, 143. *See also* Electricity
Price regulation of electricity, 134. *See also* Electricity; Regulators
Privately owned electric utility, image of, 99
Public Service Commission (PSC), 58, 61. *See also* Public Utility Commissions
Public Utilities Holding Company Act of 1935, 99, 245
Public Utility Commissions (PUCs): origins of, 245; policies of, 57; rate control, 61, 72. *See also* Regulators
Public Utility Regulatory Policy Act (PURPA) of 1978, 99, 134, 142, 246

Rate base, 68
Rate increases, 57, 58, 61, 71, 72. *See also* Electricity, demand for
Rates of return of utilities, 48, 130, 141. *See also* Costs
RD&D: allocation of resources, 114; coal-fired technology, 63, 66; end-use equipment, 70, 72; government policies, 227-29; issues, 49; levels of funding, 46, 57, 220; objectives, 113; policy, 209; reliability, 74, 247; subsidies, 60
RD&D history, 210-22
RD&D programs, 74, 242; environmental protection, 63, 82; nuclear power, 59; to promote flexibility and efficiency, 62, 110-16; types of, 222-24
Recovery of capital investments, 68, 125-26. *See also* Costs
Regional diviseness, 204-5. *See also* Increased electrification, impacts of; Population distribution
Regional electrical development, 236-42
Regional regulation, 46. *See also* Regulation
Regional trends in population migration, 193-95
Regulation, 13, 102, 121, 245; degree of, 47, 48, 75; of diversification of utilities, 47; of nuclear and coal fuel cycles, 46; regional, 46; water, 152
Regulators, 70; policies of, 58; respon-

sibility, 57, 61, 101; states' roles, 105
Regulatory lag, 131. *See also* Rate increases
Research, Development, and Demonstration. *See* RD&D
Residuals management, 158-59. *See also* Pollution
Rural Electrification Administration (REA), 88, 241

Schallenberger, O. B., 232
Seaborg, Glenn, 246
Service sector, 202-3. *See also* Increased electrification, impacts of
Shareholders vs. ratepayers, 121
"Small Coal Plants" Scenario: description of, 7, 27-28; economy, 64-65; energy consumption, 32; policy issues, 46-47, 104; population distribution, 197, 200, 202, 204; power plant cancellation, 140; power plant utilization, 137, 181, 185, 187; price regulation, 138, 141; types of appropriate RD&D, 226; utility diversification, 100, 143; utility regulation, 105, 139, 144
Small power producers, 95, 104, 144. *See also* Electricity generators
Social trends related to electric power, 82-83, 97, 113
"Soft energy" alternatives, 197, 202. *See also* Alternate sources of electricity supply
Solar photovoltaic generation. *See* Alternate sources of electricity supply
Solar satellites, 61, 75-76. *See also* Alternate sources of electricity supply
Southeastern Electric Reliability Council (SERC), 179
Southern California Edison Company, 237
Southern Power Company, 239
Southwest Power Pool (SWPP), 180
Sprague, Frank, 234
Stakeholder categories, 76-77; impacts, 74, 94
Stanley, William, 232
Steinmetz, Charles P., 235
Strauss, Lewis, 246
Super-conducting electricity generators, 86. *See also* Electricity generators
Systems' level of security, 166-67. *See also* Fuel supply vulnerabilities

Technology cycle, 85, 144-45
Tennessee Valley Authority (TVA), 88, 107, 240-42
Tesla, Nikola, 233
Third-party power producers, 56, 63, 65, 99. *See also* Electricity generators
Time-of-use electricity pricing (TOU), 11, 93, 137, 39
Trace element pollution, 161. *See also* Pollution
Transmission and distribution vulnerabilities, 185-87. *See also* Fuel supply vulnerabilities

Unconventional power generation technologies, 96. *See also* Alternate sources of electricity supply
Uranium, 181, 182
Utility insurance, 46. *See also* Catastrophic losses of third parties
Utility plant utilization, 123-24. *See also* Power plant utilization

Washington Public Power Supply Systems (WPPSS), 84
Water pollutants, 161-62. *See also* Pollution
Water regulation, 152. *See also* Regulation
Water resources, 148, 149-53, 181
Western Systems Coordinating Council (WSCC), 179
Westinghouse, George, 232-33
Westinghouse Company, 234, 238, 242
Wheeling, 68. *See also* Load management
Wild and Scenic Rivers Act, 152

TECHNOLOGY FUTURES, INC., is an educational firm that offers post-graduate courses in technology management. SCIENTIFIC FORESIGHT, INC., is a technology forecasting firm specializing in electric power and telecommunications. Both firms are located in Austin, Texas.